ROBERT E. GREEN, JR.
MATERIALS SCIENCE & ENGINEERING
MARYLAND HALL
THE JOHNS HOPKINS UNIVERSITY
BALTIMORE, MARYLAND 21218

NONDESTRUCTIVE EVALUATION:
Microstructural Characterization and Reliability Strategies

Proceedings of a symposium sponsored by the Physical Metallurgy Committee of The Metallurgical Society of AIME and the Energy and Resources Activity of the American Society for Metals at the TMS Fall Meeting, Pittsburgh, Pennsylvania, October 5-9, 1980.

Edited by
OTTO BUCK
Ames Laboratory
Iowa State University
Ames, Iowa

and

STANLEY M. WOLF
Division of Materials Sciences
Office of Basic Energy Sciences
Department of Energy
Washington, D.C.

A Publication of

The
Metallurgical Society
of AIME

A Publication of The Metallurgical Society of AIME
420 Commonwealth Drive
Warrendale, Pa. 15086
(412) 776-9000

Printed in the United States of America.
Library of Congress Card Catalogue Number 81-82177
ISBN Number 0-89520-375-8

Foreword

There are several potentially effective reliability strategies for critical components in engineering structures. "Pre-service" strategies include improved design and manufacturing, selection of samples for destructive testing, proof testing of parts, and nondestructive testing. "In-service" strategies are limited to nondestructive evaluation (NDE) for on-line monitoring and to NDE and proof testing for periodic inspection. Although proof testing has certain advantages, it could induce additional, unwanted damage to the components. Thus NDE will become increasingly important and cost-effective, and therefore attractive, as component design strives toward greater efficiency and improved performance requiring the materials to operate more reliably and closer to failure conditions.

This Proceedings reports a symposium held at the Fall Meeting of The Metallurgical Society of AIME in Pittsburgh, Pennsylvania, October 1980, to discuss the potential of NDE as a reliabilility strategy for selected advanced engineering systems and the materials to be used. Since microstructural properties of these materials can influence the nondestructive measurements, strong emphasis was placed on the role of materials science and engineering in furthering NDE technology.

The basic concept of NDE is simple, i.e., materials properties and remaining component usefulness are to be assessed from a test (or a series of tests) performed on the components; the test results are evaluated in terms of physically based models for materials performance which are sensitive to the material and the environmental conditions. Reducing this concept to practice is difficult, however, because of the complexity of application requirements (viz., geometry and load spectra) and materials (viz., their microstructure and the mechanisms which control their properties, particularly those of deformation and fracture).

NDE implies more than nondestructive inspection. Conventionally, the latter attempts to ascertain the existence of a flaw, whereas the former aims to identify and characterize the flaw and, based on this and consideration of physical mechanisms and structure, to project future performance and reliability.

The papers of this Proceedings address the need for NDE as a reliability strategy as well as define the state of the art of NDE for quantitatively determining the microstructure, the associated physical and mechanical properties of materials, and damage accumulation. This scope reflects the requirement for close interaction among the NDE, design, and materials communities if NDE is to be successful in increasing systems reliability and reducing life-cycle costs. Based on presentations at the symposium, this interaction is already developing for structural ceramics. With respect to metal alloys for rotor components of military gas turbine engines, an interdisciplinary approach has been fostered. This would clearly benefit other areas too. It is apparent from papers herein that substantive gains have been made recently. Several authors cite the need for improved techniques and analyses for both applications and basic understanding; the given examples range from the development of new inspection tools and procedures for boiler tubes in electricity generating plants to the inclusion of sound wave velocity refraction effects in determining residual stress. The editors of this Proceedings are optimistic that the technical communities can develop the necessary capabilities to advance NDE effectively. We hope this text contributes towards this result.

This Proceedings consists of three sections. The first section focuses on the reliability strategies. The opportunity to reduce component failures with NDE in commercial electricity generating stations and advanced coal conversion plants is reviewed by Armor et al. and Ellingson, respectively; both papers discuss current inspection methods and recognize the need for on-line monitoring capability. NDE also underpins the development and acceptance of the "retirement-for-cause" approach to component replacement. Thompson and Thompson introduce the methodology for this approach. Annis et al. compare it with the current practice for advanced military aircraft engines, namely, retirement based on design to lower-bound properties. Bush examines the limitations of current nondestructive inspection for ductile metallic alloys and obtains a semiquantitative ranking of flaw detectability. Katz and Lenoe add the perspective of proof testing rather than NDE of brittle materials. In contrast, Evans illustrates the potential of NDE for structural ceramics.

The second section of this Proceedings is devoted to NDE characterization of microstructure. Overviews are given on acoustic attenuation by Green, on acoustic velocity precision measurements by Shyne et al., on small-angle neutron scattering by Weertman, and on positron annihilation by Granatelli and Lynn. These techniques are, to a large extent, complementary. The acoustic techniques are able to evaluate grain size and the presence of precipitates, as well as the onset of microcracking; both papers describe the effects of texture on the measurements; the latter one also provides basic information on the acoustic measurements of residual stresses. Small-angle neutron scattering is a tool to monitor the size and size distribution of voids as well as precipitates. Positron annihilation has been effective in determining the distribution and energetics of vacancies and vacancy clusters. All three methods can "see" dislocations to some extent.

Of particular practical interest are two papers by Kupperman et al. and Simpson et al. addressing acoustic wave propagation in weldments, in particular the effects of elastic anisotropy. The presence of large columnar grain structures in austenitic weldments is of major concern in that the elastic anisotropy will affect the accurate detection and sizing of flaws. Changes in the eddy current response during solidification of castings are modeled by Siegfried et al. to indicate the presence of inclusions, hot tears, shrinkage cavities, etc., for quality control. Using eddy current probes, Swartzendruber et al. correlate the electrical conductivity response with mechanical properties of heat-treated products and note limitations on obtaining quantitative correlations. Scanning acoustic microscopy, as utilized to locate and identify voids and inclusions, is discussed by Kessler et al. for structural ceramics, and by Lee et al. for various electronic circuits.

The third section extends the second by interpreting NDE measurements with respect to physical mechanisms controlling deformation and fracture, or regions thereof. Experimental methods available for describing crack tip plastic zones are reviewed by Davidson and Lankford, and results obtained with electron channeling and optical stereoimaging are presented. Gerberich et al. evaluate acoustic emission due to thermal and athermal dislocation processes and particle fracture. Using the stress dependence of the acoustic velocity in metals, Hildebrand and Harrington describe the mapping of residual stress distributions with computer-assisted tomography. Sigler and Haworth apply optical holography to detect early stages of fatigue damage due to slip offsets and microcracks formed at surfaces. Morris et al. examine quantitatively the development and growth of microcracks initiated at different times during fatigue; the effects of surface hardening and grain boundary constraints have been included to obtain a detailed picture of the crack initiation phase of fatigue.

We appreciate the contributions of the authors in presenting these papers and the efforts of the session chairmen in enhancing discussion at the symposium. We also acknowledge the secretaries of Rockwell International Science Center and the Ames Laboratory in typing the many letters and editorial changes of the papers. Special thanks are due to Mrs. Marilyn Norman and Mr. Byron Skillings of the Ames Laboratory for their help in assembling and indexing this Proceedings.

Otto Buck
Metals and Ceramics Program
Ames Laboratory
Iowa State University
Ames, Iowa

Stanley M. Wolf
Division of Materials Sciences
Office of Basic Energy Sciences
U.S. Department of Energy
Washington, D.C.

March, 1981

Table of Contents

I. Reliability Strategies

REDUCTION OF UNSCHEDULED UTILITY OUTAGES THROUGH COMPONENT NON-DESTRUCTIVE EVALUATION

A. F. Armor
Electric Power Research Institute
Palo Alto, CA

D. N. Poole
Electric Power Research Institute
Palo Alto, CA

M. E. Lapides
Electric Power Research Institute
Palo Alto, CA

The utility industry is currently facing a future of dwindling reserve margins and, in some areas of the country, potential shortfalls in available power. This situation is being brought about by delays and cancellations of new plants (particularly nuclear units) due partly to environmental constraints and increased regulatory actions. It is fortunate that the rate of increase of electric power consumption, about 7% per year through the 1950's and 1960's, has slowed to perhaps 2-3% per year, due primarily to conservation efforts. This has put back the day of reckoning, but not eliminated the danger of future blackouts. One ominous sign has been the decreasing availability of large power plants, caused to a large degree by increasing forced outages of major components.

Non-destructive evaluation techniques offer an opportunity to improve plant availability by anticipating future component failures. This particularly applies to the on-line, continuous monitoring of operating units through newly developing diagnostic methods. The Electric Power Research Institute is addressing this aspect of plant monitoring in both nuclear and fossil units. Off-line NDE is also being studied and limitations and benefits of existing procedures in the examination of boilers, turbines, generators and other plant components will be described.

Introduction

Two factors are combining to give cause for concern regarding future generating capacity margins: slow down in new plant construction and decrease in availability of existing plants. This concern is expressed in a recent DOE report (1) which concludes that certain areas in the country may have insufficient reserve margins during the next 10 years. Specifically identified are five electric regions with current potential capability problems: Virginia-Carolinas, Gulf States Utilities Group, Northern California-Nevada, Tennessee Valley Authority and the Florida Coordinating Group. Lack of generating capacity in these areas may result in special measures including disconnection of interruptable loads.

Orders for new fossil plants totalled only just over 5000 MW for the whole of 1979 (2), down by more than 20,000 MW from a decade before. The impact of this is seen in planned capacity additions through 1990 which will average only about 10,000 MW/year according to current projections. The chance of any future improvement on this projection is small since new plants now require lead times of up to 10 years. The implication of this on plant availability is emphasized in two recent DOE draft reports (3,4). Both of these reports imply a reduction of the overall reserve margin of utilities, one by improvement of existing plants and the other (The National Power Grid Study) by increasing the physical interconnection and integrated operation of U.S. utilities.

Plant Availability Trends

In spite of the need for better availability of both fossil and nuclear plants to offset the economic and regulatory forces which are tending to erode reserve margins, the trend has been in the opposite direction. An examination of the outage data of large coal-fired plants, Figure 1, shows a loss in equivalent availability (which includes both partial and full outages) of more than 15% over the 10-year period, 1968-77 (5).

Closer examination of the causes of this loss of availability reveals that the three major contributors are, in order based on their impact, the boiler, the turbine and the generator. One aspect of downtime, which can be impacted significantly by NDE techniques, is the forced outage rate. Figure 2 shows how large units, particularly those over 600 MW in size, suffer seriously in this regard. Generally, of course, the large units are also the newest with less service life and therefore the "maturity effect" is more evident.

Plant Problem Areas

Additional studies by EPRI have identified, in more detail than is apparent from Figure 2, the areas of prime concern in plant forced outages. A series of regional meetings with representative utilities during the period 1976-77 identified those plant components contributing significantly to unavailability. For fossil fired plants, Figure 3 illustrates approximately the percentage forced outage attributed to some important areas.

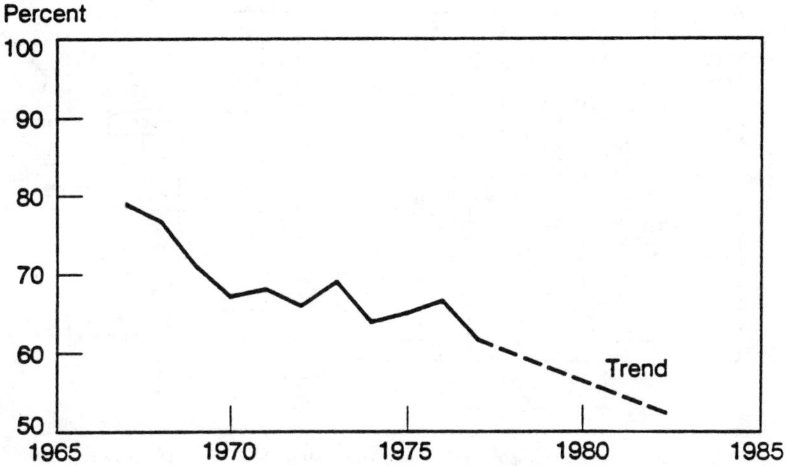

Fig. 1. Equivalent availability — coal units 400 MW and larger (EEI data)

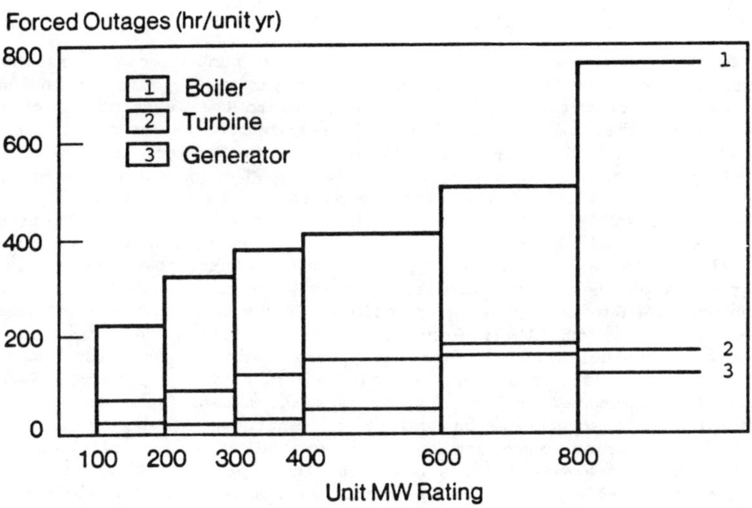

Fig. 2. Major contributors to fossil plant outages (EEI data, 1968-77)

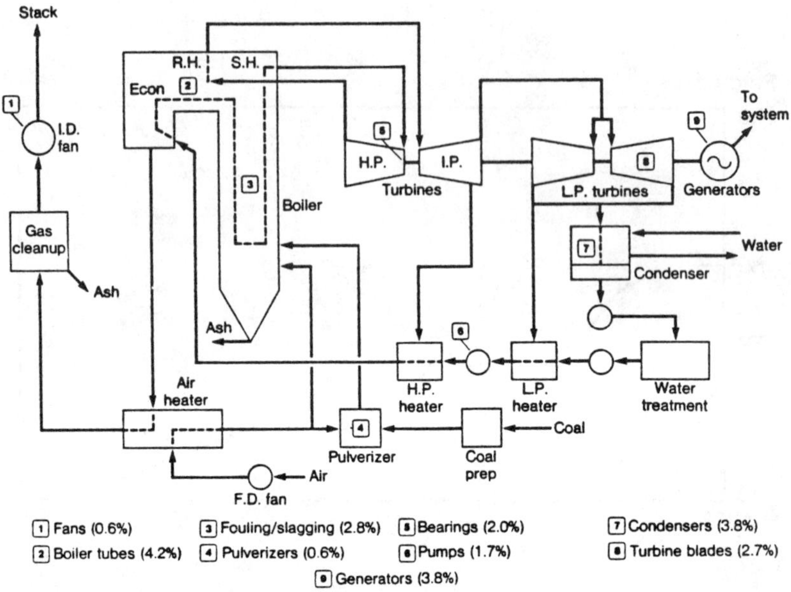

Fig. 3. Problem areas in coal plant cycle

Power Plant NDE: Approaches and Limitations

The boiler (for fossil plants) is clearly the number one contributor to plant downtime. Boiler tube damage is the major area of concern and one of the most difficult to monitor and assess, due to the large number of tubes involved and the time required for a full inspection. The turbine-generator drive train, typically more reliable than the boiler, is also of vital importance from an NDE viewpoint because of the potential catastrophic nature of any failure of the rotating system. The assessment of turbine shaft, discs and blades must be a prime target of any plant NDE program. Major failures of the generator have occurred resulting in long outages. Such incidents as retaining ring bursts, armature bar overheating or overfluxing of the magnetic stator core are some of the failure mechanisms which can result in months of downtime. Pumps and fans are of somewhat lower priority, due to the practice of installing redundancy in the system. Nevertheless, deratings and partial outages when such components fail can accumulate and significantly affect utility reserve capacity. Condensers, feedwater heaters, pulverizers and exhaust gas cleanup systems (precipitators, scrubbers, baghouses) can shutdown or derate a plant as surely as a major component and deserve the attention of preventative maintenance programs. EPRI has been studying improved NDE techniques, both on-line and off-line, in order to improve availability of these components.

Boiler NDE

A recent EPRI study (6) has evaluated utility practices in the non-destructive inspection of fossil-fired boiler tubes. The objectives of this study were to:

1. Determine the NDE tools and procedures in current use for fossil-fired boilers.

2. Identify the NDE tools found most effective for making determinations for particular types of boiler tube damage. Document the specific inspection conditions or restraints.

3. Identify areas where greater utilization of proven NDE procedures can be justified.

4. Identify types of tube failures/damage that require improved NDE tools and/or procedures for more reliable and timely assessments.

Utility experience with NDE tools for boiler tube failure/damage inspection was obtained by on-site visits and by responses to a detailed questionnaire from twenty one utilities, representing the current practices at forty-two fossil plants.

It was determined that the largest percentage of tube failures occurred in the furnace wall tube bank as Table 1 shows.

TABLE 1

DISTRIBUTION OF BOILER TUBE FAILURES ACCORDING TO LOCATION

Furnace Wall	26%
Superheater	20%
Reheater	16%
Economizer	13%
Other	20%
All	5%

At least one-quarter to one-third of the failures forcing a full outage occur in the furnace wall tube bank and reducing the number of these failures alone would result in considerable savings to the utilities. This point is particularly significant because, at least in the case of drum-type boilers, the furnace wall tubes are more accessible and amenable to both waterside and fireside inspection than are tubes in the superheater, reheater, or economizer banks (Fig. 4). Access to the waterside is readily accomplished through the feeder drums and eliminates the risk of affecting boiler tube mechanical integrity due to cutting windows for inspection access.

State-of-the-art boiler tube NDE practice is limited almost exclusively, in order of use frequency, to: (i) magnetic particle/dye penetrant, (ii) Ultrasonics (UT), (iii) visual, including fiber-optic/borescope and TV, and (iv) radiography. Boiler inspection using even these four methods is not practiced routinely but only after a forced outage has occurred. In addition, none of these state-of-the-art methods can adequately detect or size the principle boiler tube failure modes--wall thinning, damage due to overheating, or hydrogen damage. They have not been effective in preventing or reducing the number of failures causing full or partial boiler outages. In fact, at the present time there are no adequate NDE techniques for reliably detecting and sizing tube damage in fossil fired boilers.

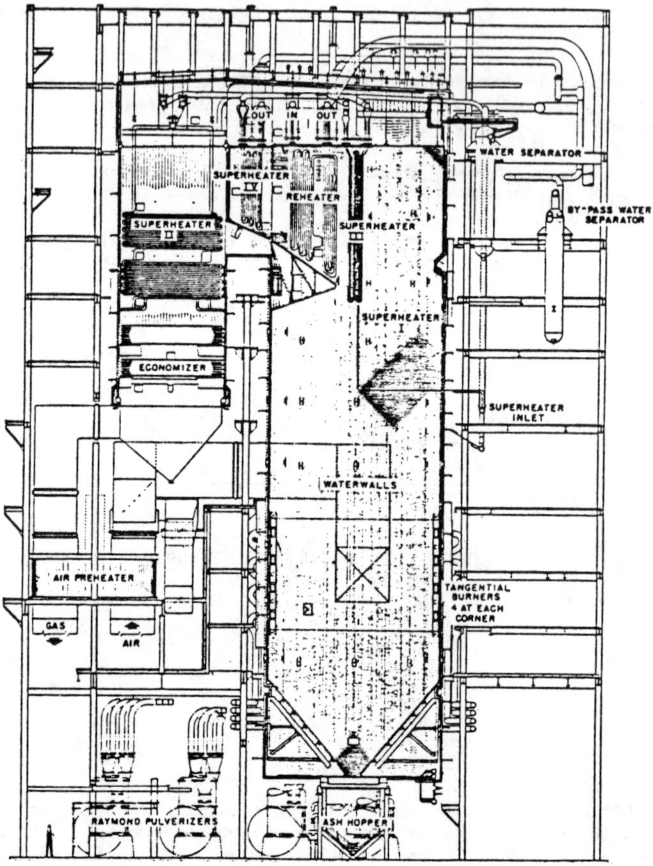

Fig. 4. Boiler Arrangement

Tube failures in utility fossil fired boilers are the single most significant cause of plant unavailability. A single tube failure in a large coal-fired unit can result in an energy cost penalty of one-million dollars even if repair time is only three days. If during the outage, similar damaged but unfailed tubes cannot be or are not located and repaired, additional failures during or after start-up can occur, substantially increasing this energy cost penalty.

There appears to be a need for developing advanced "state-of-the-art" NDE tools with the following characteristics: (a) rapid, accurate detection and sizing of boiler tube damage resulting from fireside/waterside corrosion, hydrogen damage, erosion and high temperatures (or high temperature excursions) (b) minimum interference with normal maintenance and repair activities (c) high volume coverage and (d) minimum operator interpretation.

The following inspection techniques possess some of the these capabilities and appear to be likely candidates for further development:

- Infrafred inspection has potential as a low-cost technique useful for giving rapid, high inspection coverage. While the method, by itself, is basically inadequate for sizing, it can nevertheless be used to pinpoint troublespots such as overheating, wall thinning, and excessive corrosion. The sizing problem could be addressed by eddy current and ultrasonic techniques.

- Eddy current inspection is a viable alternative to fiber-optic/ borescope and TV inspection of the waterside surfaces for pitting, wastage, changes in ID, and scale build-up. The technique is especially useful for unambiguously identifying and sizing a wide variety of defects. It is rapid, it is not limited in reach, and access from the feedwater drums eliminates the necessity of cutting windows.

- Magnetic induction type electromagnetic inspection for deep (up to 1/4-inch) penetration of the tube wall has potential for inspecting the tube wall for changes in metallurgical condition (e.g., hydrogen damage, graphitization, sigma phase formation), for assessing the fireside tube surface from the waterside, and for measuring wall thickness.

- Electromagnetic acoustic transducer (EMAT) measurement of wall thickness. Conventional UT suffers from several deficiencies: it is inadequate for detecting and sizing wastage and wall thinning; it gives variable results because of coupling problems; and it is inefficient and useful only for spot checking. EMAT, on the other hand, offers a number of improvements over conventional piezoelectric type UT transducers. It is a noncontact method which implies high inspection speed and no mechanical wear; the small size of the sensor allows inspection of tubes from the waterside and allows passage of the sensor through bends; no water or oil is required for acoustic coupling--and, EMAT gives full inspection coverage instead of spot checks of suspect troublespots.

Turbine NDE

Fractures of the turbine shaft system are few and far between, but when they happen the result is major turbine-generator damage, long outage time and possible safety implications for plant personnel. Three areas of the turbine are particularly important from the severity viewpoint: the turbine shaft, discs and blades (Fig. 5). During the last 10 years at least 14 U.S. utilities have experienced disc cracks on their turbines and this is seen a key area for improved NDE methods. Evidence of the catastrophic nature of a turbine shaft burst was given in 1974 with the fracture of the IP-LP rotor of TVA, Gallatin 2. This focused attention on methods of detecting flaws and flaw growth in turbine rotor forgings, currently performed during an inspection outage via the central bore hole of the turbine shaft. Recent turbine blade failures, particularly in the low pressure turbine, require improved methods for crack detection so that an orderly program of blade replacement can be followed.

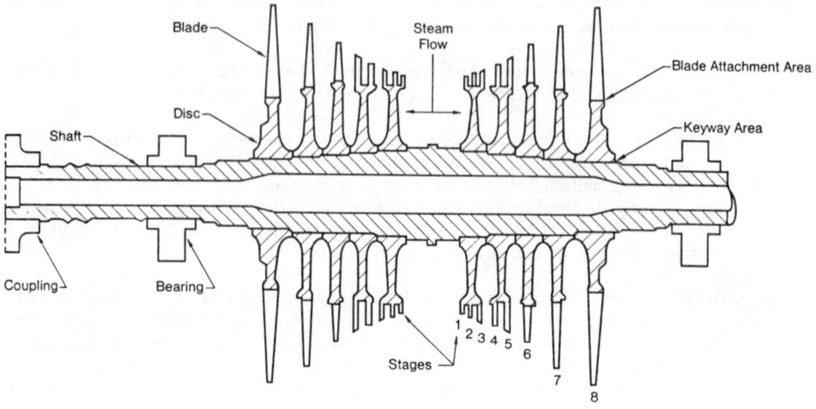

Fig. 5. Typical low pressure turbine configuration

In addition to these concerns, attention has also been given at EPRI to detecting some of the causative effects which lead to component cracking and ultimate failure. Among these effects are water induction, hard particle erosion and chemical corrosion. These areas are all the subject of on-line diagnostic monitors, currently under development.

Turbine Forgings

Research and development at EPRI has addressed the question of rotor forging integrity and has developed a computerized prediction analysis system for the quantitative evaluation of steam turbine rotor remaining life. This system, called STRAP (Steam Turbine Rotor Analysis Programs) is linked to the ANSYS structural analysis program to permit stress evaluation of Cr-Mo-V steel rotors. Using indications from NDE boresonic inspections, the program will calculate:

- Smallest crack that can produce unstable fracture.
- Critical crack size at each location of NDE concern.
- Initial crack size that can grow by low cycle fatigue to critical size.
- Potential for creep rupture and local yielding around clusters.
- Creep crack propagation.

In addition, an automated ultrasonic inspection device for turbine bores has been developed to accurately map defects in the forging.

Turbine Discs

Modern low pressure turbines are too large to forge from a single ingot. Therefore, the rotor shaft is forged with a series of diameters along the length, and separately forged discs are thermally shrunk-on the shaft. The discs are shrunk-on in sequence. Holes are drilled common to the shaft step and inner diameter of the disc and a metal pin is inserted in the hole. These holes are called keyways (the metal pins being the keys) and are used to insure that the disc won't rotate on the rotor shaft in the event of an over-speed incident. Keyways may be round, square, or other shapes such as mouse hole. Keyway cracks have been found, by non-destructive techniques, in a number of turbines, primarily nuclear units, caused apparently by stress corrosion. Inspection of these areas is carried out by ultrasonic techniques during a planned outage. Cracked rotors or discs, however, can lead to excessive vibration of the turbine generator. Ths suggests that vibration signature analysis may be applicable in some cases, to the detection of incipient cracking. The monitoring, on-line of acoustic emissions may give even earlier warning of crack initiation [(7)]. During a two-year plant monitoring project at New England Power, Brayton Point, acoustic emission sensors were placed on the bearing caps of the main turbine-generator bearings. Signals from these piezoelectric transducers appear well defined although no episodes of turbine cracking were apparent during this period. Additional acoustic emission monitoring to be carried out at TVA on the John Sevier 2 unit, will be supported by EPRI-sponsored characterization studies. These laboratory tests will accomplish the following tasks:

- Materials testing to characterize acoustic emission with rotor material, loading conditions, temperature, etc.
- Stress evaluation of large steam turbine rotors, including centrifugal and thermal stresses, particularly at startup.
- Data evaluation of the signal including spectral content, time between events and event rise time.
- Data transmission including attenuation studies and transducer mounting methods.

Turbine Blades

The failure of blades in the low-pressure sections of steam turbines is seen by EPRI as a severe and persistent problem in large power plants. The mechanisms and causes of blade failure are complex and interrelated. Turbine blades fail due to a combination of mechanical stress and the effects of the environmental to which they are subjected. A failure survey (8) indicated that the next to last blade row (L-1) of low pressure turbines had the highest incidence of failure as Fig. 6 indicates.

Non-destructive evaluation of turbine blades is usually carried out using magnetic particle inspections Typically, these inspections apply to all external surfaces including shaft, discs, blades, journals and couplings. Ultrasonic inspection of tangential entry dovetails is recommended (by one manufacturer) at 6 year intervals. The use of magnetic particle testing in the root section of blades is quite difficult because of the geometry of the region, which may cause inadequate magnetic field to be applied. Eddy current techniques, supported by finite element studies to characterize flaws, may offer improved inspection procedures for blades.

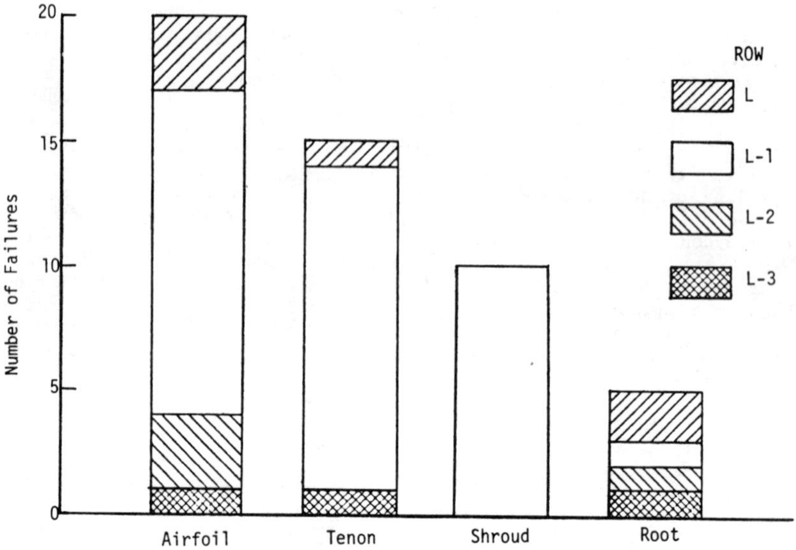

Figure 6. Location of Cracks on Blades

Generator NDE

The generator ranks closely behind the turbine in terms of impact on plant unavailability. In studying causes of generator forced outages, certain components appear to be of concern. They include retaining rings, field windings, armature windings, core and bar support systems. Inspection techniques employed routinely for generators are:

- On-line, shorted turn detector, to monitor flux patterns in the air gap.
- Core monitor, on-line, to detect the products of combustion caused by overheating of generator insulation.
- Infra-red detectors to monitor hot spots on the stator teeth. This is an off-line device which requires removal of the rotor.

Under development at EPRI are new NDE techniques to evaluate generator operational condition including:

- Fiberoptic devices to detect winding overheating.
- RF monitors to alert operators in the event of armature strand arcing.
- Torsional monitors to assess damage to generator shafts and other components in the event of system disturbances.
- Eddy current inspection methods for retaining rings.

Pumps and Fans

Boiler draft fans and boiler feed pumps can derate a power plant and in some instances shut it down. However, prior to a pump or fan failure there is often a warning signal that all is not well. This warning can take the form of excessive vibration of the pump or fan and its drive motor or turbine. Vibration signature analysis, therefore, should be a key tool in any plant predictive and preventative maintenance program. Such systems have been in operation for periods of years (7,9) and have detected incipient failure of pump or fan quite reliably. Extending this concept of monitoring component signatures, the use of acoustic emissions, which sense high frequency (into the MHz range) sounds due to micro-rupturing of metallic components, should give even earlier warning of pump or fan system failure. Studies at the Brayton Point Plant of New England Power Service Co. (10), have demonstrated the excellent potential of vibration signature analysis and acoustic emissions to provide early warning of incipient failure. This work has been further confirmed by monitoring tests on coolant pumps at the Millstone 2 Nuclear Power Station (7). At both Brayton Point and Millstone, a computer-based system was installed to monitor vibration spectra, compare spectra periodically and alert operators if significant changes occurred. One other advantage should be stated: such systems can also inform the operator that no outage is necessary for maintenance, thereby saving unnecessary downtime.

New Developments

Non-destructive evaluation techniques are utilized for confirming the condition of a piece of equipment, whether it be flawed, in which case repair or replacement is indicated, or whether it be sound, thereby confirming its suitability for further service. If this NDE can be carried out on-line, then the optimum situation is arrived at, since power is being produced at the same time as the diagnosis is being made. The obvious beneficial effect this has on plant availability, is concentrating efforts at EPRI on other NDE monitors, to diagnose adverse machine conditions previously not felt to be readily identifiable on-line.

Corrodent Detector

This on-line device is being developed to detect the presence of corrodent deposits (sodium or calcium) on turbine discs and blades. The proposed system is shown in Figure 7. The interaction of deuterium ions with the tritium target generates both neutrons and, 180° away in direction, alpha particles. An alpha detector determines the presence of a neutron headed in the appropriate direction. Approximately 10^{7} neutrons/sec. are generated which impact the sodium or calcium deposits producing gamma particles. The gamma particles impact a sodium iodide detector, producing pulses of light. The number of pulses/sec. (counts) determines the extent of the blade or disc corrosion. Appropriately calibrated, this device can indicate when a turbine needs to be removed from service due to excessive corrosion. The impact of this on the control of stress corrosion cracking and on the elimination of catastrophic turbine failure is the prime motivator for this work.

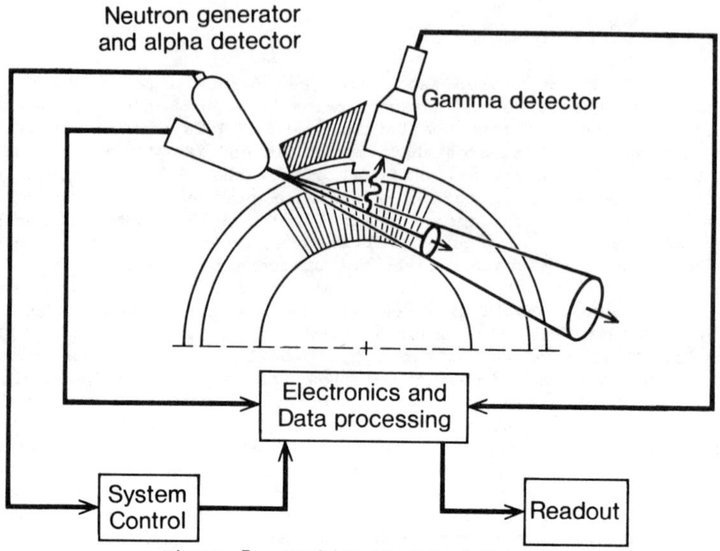

Figure 7. Turbine Corrodent Detector

Bearings

Turbine-generator bearing and lubrication problems account for a very significant proportion of plant forced outage hours. Often, the first indication of a bearing failure is a sudden rise in vibration level as monitored by the turbine shaft-riding vibration detector. In many cases the damage to the bearing, and particularly to the babbitt liner, has already been done. An ongoing study at EPRI (11) has established that contamination in the oil system is the most common cause of bearing failure. Accordingly, one diagnostic monitor being studied will detect the presence of water and particulates in lubricating oils of large units. An alternative approach recognizes the presence of tin in bearing oil as the babbitt liner deteriorates. Sensitive X-ray fluorescent devices can detect concentrations of tin down to 5 ppm, to give early warning of impending bearing failures.

The EPRI NDE Center

The Electric Power Research Institute has established an NDE Center in Charlotte, N.C.(12). The intent of this center is to:

(1) Provide for technology transfer from EPRI research projects to the utility industry.

(2) Provide training in NDE tools and techniques so that a trained corps of NDE specialists can be available, as required, to solve utility problems.

(3) Develop new approaches to NDE through feasibility studies and through some in-house research effort. This will not duplicate the efforts ongoing under separate EPRI research contracts.

As new techniques, described in this paper, are developed, effective transfer of the technology will employ the resources of this new facility.

Conclusions

The utility industry will be under increased pressure to improve plant availability during the remainder of this century. The use of improved NDE techniques offers an opportunity to reduce downtime by anticipating component failure and by optimizing preventative maintenance procedures. This particularly applies to the on-line identification of incipient failure by the use of suitable diagnostic monitoring techniques. The major contributor to fossil plant unavailability, the boiler, is seen to be deficient in terms of inspection tools and procedures. This particularly applies to the evaluation of tube damage and tube leaks. Much future attention is expected to be focused on the turbine-generator, in view of the potential seriousness of a major failure of the rotating shaft system. The on-line techniques of vibration signature analysis and of acoustic emission monitoring are seen to be a prominent part of any future diagnostic system for component cracking. Much material characterization work, relating acoustic emissions to fracture mechanics, is seen to be necessary as a precursor to a fully calibrated and operational acoustic emission monitor for turbine-generator shaft systems.

Acknowledgements

The research studies described in this paper have been contracted to a number of manufacturers, consultants and utilities under EPRI overall management. The contributions of these contractors, if not specifically identified in the text, is hereby acknowledged. The authors would also like to acknowledge the following project managers at EPRI who are involved in the management of various aspects of this work: Floyd Gelhaus, John Dimmer, Jack Parkes, Mike Kolar, Gordon Shugars and Ken Lehner.

References

1. Electric Power Supply and Demand for the Contiguous United States 1980-89, Economic Regulatory Administration, DOE, June 1980.

2. Gordon D. Friedlander, "Fossil-boiler sales rise--then plummet," Electrical World, Dec. 15, 1979.

3. Recommendations of Regulatory Actions to Promote Power Plant Productivity Improvements, National Regulatory Research Institute, Columbus, Ohio, Aug. 1979.

4. The National Power Grid Study, Economic Regulatory Adminstration, Office of Utility Systems, DOE/ERA 0056-1, Final Report, Jan. 1980.

5. Report on equipment availability for the 10-year period: 1968-77, Equipment Availability Task Force of Prime Movers Committee, EEI, July 1979.

6. Assessment of Current Non-Destructive Inspection Techniques for Determining the Type, Location and Extent of Fossil Fired Boiler Tube Damage, Final Report for EPRI project RP1266-5, Reluxtrol Co., San Francisco, CA, Nov. 1980.

7. A. F. Armor and H. G. Shugars, "On-Line Diagnostic Monitoring of Power Plants, presented at EPRI Power Plant Maintenance and Maintainability Workshop, Chicago, Illinois, Aug. 6, 1980.

8. Low-Pressure Steam Turbine Blade Failures, Workshop Proceedings, Electric Power Research Institute, WS 78-114, July 1980.

9. M. J. Nichols, "One Approach to Preventative Maintenance," presented at EPRI Power Plant Maintenance and Maintainability Workshop, Chicago, Illinois, Aug. 6, 1980.

10. P. Pekrul, et al. The Application of Computerized Vibrations and Acoustic Emission Monitoring at Fossil Power Stations, American Power Conference, Chicago, April 23, 1980.

11. Failure Cause Analysis: Turbine Bearing Systems, Draft Report for Phase I of EPRI project RP1265-3, Franklin Research Center, Philadelphia, PA, Feb. 1980.

12. G. J. Dau, "Role of the EPRI Nondestructive Evaluation Center," American Power Conference, Chicago, April 21-23, 1980.

RELIABILITY STRATEGIES FOR ADVANCED COAL CONVERSION SYSTEMS*

William A. Ellingson

Materials Science Division, Argonne National Laboratory
Argonne, Illinois

Increased usage of U.S. coal reserves to reduce the U.S. dependence on foreign oil will involve both direct combustion processes and systems based on atmospheric and pressurized fluidized-bed combustion, as well as synthetic-fuel (synfuel) production, fuel cells, magnetohydrodynamics (MHD), and other advanced technologies. Nondestructive evaluation (NDE) methods for comprehensive preservice and periodic inspection and on-line condition monitoring of components are becoming increasingly important to help assure reliable plant operation, in view of the complex mechanical designs of these systems and the growing vulnerability of industry and society to disruptions in energy supply. This paper reviews some of the NDE methods being developed for coal processes, with emphasis on plant reliability.

*Work supported by the U.S. Department of Energy.

Introduction

Coal is the nation's (and the world's) most abundant fossil fuel. Recoverable U.S. reserves amount to 6000 quadrillion Btu* (quads), which is part of a total domestic resource of about 80,000 quads and a world resource crudely estimated at about 300,000 quads. Presently, the U.S. consumes about 14 quads each year, or less than 0.3 percent of domestic recoverable coal reserves. In contrast, the U.S. extracts each year almost 10 percent of its 420-quad recoverable reserves of oil and natural gas (1).

The substitution of coal for natural gas and oil on a large scale, either through direct combustion or through synthetic coal-derived fuels, would seem to represent part of the solution to the nation's energy problems. However, even with a substantial initial price advantage, coal-based energy systems must maintain a high degree of reliability of operation in order to compete with oil and natural gas. Reliability of coal processing systems can be improved through proper design, materials selection, and implementation of appropriately rigorous nondestructive evaluation (NDE) programs. For example, in a recent reliability study, Hammelman et al. (2) recommended periodic inspection during shutdowns to improve the reliability of an atmospheric fluidized-bed boiler. This would include (a) checking the integrity of such critical components as heat exchangers, cyclone separators, and air distributors, and (b) providing a monitoring system for early detection of failures of key components such as fuel-oil pumps, air-lock valves, and air compressors. These recommendations encompass the two main aspects of NDE: Flaw detection and characterization, and on-line condition monitoring.

Coal utilization systems differ in process conditions, and this is reflected in the demands placed on materials of construction. Direct-combustion systems, for example, are less demanding on materials than MHD-based systems. Thus, materials presently used or planned for use in coal processes range from carbon steel through exotic alloys and cast stainless steels to structural ceramics and refractory concretes. Because of this wide variation in material types, comprehensive inspection programs will require the use of several NDE methods. Canonico et al. (3) have outlined the probable effects of various types of failure on plant availability, as shown in Table I. Clearly, the prediction of failure events through a comprehensive NDE program, which would encompass not only flaw detection but also on-line monitoring, would provide considerable economic benefits.

Since the reliability of a complete plant is dependent upon the reliability of each component, the potential impact of NDE on coal systems will be discussed in terms of four major components: (1) valves used in coal feed systems, (2) coal transfer lines, (3) pressure vessels, and (4) ceramic heat exchangers.

*1 Btu = 1.055×10^3 J.

Table I. Effect of Failure of Various Components on Plant Availability*

Component	Failure	Results
Valves and valve seats	Corrosion, erosion, cracking	Loss of efficiency, mild upset
Pressure, temperature and flow controllers	Metering errors, corrosion	Loss of efficiency
Lines	Fouling, plugging, cracking, stress rupture	Major upset
Compressors - reciprocating	Fatigue, erosion, corrosion, explosion	Total shutdown
Pumps - centrifugal	Wear, leakage	Mild upset
Filters	Plugging	Loss of efficiency
Absorption columns	Corrosion	Major shutdown
Steam generators	Boiler-tube failure, plugging	Major shutdown
Catalyst trays	Plugging from entrained solids	Loss of efficiency
Pressure vessels and piping	Cracking	Major outage
Structural failures	Corrosion, overload, vibration-fatigue, thermal distortions	Major upset

*From Ref. 3.

Coal Feed Valves

Coal is usually fed into high-pressure coal reactor vessels in one of two ways: (1) through a lock-hopper arrangement employing large lock-hopper valves, or (2) through slurry pumps where pulverized coal is slurried (often with water). The primary materials reliability problem is erosive wear of feed-system components induced by the abrasive coal particles. Because of excessive wear problems with slurry pumps, the lock-hopper arrangement is currently preferred; hence, this discussion will emphasize work on valves. Table II gives operating conditions for the various types of valves used in lock-hopper systems. Type I and II valves are being developed specifically for coal feed systems, as discussed by Gardner (4).

Table II. Service Conditions* for Lock-hopper Valves

	Valve Type			
	I	II	III	IV
Function	Feed valves to loading or injection hopper	Feed valves to gasifier	Char and ash removal valves	Slurry discharge outlet of ash lock hopper
Temperature (°C)				
Bulk Media	180	425 (initial) then 300	1000	300
Interface	75	425	425	300
Working Media				
Type	Any grade of coal available in U.S., and dolomite			Char and dolomite products with steam and/or process gas
Particle Size	Coal dust to 25 mm, dolomite to 12 mm			Ash-fines to 10 μm, fines to 10 mesh
Corrosive Environment	$CH_4, CO, H_2, CO_2, H_2S, SO_2, H_2O, NH_3$			CaO, Na_2O, K_2O, P_2O_5

*The following conditions are the same for all four valve types:
Pressure
Internal -- 0 to 110 atm (pressure differential in both directions)
External -- 1 atm
Maximum allowable leakage
Internal -- <10 SCFM/in. of pipe diameter
External -- <0.1 SCFM
Rate of temp. change during start-up and shutdown -- 50°C/h

Some current designs for Type I, II, and IV valves and most designs for Type III valves use some form of structural ceramic (5). Evans et al. (6) have shown that critical flaw sizes in structural ceramic components are in the neighborhood of 50-100 µm. This small flaw size poses a problem for preservice inspection. Recent work by Kupperman et al. (7) has shown that several NDE inspection methods, including holographic interferometry, acoustic microscopy, infrared imaging, and high-frequency ultrasonics, hold potential for inspection of structural ceramic components. Although the majority of this work has focused on the inspection of SiC heat-exchanger tubes (see later section), these NDE developments should also be applicable to (>99%) high-alumina materials being investigated for ceramic valves (5).

Shack et al. (8) have discussed development of noninvasive* passive acoustic systems for on-line monitoring of valve leakage. Gas leakage through valves causes a drop in pressure in the reactor vessel as the coal-feed lock hopper cycles between atmospheric pressure and the reactor pressure, which is typically 10-50 atm. Gas-valve leaks have acoustic frequency spectra between 175 and 750 kHz. Shack et al. found that quantification of leakage rate seems possible if the differential pressure across the leaking valve producers choked-flow conditions (i.e., if local sonic velocity is reached). Figure 1 shows results obtained from flow-acoustics tests conducted on several orifice configurations as well as from a full-scale valve tested at the Morgantown Energy Technology Center Valve Test Facility. Field tests carried out at operating coal conversion pilot plants have shown that mechanical background noise energies are contained in the region below 150 kHz; thus, the frequency window above 150 kHz has a high signal-to-noise ratio.

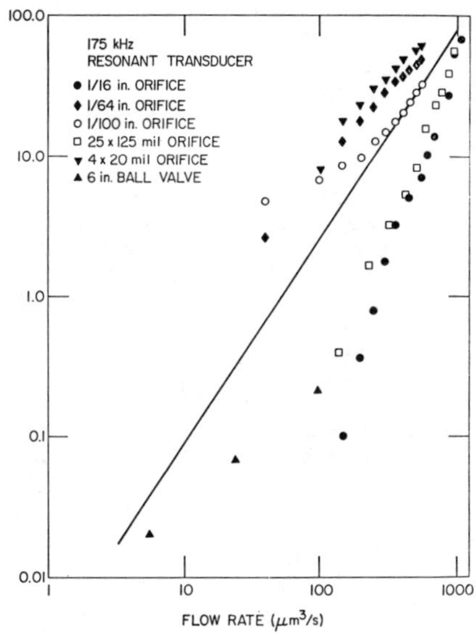

Fig. 1 - Acoustic power levels detected as a function of gas flow rate through various well-defined orifices and an actual leaking valve. All tests used air as the gas.

*I.e., not requiring penetration of the pressure boundary.

Studies on detection of internal fluid leakage past slurry blockage valves have also been reported (9). This work has shown that acoustic signal levels for fluid leaks are contained in a frequency spectrum below 80 kHz; several liquids (salt water, oil, coal-water slurries) yield similar frequency content when exiting a leakage channel (see Fig. 2). Quantification of liquid leakage rates as measured by passive acoustic systems may not be possible (10), and low mass-flow measuring instruments may be needed for quantification of liquid leaks, with passive acoustic systems in series to act only as flow/no-flow indicators.

Coal Transfer Lines

Regardless of the coal process, coal in some form needs to be transported between various vessels. A recurring materials reliability problem in coal utilization systems is erosive wear, and little is presently known about predicting erosive wear produced by 2-phase (solid-gas, solid-liquid) or 3-phase (solid-liquid-gas) flow. Prior experience and existing erosive wear theory suggest that erosive wear will be greatest at bends, elbows, and turbulent-flow regions, e.g., downstream of a protruding internal weld bead. Transport lines in coal process plants are typically of metal, either unlined or with an internal lining of refractory concrete, depending on the temperature of the material being transported. As indicated in Table I, blowout of a transfer line can cause a major upset or outage; thus, plant reliability is highly dependent upon maintaining the integrity of these lines. Recent advances have been made in NDE methods for continuous monitoring as well as periodic inspection of transfer lines.

Unlined

Youngdahl and Ellingson (11) have reported on the development of ultrasonic high-temperature systems employing acoustic waveguides for continuously monitoring erosive wear of unlined metallic transfer lines. The acoustic waveguides should be of a material as similar as possible to that being monitored; otherwise, problems involving the weldments are likely to occur, as the stud weld method used to acoustically couple the waveguide to the pressure boundary creates a shallow heat-affected zone (HAZ). More recent work by Youngdahl et al. (12) has shown that very long (>1 m) waveguides can be produced by inertial welding of two metals. With this method, only a short section of material similar to the pipe (often stainless steel, which has a high acoustic attenuation) is needed near the attachment point, and less acoustically lossy material can be used for most of the waveguide length. Long waveguides permit measurement of erosive wear at sites that are difficult to reach by more conventional methods. Figure 3 shows a current waveguide design with convective cooling fins; Fig. 4 shows several such waveguides installed on a 50 mm, Schedule XXS 321 stainless steel line at the Solvent Refined Coal coal-liquefaction pilot plant. Erosive-wear monitoring systems of this type add to the reliability of plant operation by providing information that allows planning of shutdowns; because such systems allow automatic remote data acquisition, they are also safer to use than manual inspection methods.

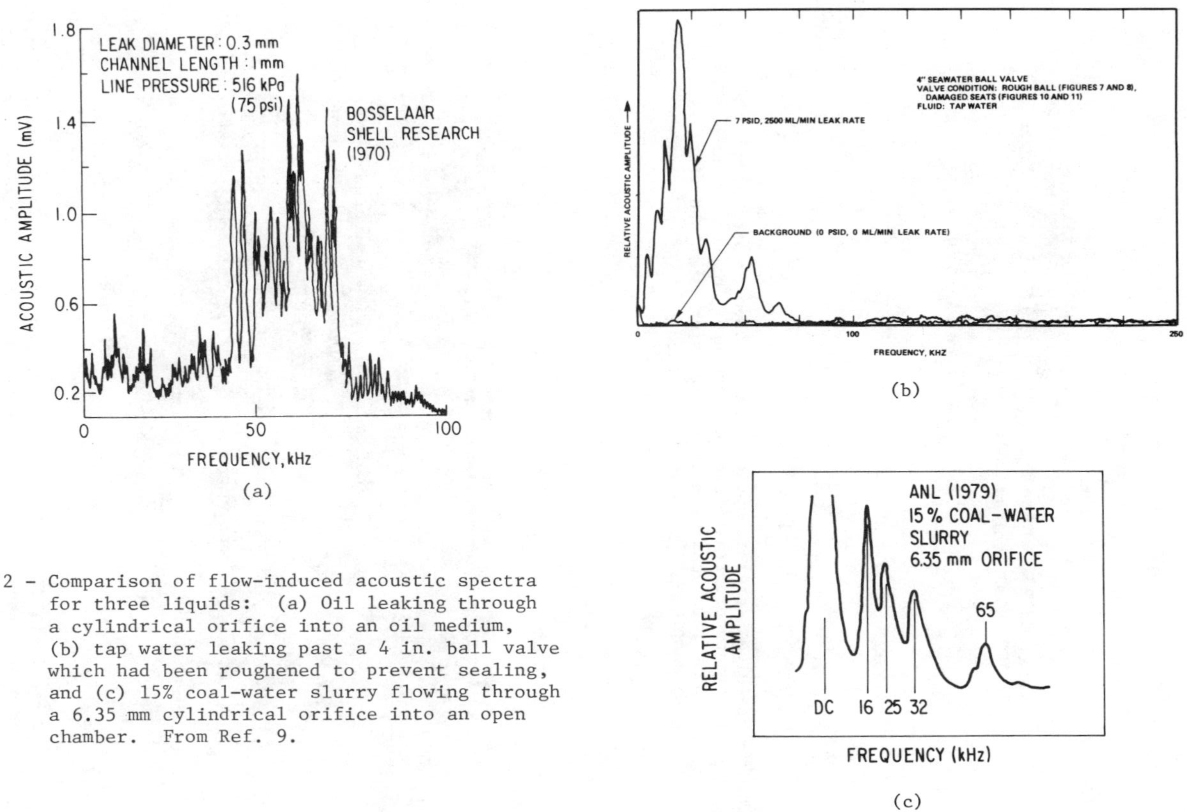

Fig. 2 - Comparison of flow-induced acoustic spectra for three liquids: (a) Oil leaking through a cylindrical orifice into an oil medium, (b) tap water leaking past a 4 in. ball valve which had been roughened to prevent sealing, and (c) 15% coal-water slurry flowing through a 6.35 mm cylindrical orifice into an open chamber. From Ref. 9.

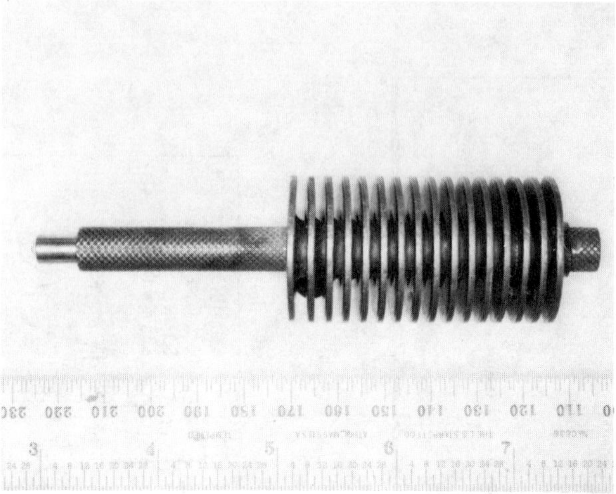

Fig. 3 - Stainless steel acoustic waveguide used as part of a high-temperature ultrasonic erosion scanner for continuous monitoring of wall thickness of metallic transfer lines and fittings. Cooling fins aid in heat dissipation.

Fig. 4 - Typical field installation of a high-temperature ultrasonic erosion scanner. Each waveguide site monitors wall thickness at a fixed position. Shown here are six sites at one 90° bend, two sites at the other 90° bend, and one point downstream of a butt weld.

Internally Lined

The inside of metal piping must be lined with refractory concrete whenever the fluid-stream temperature would cause the exposed metal to have undesirable mechanical properties. From an NDE standpoint, such layered structures present problems. Erosive wear of internal refractory linings cannot be mapped by pulse-echo ultrasonic methods because of the excessive acoustic-wave scattering caused by the lining, as well as the difficulty in acoustically coupling the metal outer wall to the internal lining. Infrared scanning has been shown (13) to be heavily dependent upon thermal contact resistance between the refractory lining and the steel outer shell, which makes quantitative measurements difficult. However, at least two NDE methods show promise of allowing inspection of such layered structures. Gamma radiography using ^{60}Co has been shown (14) to be capable of defining the internal bore of 10 cm ID, 50 cm OD piping systems. Another potentially powerful radiographic method has recently been described by Ellinger et al. (15). Using the fundamental principles of tomographic imaging (16) Ellinger has shown that cross sections of portland cement concrete cylinders can be imaged with such clarity that aggregate size distributions can be obtained (see Fig. 5). Assuming that tomographic methods can be extended to large-diameter piping systems, such images should allow complete in-situ profiling of internally refractory-lined sections. As is the case for unlined metallic lines, materials performance data as well as personnel safety would be greatly enhanced through routine employment of NDE methods to monitor refractory-lined piping.

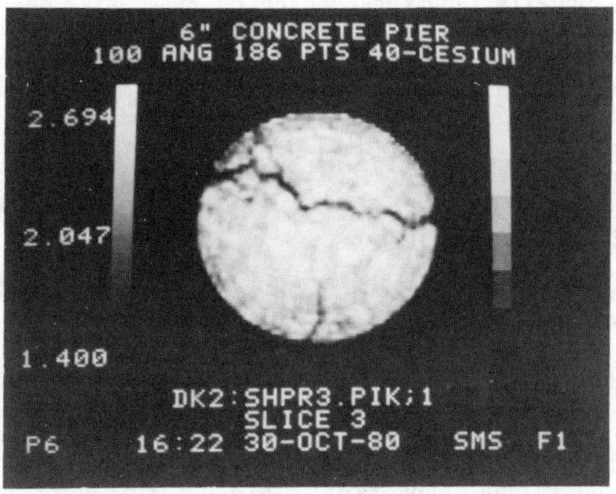

Fig. 5 - Tomographic image of a section through a 15 cm diameter portland cement concrete cylinder. A 1.6 Curie ^{137}Cs source was used to obtain this tomographic image [from Ellinger (16)].

Pressure Vessels

Because of the extremely aggressive environments produced by coal conversion processes, i.e., gases containing CH_4, H_2, H_2O, H_2S, NH_3, CO, and CO_2 at high temperatures (500-1100°C) and pressures (10-70 atm), great demands are placed on NDE for assurance of pressure vessel integrity.

Pressure Vessel Shells

Canonico et al. (3) have extensively discussed the American Society of Mechanical Engineers (ASME) boiler and pressure-vessel code with reference to coal-conversion pressure vessels and have suggested that for preservice inspection, Section VIII, Divisions I and II of the ASME code might provide acceptance criteria for pressure-vessel shells. However, this still leaves open the question of applicable codes for periodic inspection of coal-conversion pressure vessels, as well as an acceptance code for nonmetallic components. It should be kept in mind that some of the vessels proposed for synfuel plants are likely to be larger in diameter, height, and weight (by factors of 3 to 4) than any pressure vessels that have previously been fabricated (3). In addition, available information relating flaw size, stress, and temperature to the failure of candidate materials is minimal.

The large size of the proposed vessels makes field fabrication a necessity, since vessels to be transported by rail are limited to a diameter of 5 m ($\sim$ 14 ft) and a weight of 7.2×10^4 kg (800 tons) (17). Field erection will require NDE methods for inspection of large, thick-section (15-30 cm) welds. Shop-fabricated thick-walled vessels can be inspected with high-energy x-ray systems that utilize linear accelerators. Planar flaws, however, are not readily detectable by x-ray methods and supplementary ultrasonic methods are usually required. Moreover, flaw-size data obtained on thick pressure vessels by ultrasonic and radiographic methods have often been in disagreement (18), indicating that reliable inspection methods for these vessels have yet to be fully developed. Adequate field inspection will probably require the use of a mobile linear accelerator, together with advanced ultrasonic inspection methods. Techniques are also needed to measure in-place mechanical properties, e.g., residual stresses.

Pressure Vessel Linings

Pressure vessel shells of coal conversion plants operating at high temperatures must be protected by ceramic (refractory) linings. Verification of the integrity of these linings is essential for reliable operation; spalling of internal refractory linings can produce local hot spots, with potentially devastating effects on quenched and tempered regions of the shell near weldments. Bakker (19) has discussed results of research directed towards improving the mechanical reliability of thick (15-30 cm) linings of the type to be used in coal conversion systems. This work confirmed that the majority of large-crack damage to linings occurs during initial dry-out, and that cracking depends to a large extent on the heating and cooling rate used during initial firing.

Relatively little development work has been done on NDE methods for inspection of refractory concrete linings. Although infrared scanning cameras provide a qualitative means of detecting hot spots on pressure vessel shells, reliable operation requires knowledge of the refractory lining status prior to start-up. Radiography and acoustic emission detection have been discussed (20) as possible quantitative NDE methods for measuring lining integrity.

When the vessel internals are accessible, ^{60}Co radiography has been shown to be capable of detecting internal cracks on vessels with up to 50 cm of refractory lining backed by up to 9 cm of steel (14). The location and orientation of hangers, which have an important influence on lining performance (19), can also be detected by this method.

Acoustic emission detection appears to be the most promising alternate NDE method for cases where access to the vessel is not possible prior to operation (20). Refractory concrete materials are extremely heterogeneous structures and, as such, are prolific acoustic emitters when under stress. Unfortunately, this complex structure also makes detailed analysis very difficult, since it gives rise to considerable acoustic reflection and scattering. For these reasons, only comparative, qualitative tests have been conducted to date.

Detection of acoustic emissions during initial firing requires that acoustic transducers be thermally decoupled, since acoustic sensing elements cannot be used reliably for extended time periods at temperatures above 300°C. Tests at Babcock and Wilcox (21) have shown that carbon steel rods can be used as waveguides; work at Argonne has used pressure-coupled quartz waveguides.

Recent results obtained from tests on 22.9 cm (9 in.)-thick panels suggest that damage occurs during the heat-up part of the schedule for rapidly heated specimens, but during the cool-down period for slowly heated specimens. Figure 6 shows results obtained from two identically cast panels. Panel I was subjected to a slow heating schedule followed by a fast cooling schedule; Panel II was subjected to a fast heating schedule following by a fast cooling schedule similar to that used for Panel I. The total acoustic ringdown counts* from a 175 kHz transducer and a broadband (125 kHz-2 MHz) transducer are shown. Note that for Panel I, as soon as cool-down began, acoustic activity increased significantly. For Panel II, significant acoustic activity was detected during heat-up, but when cool-down began, little additional activity was noted.

A comparison of ringdown counts with modulus of rupture has shown that higher counts are correlated with lower modulus values (20). In an experiment at Babcock and Wilcox (21), a thick refractory panel, instrumented with acoustic-emission sensors, was rapidly heated with consequent explosive spalling. Figure 7 shows that prior to spalling, acoustic activity increased significantly; this increase can perhaps be considered a precursor to the failure. Such information, if quantifiable relative to refractory mechanical properties, should allow on-line surveillance of refractory status during initial firing and plant start-up and shutdown periods.

*Ringdown count is an acoustic-emission parameter which is a measure of the transducer output voltage above an electronic voltage threshold.

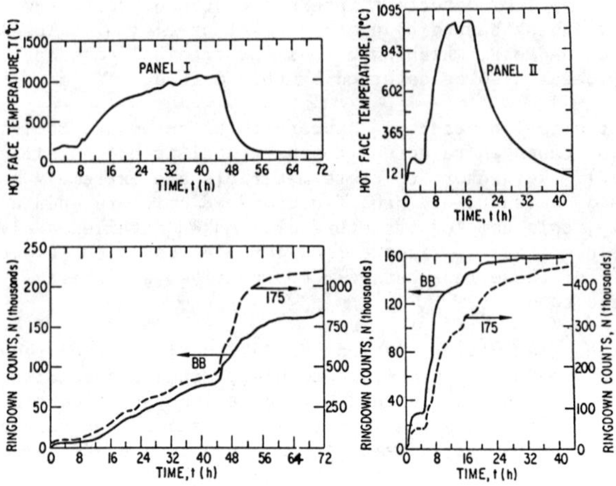

Fig. 6 - Comparison of ringdown counts as a function of time during the firing schedule for two high-alumina (95% Al_2O_3) panels heated at different rates. Top: hot-face temperature profiles. Bottom: ringdown counts from 175 kHz and broadband transducers.

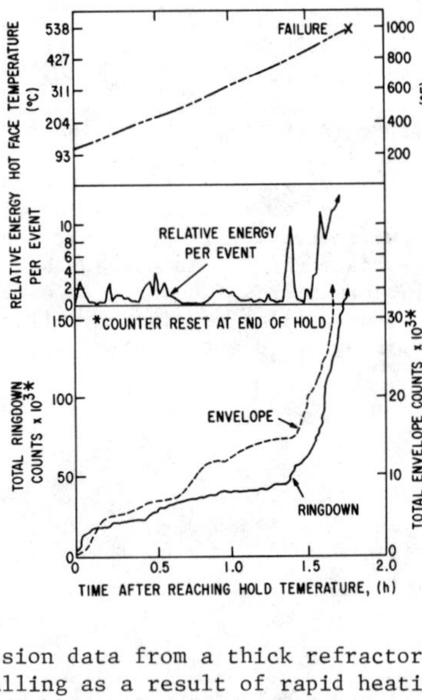

Fig. 7 - Acoustic-emission data from a thick refractory panel that underwent explosive spalling as a result of rapid heating. From Williams and Anderson (21).

Ceramic Heat Exchangers

In recent years, significant progress has been made in the use of ceramics for structural applications. Ceramic heat-exchanger components are of particular interest because they are lighter than similar metallic components, have good corrosion resistance and high-temperature strength, and can be fabricated from inexpensive and abundant elements. Since they can withstand higher temperatures than metals (> 1000°C), use of ceramics can lead to more efficient energy-conversion systems. Silicon carbide (SiC), for example, is currently being considered for heat-exchanger tubing because of its excellent thermal-shock resistance, low coefficient of expansion, high thermal conductivity, and strength at high temperature.

The reliable use of ceramics as structural components, however, requires effective failure-prediction methods. The flaws most deleterious to the lifetime of SiC components are cracks and porosity. Many ceramic fracture origins have been shown to be adjacent to the surface (22), suggesting that surface cracks are an important cause of failures. The critical size of cracks leading to fracture can be related to microstructural features such as grain size, and can be relatively small (an order of magnitude or more smaller than in comparable metallic parts). Thus, NDE techniques that are satisfactory for metals may not be for ceramics.

Inspection methods for ceramic heat exchangers must address the problems of flaw detection and characterization in (a) long tube sections (both inside and outside surfaces) and (b) ceramic butt joints. Kupperman et al. (7) have suggested several NDE methods that might be used for complete inspection of SiC heat-exchanger components and have estimated the spatial resolution limits of each, as shown in Table III. Although considerable progress has been made in the development of reliable NDE methods for ceramic heat exchangers, additional efforts are still needed in this area.

Table III. Estimated Minimum Flaw Size or Material Property Change Detectable in SiC Heat-exchanger Tubing by Various Techniques*

NDE Method	Minimum Detectable Flaw Size or Property Change
Acoustic Microscopy	100 μm or less†
Conventional Ultrasonic Testing	∿100 μm depth
Dye-enhanced Radiography	0.5% of sample thickness (surface flaws only)
Holographic Interferometry	100 μm (with special lens techniques)
Infrared Scanning	2% variation in heat-transport properties

*From Ref. 7.
†Flaw characterization also possible with this method.

Summary

Owing to the large capital investment and high operating costs involved, coal conversion and utilization processes must have a high level of reliability. The extensive use of nondestructive evaluation methods for preservice and periodic inspection as well as on-line condition monitoring should allow early failure prediction and thus provide a substantial payoff in terms of reduced downtime. Because of the wide range of ceramic and metallic materials used in coal processes, several advanced nondestructive evaluation methods will be required. Research programs are under way to address these critical issues.

References

(1) National Academy of Sciences, Energy in Transition 1985-2010, Final Report of the Committee on Nuclear and Alternative Energy Systems (1979).

(2) J.E. Hammelman, R.E. Land, J.A. Simmons, and P.J. Wood, Reliability Analysis of the Georgetown University Atmospheric Fluidized Bed Combustor, Final Report submitted to the U.S. Department of Energy (1979).

(3) D.A. Canonico, R.H. Cooper, B.E. Foster, R.W. McClung, R.K. Nansted, G.C. Robinson, and G.M. Slaughter, Assessment of Materials Technology of Pressure Vessels and Piping for Coal Conversion Systems, Oak Ridge National Laboratory Report ORNL-5238 (1978).

(4) J.T. Gardner, Program Rationale for Solids Throttling Applications in Coal Conversion and Utilization Processes, Morgantown Energy Technology Center Report METC/CR-79/14 (1979).

(5) C.F. Harper, "A Ceramic Spring for High-temperature Valves," Journal of Materials for Energy Systems, 1(1) (1979) pp. 62-67.

(6) A.G. Evans, G.S. Kino, P.T. Khuri-Yakub, and B.R. Tittman, "Failure Prediction in Structural Ceramics," Materials Evaluation, 35(4) (1977) pp. 85-96.

(7) D.S. Kupperman, D. Yuhas, W.D. Deininger and C. Sciammarella, Nondestructive Evaluation Techniques for Silicon Carbide Heat-Exchanger Tubes, Second Annual Report, Oct. 1978-Sept. 1979, Argonne National Laboratory Report ANL-79-103.

(8) W.J. Shack, W.A. Ellingson, and C.A. Youngdahl, "Development of a Noninvasive Acoustic Leak Detection System for Large High Pressure Gas Valves," pp. 391-399 in Instrumentation in the Aerospace Industry - Vol. 25/Advances in Test Measurement - Vol. 16, Part Two (Proceedings of the 25th International Instrumentation Symposium), Instrument Society of America, Anaheim, CA, 1979.

(9) W.A. Ellingson, Materials Technology for Coal Conversion Processes, Nineteenth Report, July-Dec. 1979, Argonne National Laboratory Report ANL-80-12.

(10) W.A. Ellingson, Materials Technology for Coal Conversion Processes, Progress Report for Period April-June 1980, Argonne National Laboratory Report ANL-80-93.

(11) C.A. Youngdahl and W.A. Ellingson, "Development of Ultrasonic Techniques for Remote Monitoring of Erosive Wear in Coal-Conversion Systems," pp. 305-310 in 1978 Ultrasonic Symposium Proceedings, J. deKlerk and B.R. McAvoy, eds.; IEEE, Cat. No. 78CH1344-ISU, 1978.

(12) W.A. Ellingson, Materials Technology for Coal Conversion Processes, Progress Report for July-Sept. 1980, Argonne National Laboratory Report ANL-80-118 (in press).

(13) C.K. Hsieh, W.A. Ellingson and K.C. Su, Theoretical Evaluation of Thermal Imaging for Detection of Erosive Wear of Internally Refractory-lined Transfer Lines, Argonne National Laboratory Report ANL-78-82 (1978).

(14) N.P. Lapinski, Gamma Radiography of Refractory-lined Vessels and Components, Argonne National Laboratory Report ANL-78-43 (1978).

(15) H. Ellinger, I.L. Morgan, R. Klinksiek and F. Hopkins, "Tomographic Analysis of Structural Materials," Proceedings of the Society of Photo-Optical Instrumentation Engineers, 124 (1979) pp. 179-186.

(16) W.A. Ellingson and H. Berger, "Three-dimensional Radiographic Imaging," pp. 1-38 in Research Techniques in Nondestructive Testing, Vol. IV, R.S. Sharpe, ed.; Academic Press, New York, 1980.

(17) P.B. Probert, "Design and Fabrication of the BI-GAS Reactor," pp. 51-59 in Proceedings of the Fifth Synthetic Pipeline Gas Symposium, American Gas Association, Cat. No. L51173, 1973.

(18) W.P. Johnson, S.A. Bush and R.R. Maccary, "Augmented Scope of the 1974 ASME Section XI Code, Inservice Inspection of Nuclear Power Plant Components," pp. 29-36 in Proceedings of the Conference on Periodic Inspection of Pressurized Components, Conference Publication 8, Institution of Mechanical Engineers, London, 1974.

(19) W.T. Bakker, "Refractory Applications in Gasifiers," pp. 447-463 in Proceedings of the Ninth Synthetic Pipeline Gas Symposium, American Gas Association, Cat. No. L51177, 1977.

(20) W.A. Ellingson, "Advances in Nondestructive Evaluation Methods for Inspection of Refractory Concretes," to be published in the Proceedings of the American Concrete Institute Symposium on Premature Failure in Monolithic Refractory Constructions: Preventative and Remedial Treatment, Las Vegas, NV, March 1980.

(21) R.W. Williams and E.M. Anderson, "Acoustic Emission Monitoring of Degradation in Monolithic Refractories," paper presented at Conference on Environmental Degradation of Engineering Materials, Virginia Polytechnic Institute, Blacksburg, VA, 1977.

(22) F.F. Lange, "Structural Ceramic Materials Under Development," paper 75-GT-107 presented at Gas Turbine Conference and Products Show, Houston, TX, March 1974.

NEW TECHNIQUES FOR QUANTITATIVE NDE*

R. B. Thompson and D. O. Thompson
Ames Laboratory, Iowa State University
Ames, Iowa 50011

Retirement-For-Cause, or other strategies for structural component utilization based on the expected lives of individual parts, requires nondestructive measurement techniques to evaluate flaw severity. Ideally, a technique should estimate failure related flaw parameters as accurately as possible, as well as the variances in these estimates. In this paper, the development of such a capability will be discussed for the case of ultrasonics. A review of the basic methodology used to establish such capabilities will be presented, followed by a detailed discussion of specific techniques which have been developed.

*This work was sponsored by the Center for Advanced NDE, operated by the Rockwell International Science Center for the Defense Advanced Projects Agency and the Air Force Materials Laboratories, under Contract F33615-74-C-5180.

Introduction

In major structural systems, the conservative design practices that have been adopted to ensure sufficient reliability have led to an under-utilization of components with major economic loss. One example is found in the military jet engine. Present performance standards demand that engines be designed so that no more than one in a thousand fail during the design life. After they have served for one design life, all parts are removed from service. Because of the statistical distribution of lifetimes for nominally identical components, many of the retired components have large unused remaining lives. Economic estimates (1) indicate that life cycle cost benefits of $\$250\text{x}10^6$ can be expected over a 15 year period for rotor components of the F100 engine if this loss can be recovered. This demands the use of inspection techniques which allow parts with unused remaining life to be returned to service with confidence.

Similar examples can be found in all industries where the consequences of structural failure are sufficiently large that conservative design or utilization practices are followed. For example, the cost of down time of one day in a 1000 MW nuclear reactor is $600,000 (2). In some cases, this down time can be initiated by a nondestructive test indication whose severity cannot be precisely determined. Again, techniques are needed to determine the lifetime of the flawed component so that operation can be resumed if appropriate,and further down time for unnecessary repairs can be avoided.

The prediction of remaining life requires a knowledge of flaw size and orientation, appropriate failure models, and the stress, temperature and environment in which the part must operate. The tight coupling of non-destructive measurement technique development to failure model development is an essential element in establishing a true quantitative non-destructive evaluation (NDE) capability. This ensures that the measurement techniques which are developed are primarily sensitive to specific failure related material properties. The approach thus avoids many of the problems that often plague techniques based on empirical correlations. Discussions of failure models appropriate to brittle fracture in ceramics and metal fatigue are presented in other papers (3,4) in this proceedings.

It is perhaps desirable at this point to compare the purposes and capabilities of current destructive testing technology with those envisioned for quantitative NDE as described above. The most direct way to do this is through a comparison of definitions that have come into usage. They are given below.

NDT: "...the development and application of nondestructive test methods..."

NDI: "...the performance of inspections to established specifications or procedures using NDT methods to detect anomalies."

NDE: "...the capability to assess the state of a material, a component, or a structure from a set of quantitative NDE measurements and to predict the remaining serviceability of the item in question from these measurements when evaluated in the context of appropriate failure models".

According to these definitions, non-destructive testing (NDT) refers to the development and application of various test methods. No emphasis is placed upon an interpretation of the results of the test and the nature of the interaction of the probing technique with the flaw or materials phenomenon in question. Without attention being given to this key area, it is impossible to make judgements as to the severity of the flaw in question and to the assessment of the ability of the component or structure to perform an assigned mission in the flaw's presence. The concept of non-destructive inspection (NDI) extends the definition of the NDT technology in the sense that it prescribes that inspections be performed in accordance with a specific format. This format may take the form of specifications which relate to the method, frequency, or spatial coverage of an inspection. However, the definition contains no specification for additional information that will enable the requirements for improved reliability at reduced cost to be met any more so than does the previous definition for NDT. On the other hand, the definition given for non-destructive evaluation (NDE) represents an enormous extension in scope, scientific content, and ultimate payoff from either of those given for NDT or NDI. There are essentially three new and important concepts contained within this definition. They are the concepts of quantitative NDT measurements, the assessment of a flaw's criticality in a material or component, and the prediction of the remaining serviceability of the component because of the presence of the flaw. The introduction of quantitative test methods is an essential ingredient of the NDE test technology. Simply, the words mean that the test techniques are sufficiently advanced so that quantitative, rather than qualitative, information is available concerning the target of interest, or flaw.

The weak link in today's technology which has prevented achieving prediction of serviceability via the NDE concept is the inability of the presently used NDT techniques to provide quantitative flaw information. Operationally, the lack of quantitative prediction capability is described by a probability of rejection curve, Fig. 1, which is broader than desired. The ideal technique, indicated by the dashed line, would have a zero probability of rejecting all flaws less than the critical size and a unity probability of rejecting those larger than this size. A real technique will make errors, as indicated by the solid curve. Some flaws less than the critical, failure producing size will be rejected (false rejects) and some greater than this critical size will be accepted (false accepts).

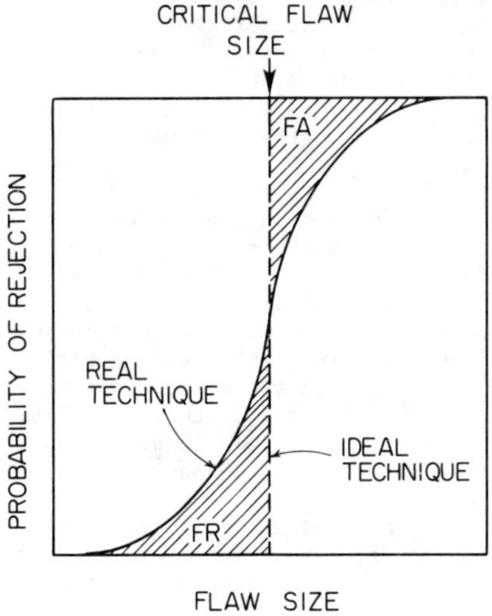

Fig. 1. Probability of rejection curve showing ideal techniques (dashed curve) and realistic technique (solid curve). Shaded areas indicate total probabilities of falsely accepting bad parts (FA) and falsely rejecting good parts (FR).

Three approaches can be used to determine such probability of detection curves. Laboratory experiments can be performed on samples containing fabricated flaws, the results of field or production data obtained on real parts can be analyzed, or predictive analytical techniques can be employed. The former two approaches are the ones which have been employed in the past. However, predictive, analytical approaches would offer several advantages. They would cost less to implement. They would provide a computational basis for establishing extremely high reliabilities in cases for which the large number of samples required for experimental validation would be prohibitive. They would provide a foundation for measurement technique optimization which would lead to performance closer to the ideal curve. Finally, they would serve as a computational basis for accept/reject criteria determination to minimize total system costs.

In recognition of the need for more quantitative measurement techniques and improved reliability, the Interdisciplinary Program for Quantitative NDE was established in 1974 by the Defense Advanced Research Projects Agency (DARPA) and the Air Force Material Laboratory (AFML). The purpose of this paper is to review a few of the results of that program which have led to improved ultrasonic techniques for quantitative flaw characterization.

Methodology

Figure 2 summarizes the basic approach that has been followed in developing this capability for quantitative NDE. The foundation required is an understanding of the energy (in this case ultrasound) flaw interaction. This requires the development of theoretical scattering models and their confirmation by experiments on samples containing flaws of controlled shape. Because of the great complexity of elastic wave scattering problems, exact solutions are not possible for any but the most simple flaw shapes. Hence, approximate theories have been emphasized which apply to more generalized flaw shapes. The development of these theories, and their comparison to experiment is discussed (5). In many cases, these experiments guided the development of the theories since particular approximations were often suggested by the measurement results.

The scattering models predict the elastic wave fields scattered by a flaw of known dimensions and material. However, of greater practical interest is the problem of identifying these characteristics of an unknown flaw from measurements of the scattered fields. This problem is often known as the inverse scattering problem. The forward scattering solutions developed in the DARPA/AFML program have been used as the "kernel" for developing solutions to this inverse problem.

Figure 3 summarizes the general classes of inversion techniques that have been developed. At the bottom, a plot of the backscattered signal from a spherical void is presented as a function of ka, where k is equal to 2π/wavelength. This defines several scattering regimes of interest. For ka $<<1$, the sphere is small with respect to a wavelength and one is in the Rayleigh regime. Here the scattered amplitude increases quadratically with frequency. When ka ~ 1, the resonance regime occurs and the scattering oscillates due to interference of rays following different paths. Finally, when ka $>>1$, one enters the geometrical cross-section regime in which the scattering is determined by the cross-sectional area of the sphere.

Above the graph, a number of inversion techniques are identified and compared. These are ordered in decreasing values of ka. In general, as ka decreases, the ultrasonic signals contain less detailed information about

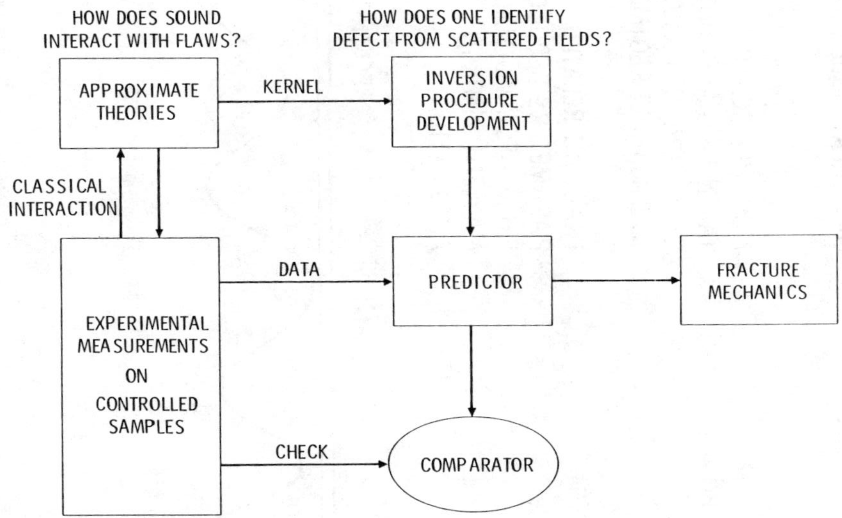

Fig. 2. Methodology for development of quantitative non-destructive evaluation capability.

TYPE	REGIME	ADVANTAGES
IMAGING	$ka > 6.3$ ($\lambda < a$)	HIGH INFORMATION CONTENT EASILY INTERPRETED DISPLAY RESOLVES MULTIPLE FLAWS
MODEL BASED RECONSTRUCTION	$ka \geqslant 3$	PHYSICAL PRINCIPLES USED TO IMPROVE RESOLUTION AND TREAT MODE CONVERSION
MODEL BASED ADAPTIVE LEARNING NETWORKS	$0.4 < ka < 3$	MULTIPLE SCATTERING TAKEN INTO ACCOUNT GAIN MORE INFORMATION IN DIFFICULT REGIME
LONG WAVELENGTH SCATTERING	$ka < 0.5$	FRACTURE RELATED PARAMETERS DEDUCED FROM A FEW MEASUREMENTS MAY BE USEFUL IN AUTOMATION

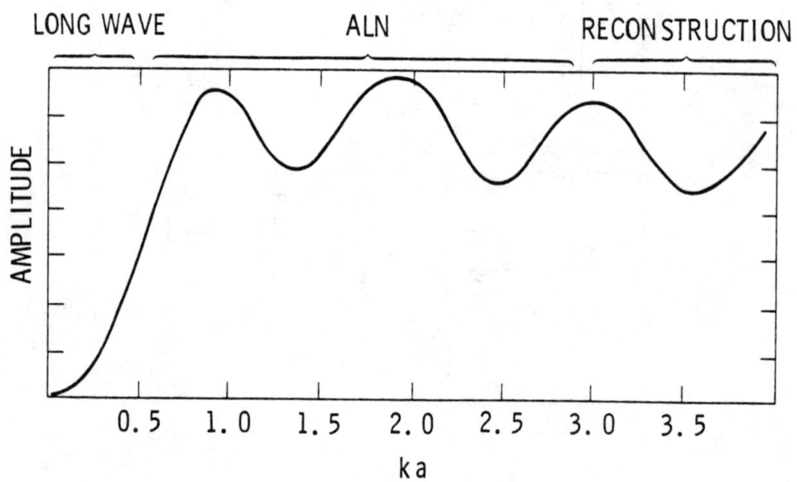

Fig. 3. General classes of inversion techniques.

the shape and size of the flaw. More independent assumptions, or a priori information, must be used to obtain a size estimate.

When ka >>1, the wavelength is small with respect to the flaw size and imaging techniques can be used. The techniques for processing the signals are equivalent to those used in optical imaging. These are generally based on the assumption that the flaw is a diffuse reflector, and that the reflection from various points on the object are independent of one another. Thus lenses, or equivalent electronic means, are used to gather signals scattered over a wide aperture. At each point in the reconstructed image, these are summed with appropriate phases so that the signals leaving the equivalent point on the object interfere constructively, and those leaving all other points on the object interfere destructively.

When ka ~ 6 ($\lambda \sim a$), imaging systems are no longer able to determine flaw shape details due to resolution limits. Furthermore, the existence of a variety of physical phenomena, such as Rayleigh waves which are excited by the incident wave and propagate around the periphery of flaw and reradiate at a second point, add complications which lead to distortion or the appearance of ghosts in images. For this regime, a set of techniques known as model based reconstruction procedures have been developed. These are similar to imaging approaches in that one attempts to reconstruct a detailed picture of the flaw shape. However, they differ in that the algorithms are based upon more complete models of the ultrasound-flaw interaction.

When the wavelength becomes larger so that ka $\lesssim 1$, there is no longer adequate information available to reconstruct any detailed picture of the flaw. Instead, one must either make use of independent information about the possible classes of flaws that are expected or in some way parameterize the limited information recovered. The adaptive learning approach is an empirical way of doing the former, while long wave measurements are an example of the latter which have a very important relationship to failure related parameters in certain cases.

Examples of each of these techniques will be presented in the following section. These have been selected for the most part from results obtained in a program to characterize flaws in ceramic materials. Aspects of a closely coupled failure modeling program are discussed elsewhere in this proceedings (3).

Inversion Techniques

Imaging

Several approaches to imaging are under investigation in the DARPA/AFML program as well as under other support. Focused beams from a single transducer can be used to produce a high quality image whose resolution is limited only by the diffraction limited spot size of the transducer (equal to $\lambda/2 \sin\theta$ where λ is the ultrasonic wavelength and θ is the half-angle of the aperture). Some images have been obtained using a variant of this technique suitable for high frequency applications (6). The primary problem with this approach is that it is very time consuming, since the transducer must be mechanically scanned in a three-dimensional pattern corresponding to the volume of material to be inspected.

To overcome this problem, a number of techniques are under development for the high speed formation of images. Synthetic aperture (7) systems and wavefront reconstruction systems (8) share the common characteristic that the ultrasonic fields are sampled in a finite aperture by an array of piezoelectric transducers and an image is reconstructed by the use of high speed electronic circuitry. Some degradation in image quality occurs since the fields are discretely rather than continuously sampled, so this may be thought of as trading some image quality for increased inspection speed.

Another approach uses a scanned laser beam as the transducer which detects the ultrasound. This has the advantage that the beam may be moved very rapidly over the part and can continuously sample the fields. Its main limitation is a considerably lower efficiency than that of piezoelectric transducers. Images of a number of flaws in ceramic materials have been obtained using this approach (9).

Model Based Reconstruction

The basic premise upon which image formation algorithms are based is that the signal is diffusively and instantaneously reflected from the flaw surface without any delayed signals arising from internal reverberations or creeping waves propagating around the periphery of the flaw. In many ultrasonic situations, these assumptions are strongly violated. For example, Fig. 4 presents the theoretically predicted longitudinal ultrasonic backscattered signal as a function of time when the wave is backscattered from a spherical Si inclusion in Si_3N_4 (10). It can be seen that the signal consists of two parts. The first arrival is produced by the reflection from the front surface of the inclusion. It is this signal which imaging systems are intended to process. The signal train occurring later is actually the superposition of several events whose nature are indicated in the sketches. These features can cause distortion and ghosts in images and hence are viewed as spurious from the imaging point of view. However, they contain valuable information that can be [and has been (10)] used to identify the flaw material and size.

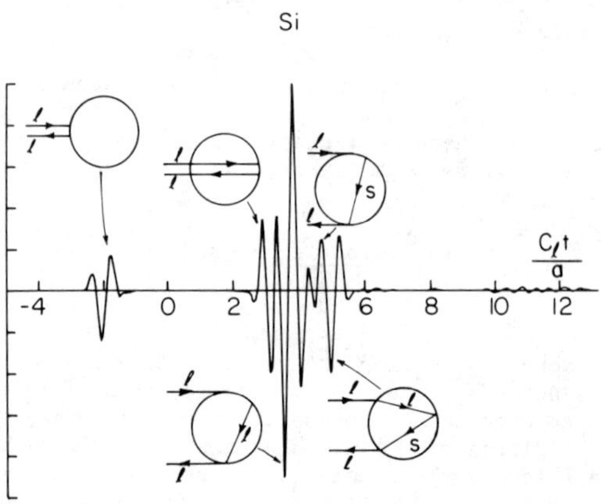

Fig. 4. Longitudinal ultrasonic signal backscattered from spherical silicon inclusion in silicon nitride.

An alternate, somewhat more formal approach makes explicit use of the low frequency components of the scattered signal (11). Elastic wave scattering shows that, for either a flaw with a center of inversion symmetry, or for a weakly scattering flaw, the Fourier transform of the scattered signal can be written in the form

$$A(\omega) = A_2\omega^2 + O(\omega^4) \tag{1}$$

when the origin is chosen at the centroid of the flaw. In this expansion, there is no cubic term. If, however, the origin is not chosen at the center of the flaw, a cubic term arises. This may be evaluated from Eq. (1) by noting that, if the origin is shifted a distance Δx, so that the signals are delayed by a time $\tau=2\Delta x/v$ where v is the longitudinal wave velocity, then the Fourier integral shift theorem predicts that,

$$A(\omega,\tau) = A(\omega)e^{i\omega\tau} \tag{2a}$$

$$= A(\omega)\ [1 + i\omega\tau - \frac{\omega^2\tau^2}{2} + \ldots] \tag{2b}$$

$$= A_{2\omega}^{\ 2} + A_{3\omega}^{\ 3} + O(\omega^4) \tag{2c}$$

where

$$A_3 = i\tau A_2. \tag{3}$$

τ may then be experimentally determined from measurements of A_2 and A_3 and use of Eq. (3). This provides a direct measure of the time at which a signal would have been reflected from the center of the flaw, were there a discrete reflector at that point. If we define T as the time at which the front surface reflection arrives (which is clearly visible in Fig. 4), then the distance from the front surface to the flaw centroid, d, is given by

$$d = \frac{\tau - T}{2v} \tag{4}$$

where v is the velocity of ultrasound in the host medium.

This approach is different from an imaging approach in that explicit use is made of the low frequency components of the scattered wave. This allows a direct determination of the radius of the flaw along the direction of measurement, as distinct from the lateral dimensions of a flaw that are usually found in an imaging approach. If a variety of observation angles are possible, then the flaw shape can be reconstructed. For non-spherical shapes, the distance d must be interpreted as the distance from the flaw center to a plane tangent to the surface of the flaw and perpendicular to the line of observation (12). The flaw surface can be deduced from a set of these front surface tangent plane distances.

In the reconstruction of flaw shapes, this approach has an important advantage over imaging approaches when the sound velocity in the host medium is anisotropic. The formation of an image usually depends upon time delay information to determine the position of the flaw surface. This is usually measured with respect to the time at which the signal left the transducer. Small errors in the velocity in the host material such as are produced by anisotropy, cause errors in the position assigned to the flaw surface. These errors can easily be on the order of the sizes of a flaw when the velocity errors are integrated over a long path length. If the material is isotropic, this is equivalent to a slight error in flaw location. However, if it is anisotropic, serious shape and size distortions can occur. The problem is greatly reduced in the technique described above since all times are referenced to the center of the flaw rather than the surface of the part.

From a historical perspective, the above arguments were developed as an extension of the Inverse Born approximation. This approximation, which is only rigorously required to give accurate results for flaws whose elastic constants were close to those of the host material, has been found to be surprisingly accurate for a wide range of flaw properties. For example, Fig. 5 shows the successful use of the technique to determine the size of a cavity. The fundamental reason is explained by the previous discussion, whose details are independent of the magnitude of the difference between the properties of the flaw and the host.

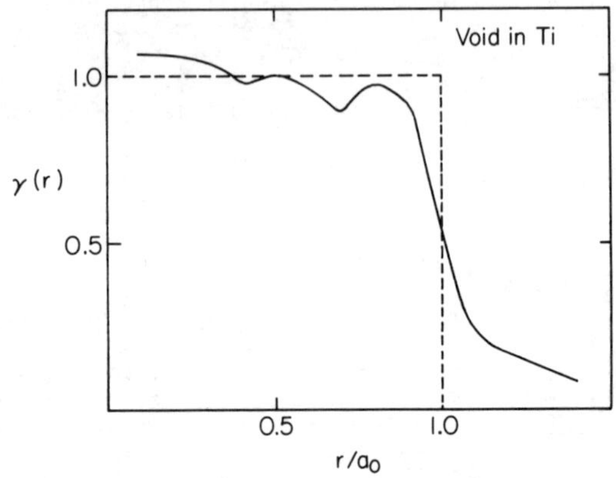

Fig. 5. Characteristic function for a spherical void (γ=1 inside void, γ=0 outside void) in titanium based on Inverse Born approximation.

Other approaches have also been developed for making use of physical models in the reconstruction of flaw size and shape. Included are the physical optics far field inverse scattering approach (13) for volumetric flaws such as voids and inclusions, and related inverse scattering integrals (14) for two-dimensional flaws, i.e., cracks. In each case, these are based on Fourier-type inversion integrals which are used to recover size, shape, and orientation from far-field data.

Adaptive Learning

As the wavelength becomes larger with respect to the flaw size, there is no longer sufficient information to allow a detailed reconstruction of the flaw shape to be made. In general, size parameters can only be deduced if certain assumptions are made about the flaw shape. Fortunately, however, there are certain cases in which these assumptions need not be too restrictive, and quite useful results can be obtained in this regime.

When ka $\sim$1, the ultrasound-flaw interaction is extremely complex, and no satisfactory analytical inversion algorithms have been developed. However, an empirical adaptive learning procedure has shown considerable promise (15). This technique develops a fourth order polynomial which predicts the size of a flaw in terms of experimental observables such as the relative ultrasonic power scattered in various directions. The polynomial is developed by fitting it to the known sizes and measurements on a set of "training" flaws. It is then confirmed by measuring the accuracy of the predictions on a second, independent set of "test" flaws. Normally, both the training and test flaw data are obtained experimentally. However, it has recently been shown that the analytical scattering models can be used in the training phase, and that the networks so derived make satisfactory predictions of the sizes of machined flaws when the experimental data are used as input parameters. Table I presents the results of one such comparison for the case of ellipsoidal flaws.

Long Wavelength

When the wavelength becomes large with respect to the flaw size, $ka \ll 1$, analytical techniques can again be employed. In this Rayleigh scattering regime, the scattered signal is proportion to ω^2, and flaw size and shape information are deduced from the A_2 coefficient in Eq. (1). The variation of A_2 with flaw parameters can be computed using the quasi-static approximation (16). Experience from problems involving a scalar wave equation suggests that measurements of the coefficient A_2 provides little information about the flaw parameters. However, when the scattering is governed by the elastic wave equation, which relates tensor fields, the situation is somewhat different. In the most general case, A_2 is a function of 22 independent parameters of the flaw (17). Hence, it could be hoped that measurements of A_2 at a sufficient number of angles would allow one to deduce these parameters. In practice, such an ambitious undertaking has not yet been performed. However, the technique has been employed to measure the major axis of an ellipsoidal void, as shown in Fig. 6. In this experiment, it was assumed that the flaw was an ellipsoidal cavity, and the measured values of A_2 at several angles were compared to the theoretical values with the flaw size as a variable parameter. An estimation theory approach was used to produce the probability density function of the flaw's major radius as shown.

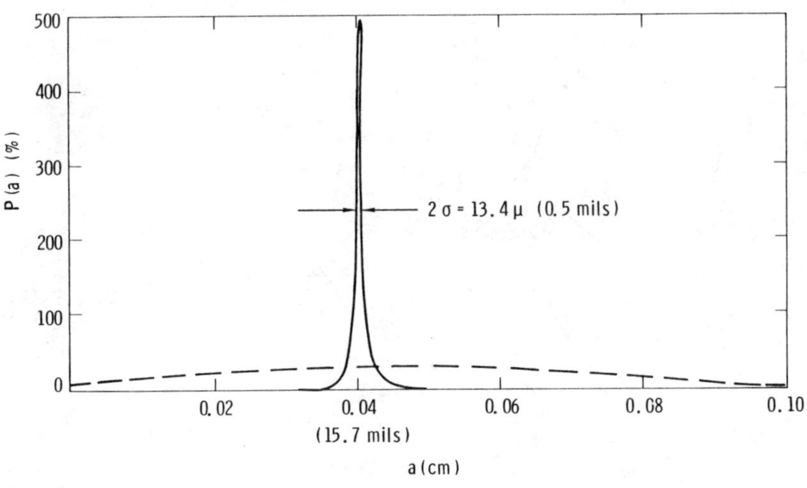

Fig. 6. Probability density of major radius of ellipsoidal void as determined from experimental measurements of long wavelength scattering coefficients.

Table I. Size and orientation estimates for nine experimental spheroid defect samples determined from theoretically trained ALN models. Dimensions defined in the sketch below.

Exp. No.	A (μm) True	A (μm) Estimated	B (μm) True	B (μm) Estimated	α (degs) True	α (degs) Estimated	β (degs) True	β (degs) Estimated
1	200	123	400	413	0	4	-	-
2	200	180	400	381	30	23	225	222
3	100	105	400	415	80	79	160	167
4	100	72	400	372	0	12	-	-
5	200	127	400	341	80	89	160	174
6	100	94	400	406	30	29	180	190
7	200	121	400	370	30	46	180	189
8	100	97	400	393	30	36	225	227
9	200	216	400	453	15	14	180	184

Error Summary

Parameter	Relative Error
A (μm)	23%
B (μm)	6%
α (degs)	7%
β (degs)	4%

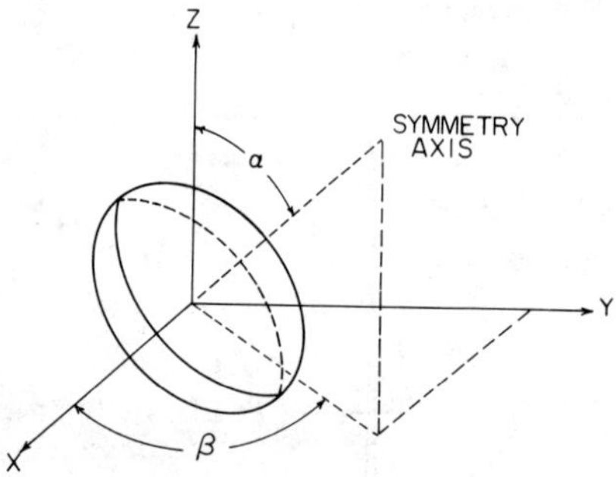

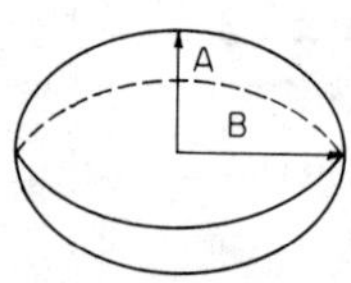

This result illustrates one important property of the long wavelength technique, high leverage on flaw parameters. This is primarily a consequence of the fact that $A_2 \alpha a^3$, where a is the flaw radius. Thus, a 30% accuracy in the measurement of A_2 is equivalent to a 10% accuracy in the measurement of a. The small width of the probability density shown in Fig. 6 is a consequence of this effect. Another important advantage of the long wavelength technique is its insensitivity to minor perturbations in flaw shape. Average flaw parameters will be obtained if the flaw does not coincide exactly to the initially assumed shape.

Perhaps the most striking and important property of the long wavelength measurement is its relationship to the reduced stress intensity factor of a flaw. It has been shown (18) that a measurement of A_2 perpendicular to the face of an elliptical crack can be used to directly determine the reduced stress intensity factor, essentially independent of the ellipticity of the flaw. This result, discussed in greater detail elsewhere in this proceedings[3], arises because both A_2 and the reduced stress intensity factor are computed from related integrals in which the static deformation of the flaw under an applied load appears in the integrand. Although only a limited set of flaw parameters can be measured, one of them can be directly used in the prediction of failure. Thus, the value of A_2 measured along an axis perpendicular to the face of an idealized elliptical crack contains all of the information necessary to predict failure loads under conditions of brittle fracture.

As initially derived, this result was only rigorously valid for the case of an interior flaw. It was then generalized to the more common case of a surface crack. Figure 7 illustrates an experimental verification of the results on a glass sample (19). Here an indenter was used to initiate a small crack in the specimen, and the A_2 coefficient for Rayleigh wave backscattering was then measured. From this, the reduced stress intensity factor was estimated, a fracture stress was predicted, and this stress was compared to the actual stress at failure. The figure shows that there was a linear relation between the predicted and measured fracture stresses with the latter being systematically low by about 15%. This discrepancy was later explained by noting that the failure was not purely brittle and that a slow crack growth occurred during stressing which caused the actual crack size at fracture to be somewhat greater than at the time of the ultrasonic measurements.

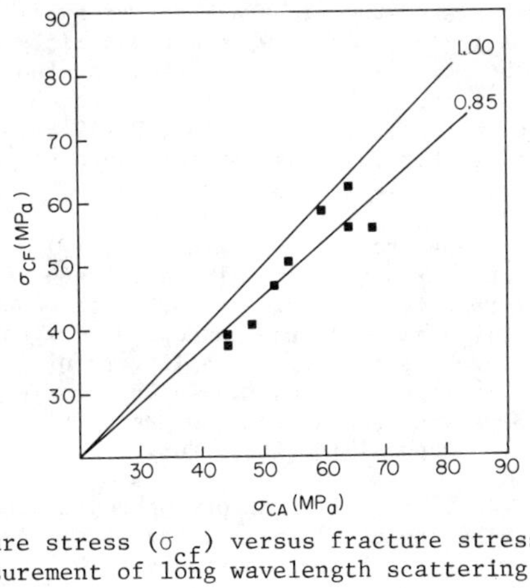

Fig. 7. Actual fracture stress (σ_{cf}) versus fracture stress predicted from acoustic measurement of long wavelength scattering coefficient (σ_{ca}).

Statistical Estimation Theory

It must be recognized that none of the above techniques will ever determine flaw parameters without possible error or ambiguity. The techniques should therefore be asked to produce not only a best estimate of the flaw parameters, but a measure of the variances of these estimates as well. This is particularly important when the NDE is coupled to probabilistic failure models. The variances clearly will depend upon such parameters as the signal to noise ratio of the data and the sensitivity of the measured signals to flaw parameters. One approach that has been developed to make such estimates is summarized in simplified form below.

The technique will be discussed in the context of a particular problem of current interest, the identification of both the material and size of inclusions in ceramics (20). The motivation for this problem, and its relationship to the development of accept/reject criteria, are discussed in greater detail elsewhere in this proceedings (3).

It is assumed that the non-destructive measurement can be described by a model of the form

$$y = h_g(x) + \nu \tag{5}$$

where y is the measurement result, x is the flaw size, $h_g(x)$ is the scattering model relating the size of a flaw of material g to the measurement, and ν is the noise, or measurement error. Clearly, y and ν are random variables. In most actual cases, there are multiple flaw parameters and measurements of interest and the analysis takes a matrix form. However, this multidimensionality will be suppressed here for simplicity.

What one wishes to determine is the probability that the flaw has a size x and material g, given the measurement result y. In other words, one seeks the conditional joint probability $P(x,g|y)$. Elementary probability analysis says that this is given by the identity.

$$P(x,g|y) = P(y|x,g)\ P(x,g)\ P(y)^{-1} \tag{6}$$

where the three factors on the right are the probability distribution of measurement results for a given flaw, the probability that this type of flaw is found in the material (sometimes known as the prior distribution) and the reciprocal of the probability that a particular measurement result is obtained. Note that this latter factor, $P(y)^{-1}$, is only a normalization factor and affects neither the peak nor the width of the desired function $P(x,g|y)$.

The brute force approach to determine $P(x,g|y)$ would be a numerical evaluation of Eq. (6) for a sufficiently dense mesh of x and g values. This unfortunately requires extensive computer time and is not practical. It is philosophically similar to the concept of "signature analysis" in which a measured signal is compared to a library of signatures from known flaws to see which it is most like. The problem of extensive computer time requirements is in many ways analogous to the problem of gathering a sufficiently complete set of possible signatures.

In the statistical estimation theory approach, one starts with an assumed prior distribution at flaw sizes and materials, $P(x,g)$. These are then weighted according to their consistency with the measurement results, given the measurement noise. In other words, those members of the prior distribution who are inconsistent with the measurements, given the noise, are eliminated. Clearly, the lower the noise and the more sensitive the measurements to the flaw parameters, the sharper the resulting distribution. The predicted flaw size and variance is obtained from the properties of this weighted distribution. In the development of a unified inversion algorithm for the ceramic problem (20), it was assumed that the measurement noise was Gaussian so that one could represent $P(y|x,g)$ by

$$P(y|x,g) \propto \exp \{-\frac{1}{2} [y-h_g(x)]^2/C_\nu\} \tag{7}$$

where C_ν is the noise covariance. This, coupled with a Taylor series expansion about the most likely flaw size, allows one to analytically find the form of the desired $P(x,g|y)$ which is

$$P(x,g|y) \propto \exp \{-\frac{1}{2} (x-\hat{x})^2 H_g\} \tag{8}$$

where $\hat{x}$ is the most likely flaw size and H_g determines the width of the probability density function, i.e., the variance about this prediction. The technical problem is to determine $\hat{x}$ and H_g. From a formal point of view, these are given by

$$x = h_g^{-1} \{y + \frac{[\partial P(x,g)/\partial x]C_\nu}{2P(x,g)[dh_g(x)/dx]}\} \tag{9}$$

The width of distribution is

$$\Delta x = H_g^{-1/2} \tag{10}$$

where

$$H_g = \frac{\partial}{\partial x} \left\{\frac{-2[y-h_g(x)][\partial h_g(x)/\partial x]}{C_\nu} - \frac{[\partial P(x,g)/\partial x]}{P(x,g)}\right\} \Bigg|_{x=\hat{x}} \tag{11}$$

Inspection of Eqs. (9) and (11) reveals the relative importance of noise, prior distribution information, and measurement model leverage. Consider first the case of zero noise, $C_\nu = 0$. In this case, the best estimate of flaw size, $\hat{x}$, is just the value $h_g^{-1}(y)$ that would normally be expected, and the width of the probability distribution is zero. In other words, the flaw parameters are determined with zero error. When $C_\nu \neq 0$, an uncertainty exists and two other factors must be considered. Recall that $P(x,g)$ is the prior distribution of flaw size and type. In other words, this is a function which incorporation any <u>a priori</u> information that might be available regarding the possible flaws that could be expected. This distribution is, of course, strongly influenced by the fabrication and service history of the part. If no such information is available, the distribution is flat, $\partial P(x,g)/\partial x = 0$. Again, the best estimate of flaw size, $\hat{x}$, has the value $h_g^{-1}(y)$. When $\partial P(x,g)/\partial x \neq 0$, there is a shift in this estimate, with the magnitude of the shift being determined by the noise, the derivative of the prior distribution, and the leverage of the measurement, $\partial h_g/\partial x$. This latter factor is large when the measurement y changes by a large amount of small changes in the flaw size $\hat{x}$. In this case of large measurement leverage, the shift in the estimated flaw size produced by a fixed amount of noise and non-constant prior distribution information is relatively small. Here the estimate is dominated by the measurement result. Conversely, when $\partial h_g/\partial x$ is small, the shift in $\hat{x}$ from the noise free value can be large. In this case, one is said to be dominated by the prior distribution.

The same factors influence the variance of the flaw estimate. As noted above, small values of noise variance, C_ν, produce small variances in the flaw size estimate. Similarly, if the measurement leverage or the prior distribution leverage is large, the variance in flaw size estimate is again small.

In the above equations, h_g^{-1} symbolizes the solution to the inverse problem. In actual cases, these will often not be available in closed form, and the above formal procedures cannot be directly implemented. Instead, more complex numerical approaches are required (20).

Figure 8 illustrates the result of applying the unified inversion algorithm based on these principles to the measurement of inclusion size and material in Si_3N_4 ceramic. As indicated in the sketch at the top of the figure, the measured parameters were A_2 and d as discussed in previous sections. The algorithm assumed that the inclusion was either a sphere or an ellipsoid composed of one of six candidate materials (Void, BN, Si, SiC, WC, Fe). The output for one case is shown at the bottom of the figure in the form of the distribution function $P(x,g|y)$. The probability that the flaw is composed of material g is, of course, given by $P(g|y) = \int P(x,g|y)dx$. For this case, the correct material, iron, was predicted with a 71% probability and, given this prediction, the correct size was obtained within 4μ.

Summary

A variety of techniques have been presented for quantitatively deducing flaw parameters from ultrasonic measurements. Ranging from the short to long wavelength limits, these include imaging, model based reconstruction, adaptive learning, and long wavelength techniques. The former provide the most detailed information. However, the latter can be used to deduce specific failure related material parameters, the most striking example being the determination of reduced stress intensity factors from long wavelength scattering data.

One in general wishes to know not only the best estimate of flaw parameters, but the variances in these parameters as well. Statistical estimation theory has been used to demonstrate such a capability for the particular problem of identifying inclusion material and size in ceramics. It is believed that methodologies of this type will be required in the future as NDE is coupled to probabilistic fracture mechanics to produce probabilistic life predictions and optimum selection of accept/reject criteria.

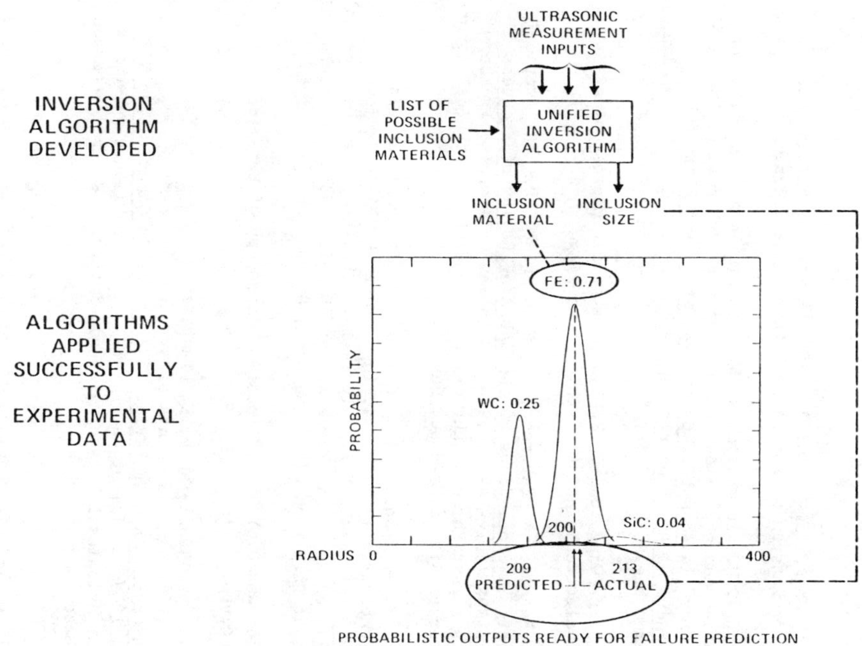

Fig. 8. Joint probability distribution of flaw size and material for inclusions in Si_3N_4 based on experimental measurements processed by unified inversion algorithm.

References

1. C. G. Annis, J. J. Cargill, J. A. Harris, Jr., M. C. VanWanderham, "Engine component retirement for cause: a nondestructive evaluation (NDE) and fracture mechanics based maintenance concept", this proceedings.

2. G. Dau, "NDE: Indepth search for flaws", EPRI Journal 5, July/Aug. (1980).

3. A. G. Evans, "Non-Destructive Failure Prediction for Brittle Solids," this proceedings.

4. W. L. Morris, M. R. James, and O. Buck, "Remaining Fatigue Lifetime Prediction for Retirement-for-Cause," this proceedings.

5. R. K. Elsley, J. M. Richardson, R. B. Thompson, and B. R. Tittmann, "Comparison between experimental and computational results for elastic wave scattering" in Acoustic, Electromagnetic, and Elastic Wave Scattering-Focus on the T-matrix Approach, V. K. and V. V. Varadan, eds.; Pergamon Press, New York, 1980, p. 57.

6. C. C. Lee, J. K. Wang, and C. S. Tsai, "Acoustic Microscopy of Hybrid Microelectronic Joints," this proceedings.

7. D. Corl, S. Bennett, K. Peterson, and G. S. Kino, "Digital synthetic aperture system", Proceedings of the DARPA/AF Review of Progress in Quantitative NDE (July 14-18, 1980, La Jolla, California) (to be published as an Air Force Wright Avionics Laboratory technical report, Wright-Patterson Air Force Base, Dayton, Ohio).

8. K. M. Lakin, William Sheppard, and K. Tam, "Ultrasonic imaging by wavefront reconstruction", Proceedings of the DARPA/AF Review of Progress in Quantitative NDE (July 14-18, 1980, La Jolla, California) (to be published as an Air Force Wright Avionics Laboratory technical report, Wright-Patterson Air Force Base, Dayton, Ohio).

9. L. W. Kessler and D. Yuhas, "Acoustic Microscopy of Ceramics," this proceedings.

10. C. H. Chou, B. T. Khuri-Yakub, K. Liang, and G. S. Kino, "High frequency bulk wave measurements of structural ceramics", Proceedings of the ARPA/AF Review of Progress in Quantitative NDE (July 8-13, 1979, La Jolla, California) (to be published as an Air Force Wright Avionics Laboratory technical report, Wright-Patterson Air Force Base, Dayton, Ohio).

11. J. H. Rose and J. Richardson, "Determination of flaw characteristics in the time domain", Proceedings of the DARPA/AF Review of Progress in Quantitative NDE (July 14-18, 1980, La Jolla, California) (to be published as an Air Force Wright Avionics Laboratory technical report, Wright-Patterson Air Force Base, Dayton, Ohio).

12. J. M. Rose, R. K. Elsley, B. Tittmann, V. V. Varadan, V. K. Varadan, "Inversion of ultrasonic scattering data", in Acoustic, Electromagnetic, and Elastic Wave Scattering-Focus on the T-matrix Approach, V. K. and V. V. Varadan, eds.; Pergamon Press, New York, 1980, p. 605.

13. N. Bleistein and J. K. Cohen, "Progress on a mathematical inversion technique for non-destructive evaluation", Proceedings of the ARPA/AF Review of Progress in Quantitative NDE (July 8-13, 1979, La Jolla, California) (to be published as an Air Force Wright Avionics Laboratory technical report, Wright-Patterson Air Force Base, Dayton, Ohio).

14. J. D. Achenbach, A. Norris and K. Viswanathan, "Inversion of crack scattering data in the high-frequency domain", in Acoustic, Electromagnetic, and Elastic Wave Scattering-Focus on the T-matrix Approach, V. K. and V. V. Varadan, eds.; Pergamon Press, New York, 1980, p. 591.

15. M. F. Whalen, L. J. O'Brien, and A. N. Mucciardi, "Application of adaptive learning networks for the characterization of two-dimensional and three-dimensional defects in solids", Proceedings of the ARPA/AF Review of Progress in Quantitative NDE (July 8-13, 1979, La Jolla, California) (to be published as an Air Force Wright Avionics Laboratory technical report, Wright-Patterson Air Force Base, Dayton, Ohio).

16. J. E. Gubernatis, "Long wave approximations for the scattering of elastic waves from flaws with application to ellipsoidal voids and inclusions", J. Appl. Physics, 50, 4046 (1979).

17. J. M. Richardson, "The inverse problem in elastic wave scattering at long wavelengths", 1978 Ultrasonics Symposium Proceedings (IEEE, New York, 1978), p. 759.

18. B. Budiansky and J. R. Rice, "On the estimation of a crack fracture parameter by long wavelength scattering", Trans. ASME, J. Appl. Mech., 45, 453, 1978.

19. B. T. Khuri-Yakub, G. S. Kino, J. C. Shyne, M. T. Resch, and V. Domarkas, "Surface crack characterization: Geometry and stress intensity factor measurements", in Proceedings of the ARPA/AFML Review of Progress in Quantitative NDE, Air Force Materials Laboratory technical report AFML-TR-78-205 (Wright Patterson Air Force Base, Dayton, Ohio 45433), p. 505.

20. K. W. Fertig and J. M. Richardson, "Inverse scattering at low and intermediate frequencies", Proceedings of the ARPA/AF Review of Quantitative NDE (July 8-13, 1979, La Jolla, California) (to be published as an Air Force Wright Avionics Laboratory technical report, Wright-Patterson Air Force Base, Dayton, Ohio).

ENGINE COMPONENT RETIREMENT-FOR-CAUSE: A NONDESTRUCTIVE EVALUATION (NDE) AND FRACTURE MECHANICS BASED MAINTENANCE CONCEPT

C.G. Annis, J.S. Cargill, J.A. Harris, Jr.
and M.C. Vanwanderham

Materials & Mechanics Technology
Pratt & Whitney Aircraft Group
Government Products Division
West Palm Beach, Florida 33402

ABSTRACT

Historically, cyclic life limited gas turbine engine components have been retired when they reach an analytically determined life where the first fatigue crack per 1000 parts could be expected. By definition, 99.9% of these components are being retired prematurely as they have considerable useful life remaining. Retirement-for-cause is a procedure which would allow safe utilization of the full life capacity of each individual component. Since gas turbine engine rotor components are prime candidates and are among the most costly of engine components, adoption of a RFC maintenance philosophy could result in substantial engine systems life cycle cost savings. Two major technical disciplines must be developed and integrated to realize those cost savings: Fracture Mechanics and Nondestructive Evaluation. This paper discusses the methodology, and development activity required, to integrate these disciplines to provide a viable RFC system for use on military gas turbine engines, and illustrates potential benefits of its application.

Introduction

Historically, methods used for predicting the life of gas turbine engine rotor components have resulted in conservative estimation of useful life. Most rotor components are limited by low cycle fatigue, generally expressed in terms of mission equivalency cycles. When some predetermined cyclic life limit is reached, components are retired from service. Currently, engine rotor components life limits are analytically determined using lower bound low-cycle fatigue (LCF) statistics: If the data indicate, e.g. that one in one thousand components (such as disks) will have a fatigue induced crack of approximately 0.03 inches in length, all parts are removed from service. The life prediction criterion is established by considering uncertainties in any design system (e.g. stress analysis errors, field mission definition, fabrication deviation, temperature profile uncertainty, etc.). This procedure has been very successful in preventing the occurrence of catastrophic failure of disks in the field. However, in retiring one thousand disks because one may fail, the remaining life of the 999 unfailed disks is not utilized. The amount of usable life remaining can be significant, as shown in Fig. 1, where over 80% of the disks have at least 10 lifetimes remaining.

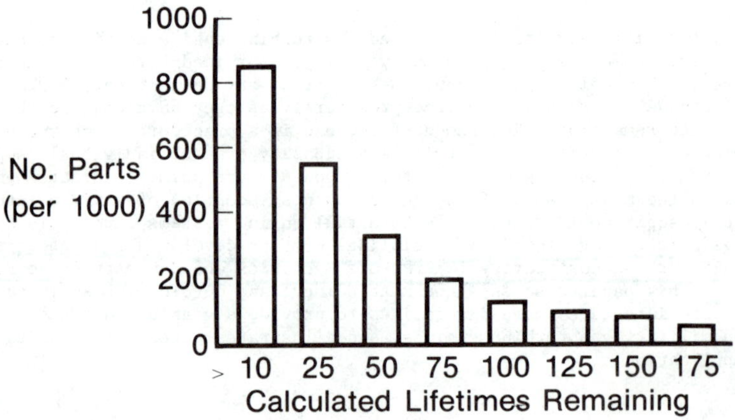

Figure 1. The Majority of Disks Have Useful Life After Retirement

Retirement for Cause (RFC) is a desirable alternative to that procedure. RFC would allow each component to be used to the full extent of its safe total fatigue life, retirement occurring when a quantifiable defect necessitates removal of the component from service. The defect size at which the component is no longer considered safe is determined through nondestructive evaluation (NDE) and fracture mechanics analyses of the disk material and the disk fracture critical locations, the service cycle and the overhaul/inspection period. Realization and implementation of a RFC Maintenance Methodology will result in system cost savings of two types:

(i) direct cost savings resulting from utilization of parts which would be retired and consequently require replacement by new parts, and

(ii) indirect cost savings resulting from reduction in use of strategic materials, reduction in energy requirements to process new parts, and mitigation of future inflationary pressure on cost of new parts.

Retirement-for-Cause Methodology

Philosophy - The fatigue process for a typical rotor component such as a disk can be visualized as illustrated in Fig. 2. Total fatigue life consists of a crack initiation phase followed by growth and linkup of microcracks. The resulting macrocrack(s) would then propagate subcritically until the combination of service load (stress) and crack size exceeded the material fracture toughness. Catastrophic failure would result had not the component been retired from service. To preclude such cataclysmic disk (and possibly engine) failures, disks are typically retired at the time where 1 in 1000 could be expected to have actually initiated a short (0.03 in.) fatigue crack. By definition 99.9% of the retired disks still have useful life remaining at the time they are removed from service. Under the Retirement-for-Cause philosophy, each of these disks could be inspected and returned to service. The return-to-service (RTS) interval is determined by a fracture mechanics calculation of remaining propagation life from a crack just small enough to have been missed during inspection. This procedure could be repeated until the disk has incurred measurable damage, at which time it is retired for that reason (cause). Retirement-for-Cause is a methodology under which an engine component would be retired from service when it had incurred quantifiable damage, rather than because an analytically determined minimum design life had been reached. Its purpose is not to extend the life of a rotor component, but to utilize safely the full life capacity inherent in that component.

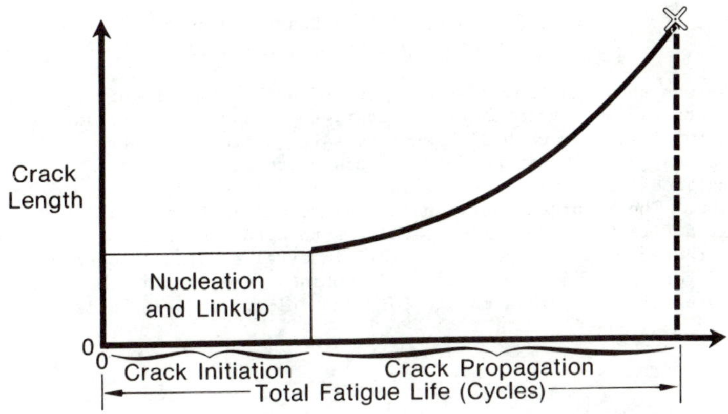

Figure 2. Total Fatigue Life Segmented Into Stages of Crack Development, Subcritical Growth, and Final Fracture

Residual Life - For simplicity this paper will consider component life limits which have been determined from the crack initiation characteristics of the specific disk material, and will not address the problem of intrinsic crack-like defects. While damage tolerant concepts are utilized in some instances to establish life limits, the majority of components in current gas turbine engines have life limits set by an initiation criterion.

All fatigue data have inherent scatter. The data base used for design life analyses purposes must be applicable to all disks of a given material, and therefore includes test results from many heats and sources. Data are treated statistically as shown schematically in Fig. 3. The distribution of life, defined as the number of cycles necessary to produce a crack approximately 0.03 in. long, is obtained for a given set of loading conditions (stress/strain, time, temperature). As can be seen, the $\pm 2\sigma$ bounds which contain 95% of the data, may span two orders of magnitude in fatigue initiation life.

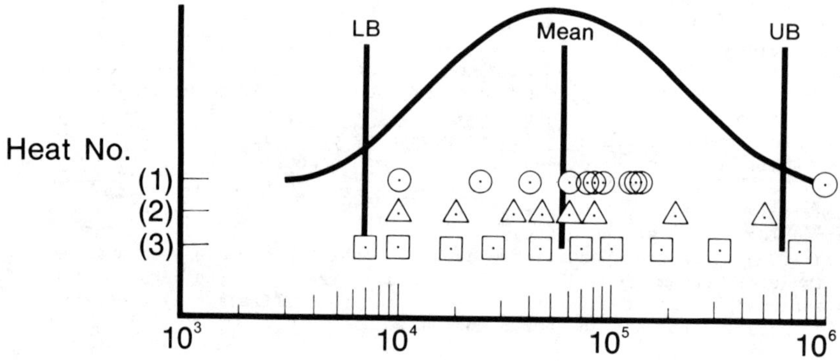

Figure 3. Material Data Scatter Results in Conservative Life Prediction

As was shown in Fig. 1, the amount of usable life remaining in retired components can be significant. However, the means of extracting the remaining useful life from each disk must be safe to avoid catastrophic failure. This is done by determining the disk crack propagation life (N_p) (at every critical location) from a defect barely small enough to be missed during inspection. The Return-to-Service (RTS) interval is then calculated by conducting a Life Cycle Cost (LCC) analysis to determine the most economical safety factor (SF) to apply to N_p (RTS interval = N_p/SF). Cost vs SF is plotted for each individual disk and combined to determine the most economical interval to return a module for inspection. An example is shown in Fig. 4.

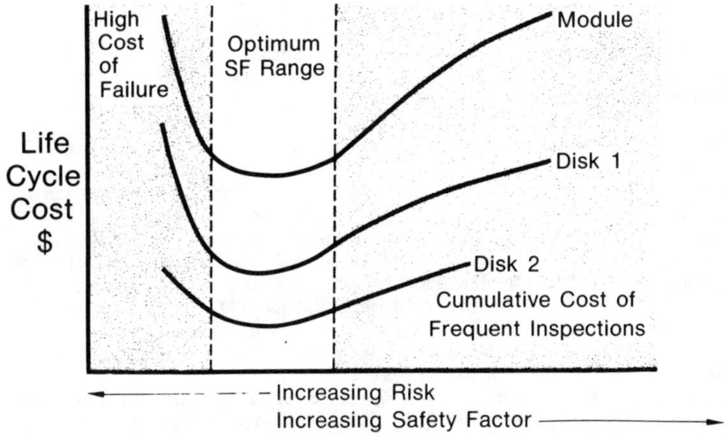

Figure 4. Safety Factor Is Determined from an Economic Balance between High Cost of Failure vs Cumulative Costs of Frequent Inspections

The first required disk inspection is near the end of the analytically determined crack initiation life. Only one disk in 1000 inspection should have a crack and be retired. The other 999 will be returned to service for the calculated RTS interval. This inspection is repeated at the end of each RTS interval with the cracked disks being retired and all others returned to service. Fig. 5 illustrates how the residual life is extracted from each disk after the crack initiation life has been used.

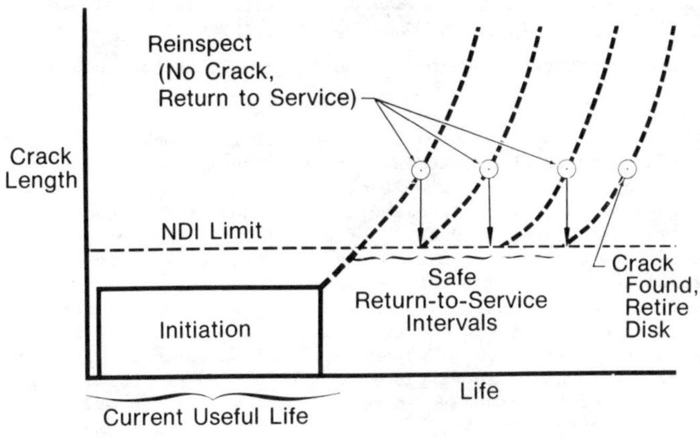

Figure 5. Base Retirement-for-Cause Concept

Technology Development Required

As Figs. 4 and 5 illustrate, Return-to-Service (RTS) intervals are based on two broad technologies: Nondestructive Evaluation (NDE) and applied Fracture Mechanics, and evaluated based upon economic factors.

Fracture Mechanics must provide an assessment of the behavior of a cracked part should it pass NDE with a defect just below an inspection limit. To assure safe return to service of a part which may contain a small crack, an accurate crack propagation prediction is imperative. Recent strides in applied elevated temperature fracture mechanics (1, 2, 3, and 4) have provided the necessary mathematical description (models) of basic propagation, i.e., crack growth under conditions of varying loading frequency (ν), stress ratio (R), and temperature (T). Further work (5 and 6) has expanded this capability to include loading spectra synergism, i.e., crack growth subjected to (frequent) periodic major load excursions separated by a small number (10-50) of varying subcycles. It is important to note that a typical mission loading spectrum to which gas turbine engines are subjected bears little resemblance to that experienced by air frames, and therefore different predictive tools are required for each (7).

Referring again to Fig. 5, it is seen that accurate propagation predictions constitute a necessary, but not sufficient, condition for the implementation of Retirement for Cause. The other requisite technology is high reliability nondestructive evaluation (NDE).

NDE must provide the means of screening disks with flaws that could cause component failure within an economically feasible RTS interval. Insufficient NDE reliability has been a major argument against implementation of an RFC maintenance program. NDE capability with acceptable flaw detection resolution has been available for some time (8, 9), but adequate reliability of flaw detection has been lacking (10). Complementary inspections and improvements in NDE single inspection reliability (by automation), can provide the required reliability for many gas turbine engine components to economically utilize the RFC maintenance concept.

These two requisite technologies are integrated in the component life analysis.

The fracture mechanics approach to estimating component service life is based on the assumption that materials may contain intrinsic flaws, and that fatigue failure may occur as a result of progressive growth of one or more of those flaws into a critically sized crack. Thus, the prediction and monitoring of crack growth as a function of time (or cycles) becomes one of the basic requirements of the analysis system. To utilize such an approach in practice requires quantitative information on component stress, materials characteristics, and nondestructive evaluation (NDE) capabilities. Much of this information cannot be defined as a single value, but must be described by a probability distribution. Two examples are the probability that a flaw of a given size will exist in virgin material, or the probability of finding a given flaw size with a standard inspection procedure. In order to obtain a deterministic fracture mechanics life prediction (given these distributions), the conventional approach has been to use worst case assumptions for all parameters. Employing all worst case assumptions (deterministic) necessarily results in a conservative estimate for the service life of the component.

To circumvent this difficulty, the problem can be treated probabilstically. A closed-form solution, which takes into account all the required probabilities, is far too complex to be practicable. An alternative solution is to employ computer simulation techniques. A probabilistic life analysis (11, 12) would use a distribution of flaw sizes. This type of analysis results in failure probability as a function of time, includes NDE reliability, and allows selection of an RTS interval to obtain an acceptable (low) failure probability with realistic NDE reliability.

Both deterministic and probabilistic methods could provide some of the NDE reliability through multiple inspections and/or through higher NDE limits due to shorter RTS intervals. Since many NDE errors are the result of human frailty, multiple inspections and automation can enhance detection reliability. The probabilistic system , however, has the ability to accommodate NDE reliability (probability of detection versus crack length) distributions and assess their effect upon RFC efficiency. Obviously, high reliability NDE is desired to optimize the economic benefits of Retirement-for-Cause.

The RFC Procedure - The Retirement-for-Cause (RFC) flow chart (Fig. 6) illustrates a simplified view of how this maintenance concept can be utilized. When an engine is returned for maintenance, an economic analysis is performed on the engine component/module identified as a participant of the RFC maintenance program. If the module has already been in service for several inspection intervals, the probability of finding cracked parts may be great enough to make reinspection economically undesirable and specific components of that module are retired without being inspected. This is determined by the economic analysis at decision point one and is one of three possible decisions. An unscheduled engine removal (UER) may bring a module out of service that is more economical to return to service for the remainder of its inspection interval than to inspect and recertify it for a new full interval (the second possible decision at point one). The remaining choice at point one is to tear down the module and inspect the parts. During inspection there again are three possibilities (decision point two). If no crack is found, the part is returned to service. If the disk is found to be unsafe, it is retired. The third choice is to investigate modification or repair of a flawed part. An economically repairable part may be repaired and returned to inspection (decision point three).

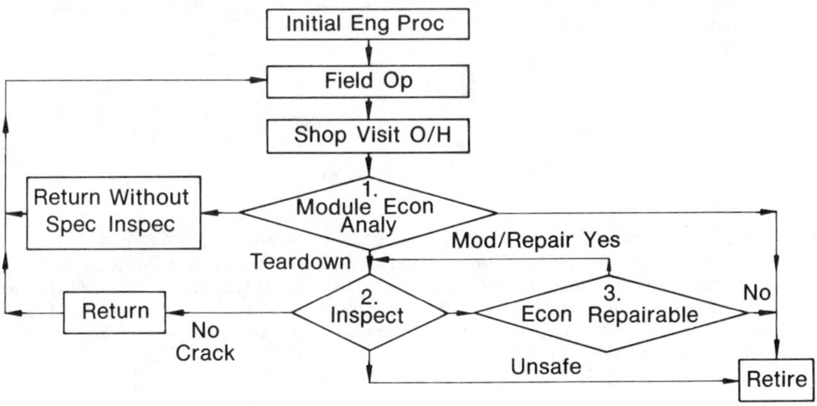

Figure 6. Retirement-for-Cause Procedure Flow Chart

An Application of Retirement-for-Cause

The application of a Retirement-for-Cause maintenance approach to a military gas turbine system has been studied* and technology development programs are currently underway. The objective of the program was to determine the feasibility of applying a Retirement-for-Cause (RFC) maintenance approach to the USAF F100 engine. The study was directed primarily toward rotating components of that engine, specifically the various disks and air-seals/spacers that comprise the prime rotor structure. The technical effort consisted of five tasks:

- Define an RFC methodology and a means of assessing the ROl for its application
- Evaluate the disks of the F100 engine plus other appropriate engine rotor components for RFC applicability
- Assess nondestructive evaluation (NDE) requirements for implementation
- Establish a ranking of components for development priorities
- Establish development plans leading to implementation.

The methodology has been discussed in the preceding section of this paper. The component evaluations used a deterministic life analysis. Based upon this life analysis and life cycle cost analysis, 21 candidate components were selected, and development priorities established. Nondestructive evaluation requirements were determined for each of the components, and a development plan established, which when followed should enable implementation of RFC for the F100 engine in the 1985 time period.

The development plan is shown schematically in Fig. 7, and addresses the technology development requirements discussed previously. At the present time, many of these areas are the subject of Government contract procurement activities, and cannot be discussed in detail. However, to illustrate some of the items which must be addressed, typical NDE requirements can be used as an example.

*The demonstration engine system is the USAF F100 engine. Under Defense Advanced Research Projects Agency (DARPA) and Air Force Wright Aeronautical Laboratories (AFWAL) sponsorship, a study entitled "Concept Definition: Retirement-for-Cause of F100 Rotor Components" Contract F33615-76-C-5172 (as modified in 1979) was performed. The results of this study are documented in AFWAL-TR-80-4118, the final report for that program.

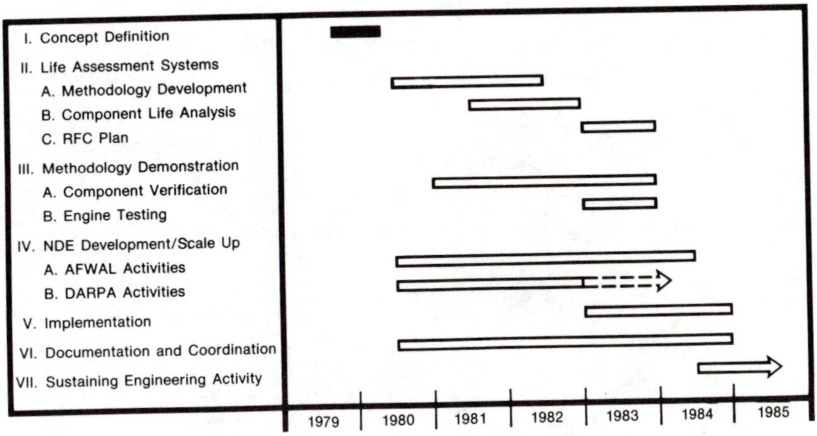

Figure 7. Development Plan Engine Component Retirement-for-Cause

A composite sketch of a typical gas turbine engine rotor component is shown in Fig. 8. As can be seen, the configuration of these components is complex and will require innovative techniques to enable reliable inspection. Examples of the types of defects which will require detection on these components are shown in Fig. 9. Techniques currently exist, such as eddy current, ultrasound and/or penetrants with the capability to detect these flaws. The major problem to be solved is detecting these flaws in a high volume maintenance environment, and integrating other techniques, such as proof testing, into a system with sufficient reliability to enable Retirement-for-Cause to be economically viable.

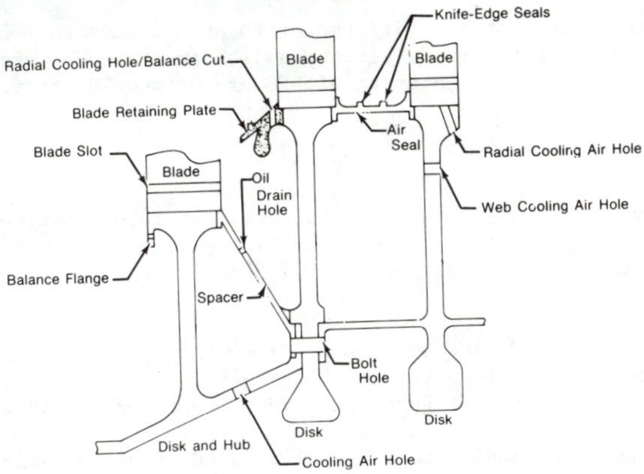

Figure 8. Composite Sketch of Typical Gas Turbine Engine Rotor Components (Not All Features Are on All Parts)

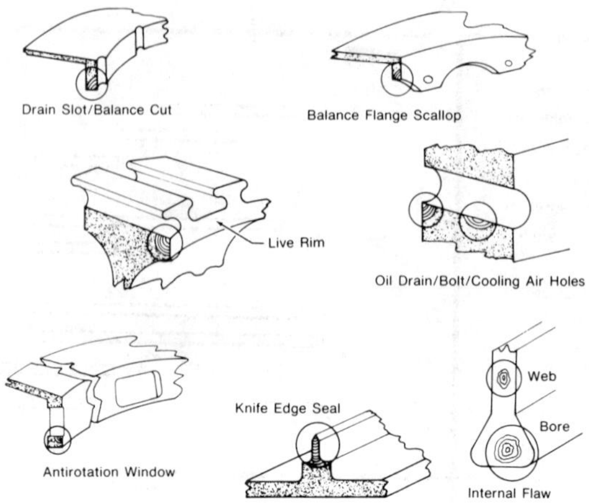

Figure 9. Sketch of Typical Gas Turbine Engine Rotor Component Flaw Types

Benefits - The assessment of the benefits of a Retirement-for-Cause maintenance approach to a gas turbine engine is contingent upon many assumptions. These assumptions include: fleet size, anticipated usage rates, usage life, inspection interval, labor costs, parts cost and many others.

To quantify the benefits of this maintenance concept for the USAF F100 engine, life cycle cost analyses were conducted. These analyses determined the change in life cycle costs of the F100 engine that could accrue based upon implementation of an RFC maintenance procedure in January 1985 as opposed to a continuation of current or baseline maintenance practices.

The life cycle cost benefits amount to an approximate $250 million savings over a 15-year period. In comparison to the investment required, the development and implementation of RFC are extremely attractive.

References

1. C.G. Annis, R.M. Wallace and D.L. Sims, "An Interpolative Model for Elevated Temperature Fatigue Crack Propagation," AFML-TR-76-176, Part I, November 1976, presented at 1977 SESA Spring Meeting, Dallas, TX, May 1977.

2. R.M. Wallace, C.G. Annis and D.L. Sims, "Application of Fracture Mechanics at Elevated Temperatures," AFML-TR-76-176, Part II, November 1976, presented to Air Force Materials Laboratory, WPAFB, OH, May 1977.

3. D.L. Sims, C.G. Annis and R.M. Wallace, "Cumulative Damage Fracture Mechanics at Elevated Temperature," AFML-TR-76-176, Part III, November 1976.

4. D.L. Sims, "Evaluation of Crack Growth in Advanced P/M Alloys," AFML-TR-79-4160, September 1979.

5. J.M. Larsen, C.G. Annis, Jr., "Observation of Crack Retardation Resulting from Load Sequencing Characteristic of Military Gas Turbine Operation," presented at ASTM Symposium on Effects of Load Spectrum Variables on Fatigue Crack Initiation and Propagation, San Francisco, CA, May 1979.

6. J.M. Larsen, B.J. Schwartz, C.G. Annis, Jr., "Cumulative Damage Fracture Mechanics under Engine Spectra," AFML-TR-76-4159, September 1979.

7. C.G. Annis, Jr., "An Engineering Approach to Cumulative Damage Fracture Mechanics in Gas Turbine Disks," presented at ASME Gas Turbine Conference, San Diego, CA, March 1979.

8. J.M. Hyzak, J.E. Allison and W.H. Reimann, "Development of Quantitative NDI for Retirement-for-Cause," AFML-TR-78-198, February 1979.

9. J.S. Cargill, J.K. Malpani and Y.W. Cheng, "Disk Residual Life Studies," AFML-TR-79-123, September 1979.

10. W.H. Lewis, et al., "Reliability of Nondestructive Inspections," SAALC/MME 76-6-38-1, December 1978.

11. C.G. Annis, F.K. Haake and D.L. Sims, "Probabilistic Fracture Mechanics and Retirement-for-Cause," P&WA internal communication, to be submitted for external publication.

12. C.A. Rau, Jr., "The Impact of Inspection and Analysis Uncertainty on Reliability Prediction and Life Extension Strategy," presented at ARPA/AFML Quantitative Nondestructive Evaluation, San Diego, CA, July 1978.

RELIABILITY STRATEGIES IN FLAW EVALUATION

Spencer H. Bush

Battelle-Pacific Northwest Laboratories
Richland, Washington 99352

A sensitivity index is proposed as a tool to fix flaw detection limits appropriately below the critical flaw size without setting unrealistically low values. The index could be used to establish unreliability of UT as well as determining conditions where crack growth rates are too high to set realistic values for flaw detection reliability with UT.

Introduction

The issue of flaw detection and sizing in structures by techniques such as ultrasonics often is beclouded by its adherents and opponents. The adherents often make unrealistic claims as to sensitivity of detection and accuracy of sizing of flaws while the opponents rail against the total inadequacy of UT. If one approaches the problem objectively it becomes apparent that the truth is somewhere between the extremes.

If we relegate UT to its proper niche -- as a tool to assist us in establishing the failure probability of a structure or component, we can better assess its strengths and weaknesses. Flaw detection and sizing is only one input into a failure evaluation. We must know the loads and the material properties to complete such an evaluation. Unrealistic claims of the accuracy of detecting small flaws in complex structures is a disservice to the owner and to the public when the critical flaw size lies within the region of marginal detectability. Similarly, undue concern with detection of flaws below, for example, five percent of the through-thickness is unrealistic when a 50 percent through-wall flaw is not expected to fail.

This paper will review some of the factors influencing reliability of detection and sizing of flaws with UT in the context of what is required, based on fracture mechanics analyses. One possibility advanced is a sensitivity index where UT reliability is set on the basis of failure assessment.

Some "Horrible" Examples

Three examples will be discussed relating to the unreliability of NDE(UT) in components; namely:

- Detection and sizing of flaws in thick sections typical of nuclear reactor pressure vessels;
- Detection and sizing of stress corrosion cracks in nuclear piping;
- Detection and sizing of flaws in turbine rotors and discs.

A program to assess the reliability of detection and sizing of fabrication flaws by UT in thick sections typical of reactor pressure vessels has been underway since the late 1960's under the auspices of the Pressure Vessel Research Committee (PVRC). In the late 1970's the program was expanded to include countries within the European Community under the Plate Inspection Steering Committee (PISC) and in Japan under the Japan Pressure Vessel Research Committee (JPVRC). The so-called PISC-I results will be briefly reviewed. Figure 1 illustrates one typical test geometry, a nozzle-in-shell weldment. PISC UT equipment and procedures fell into three categories.

(A) Those complying with the PISC procedure (1), which in turn, closely followed the requirements of ASME XI(2);
(B) Relatively minor modifications of the PISC procedure;
(C) Markedly different techniques providing a truly alternative procedure.

Using the different procedures three questions were to be answered:

(1) How well can flaws be detected;
(2) How well can they be sized to establish acceptability;
(3) How well can they be sized to establish that they are rejectable.

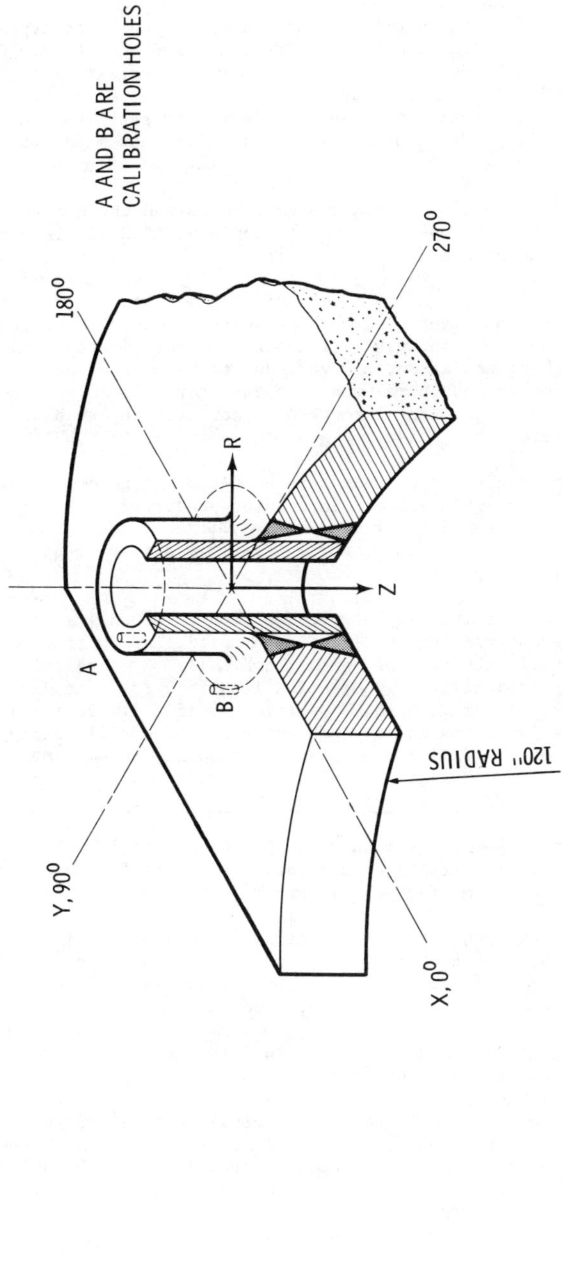

Fig. 1 - Typical Nozzle-Plate Weldment Geometry

The flaws can be grouped as follows:

(a) Small acceptable defects,
(b) Small rejectable defects,
(c) Large continuous defects,
(d) Composite (multiple type) defects.

Figure 2(A), (B), (C) permits a comparison of the three types of techniques with regard to defect detection (DD), correct acceptance (CA) and correct rejection (CR) for types a, b, c and d defects with A being the PISC procedure, B the National Standard or procedures in the spirit of the ASME code, and C alternative techniques. As can be seen the track record is good only for large continuous defects with PISC and modified PISC techniques. It becomes better for more sophisticated techniques.

After examination of the data one can understand the somewhat restrained comment from Europe that the conventional ASME XI UT technique did not appear to be as good as might be desired.

Another example pertains to a case of stress corrosion cracking in austenitic piping (3). Eight similar pipes were examined after leakage was detected in one due to a through-wall crack. In this instance the specific location of cracking was known; however, due to a combination of tight cracks, and use of an unduly relaxed accept-reject procedure of cracks that proved to be up to 70 per cent through-wall were not reportable. The reasons will be discussed later.

A third case occurred in steam turbine discs and rotors. In one instance UT reported a surface crack in a rotor predicted to be one to two inches deep. After removing the crack only about one inch of material remained until reaching the inner bore. Result - a rotor requiring over two hundred days to replace.

Turbine disc cracking can be more insidious. Depending on material, heat treatment and operating conditions the critical flaw size for failure may be approximately the size of one's thumb nail (4). Needless to say one must have a high reliability of detection if a flaw is to be detected prior to reaching critical size. With UT systems specifically designed for examining discrete locations one may achieve the needed reliability; however, there have been enough failures so that I personally require more convincing.

Factors Influencing UT Reliability

At least three sets of variables influence the reliability of detection and sizing of flaws in metallic components. I classify them somewhat arbitrarily into external factors, internal factors and procedures.

Examples of external factors affecting UT reliability include equipment, transducers, search angles, couplant and one could arbitrarily include examination surface and component geometry. Equipment can be further divided with regard to type, frequency, band width, etc.

Internal factors include type of material, condition of material (wrought, cast, etc.), flaw location, flaw orientation, flaw tightness, etc. Procedures usually relate to calibration and accept-reject reporting. Table I lists some of the external and internal factors. Figure 3 illustrates how different procedures may influence rejection (5). For example ASME Section XI at one time or another has required 20%, 50% and 100% reporting and/or rejection using the Distance-Amplitude-Curve (DAC). The

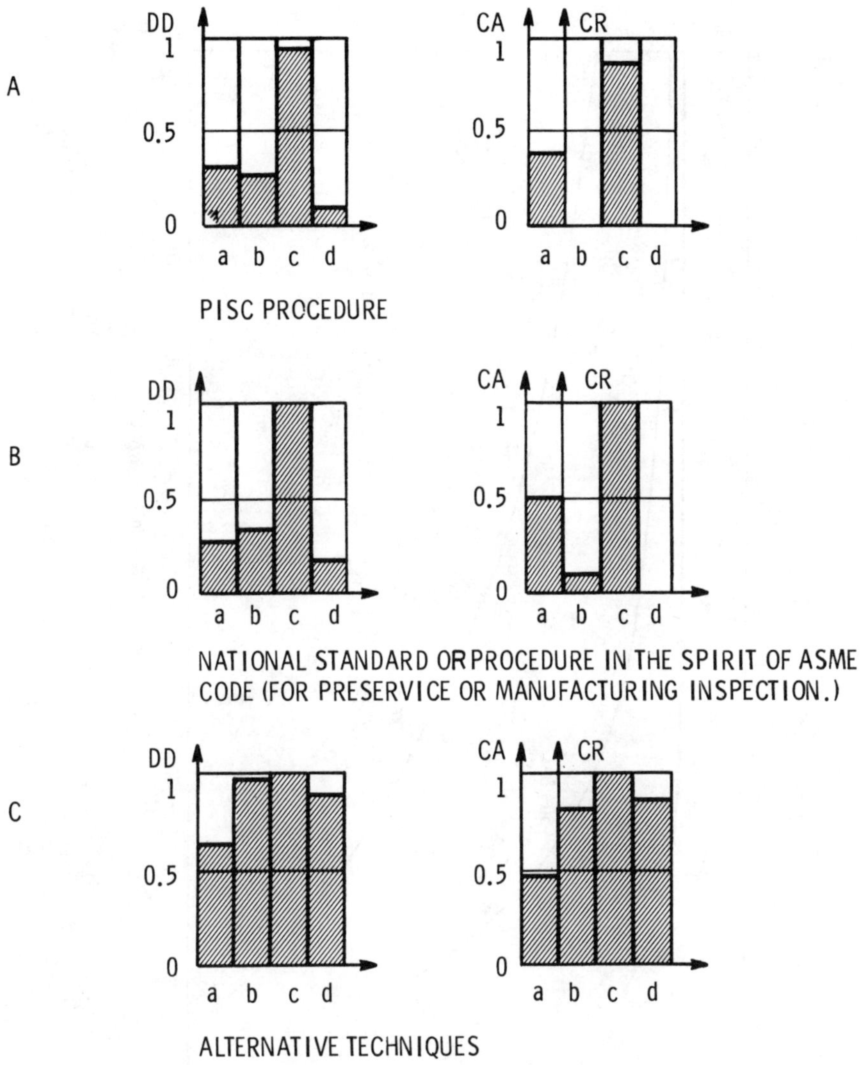

Fig. 2 - Average Cumulative Diagrams of Defect Detection (DD) and Correct Rejection or Acceptance [CR(CA)] as a Function of the Defect Categories a,b,c,d.

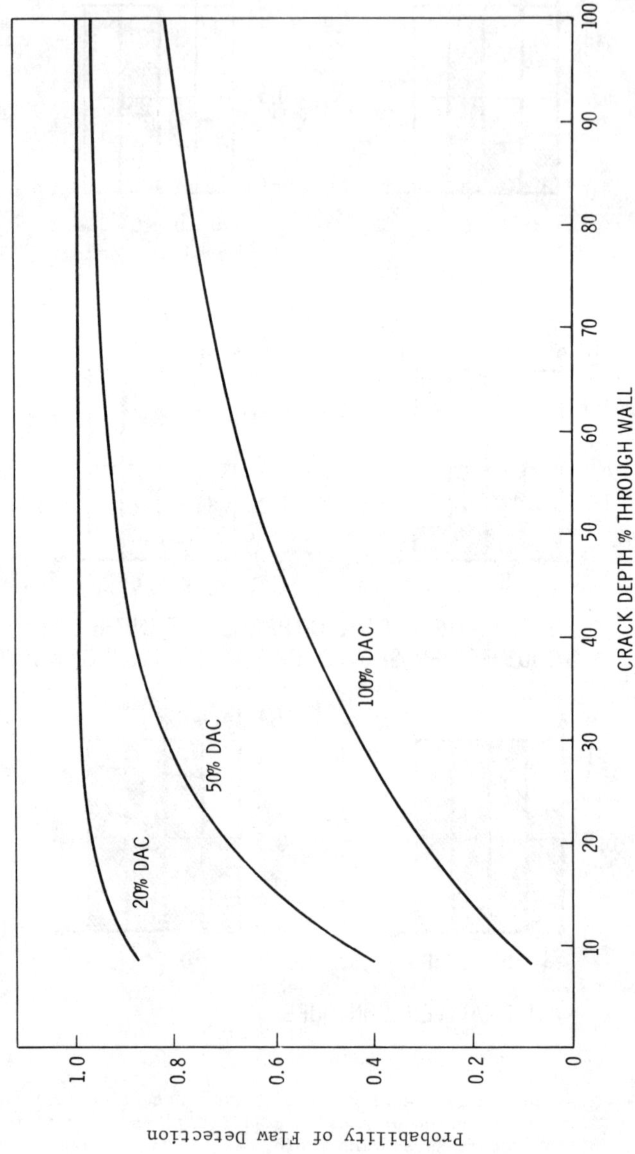

Fig. 3. - Predicted Probability of Detection (POD) Curves For 20, 50, and 100% DAC Reporting Levels

results after examining the same flaw speak for themselves. "This was the factor with regard to IGSCC in Nuclear Piping" (3).

Table I. External and Internal Factors Influencing Ultrasonic Reliability

- Theoretical Considerations
  - Limits of Resolution
- External Factors
  - The Operator
    - Same Operator
    - Operator to Operator
  - Equipment
    - Transducers
    - Frequency
    - Signal-to-Noise
  - Calibration Block
    - Acoustic Properties
    - Type Notch or Hole
    - Natural Flaws
- The Interface
  - Couplant Properties
  - Surface Condition
- Internal Factors
  - Component Geometry
    - Weld Joint Design
    - Cladding
  - Material Variables
    - Grain Boundary Scattering
    - Beam Attenuation
  - Flaw Characteristics
    - Size
    - Location
    - Orientation
    - Roughness
    - Transparency

Another way of expressing external and internal variability is in terms of standard deviations of specific parameters expressed in decibels (dB) of equipment response. Table II from Becker (5) compares his data and that of Silk (6) and Forli (7). One could add material variables to this listing. For example, cast stainless steels may severely attenuate the ultrasonic beam.

Since this is not a treatise on UT reliability, we shall limit further discussion to one parameter affecting the reliability of detection of a crack. Figure 4 illustrates the effect of bending loads on a flaw as such tensile or compressive loads influence the UT signal amplitude. As noted, the amplitude difference may exceed 15 dB; this is a real effect because of transparency to UT of a tight crack. The incident of poor detectability of the IGSCC crack at Duane Arnold was due primarily to this effect.

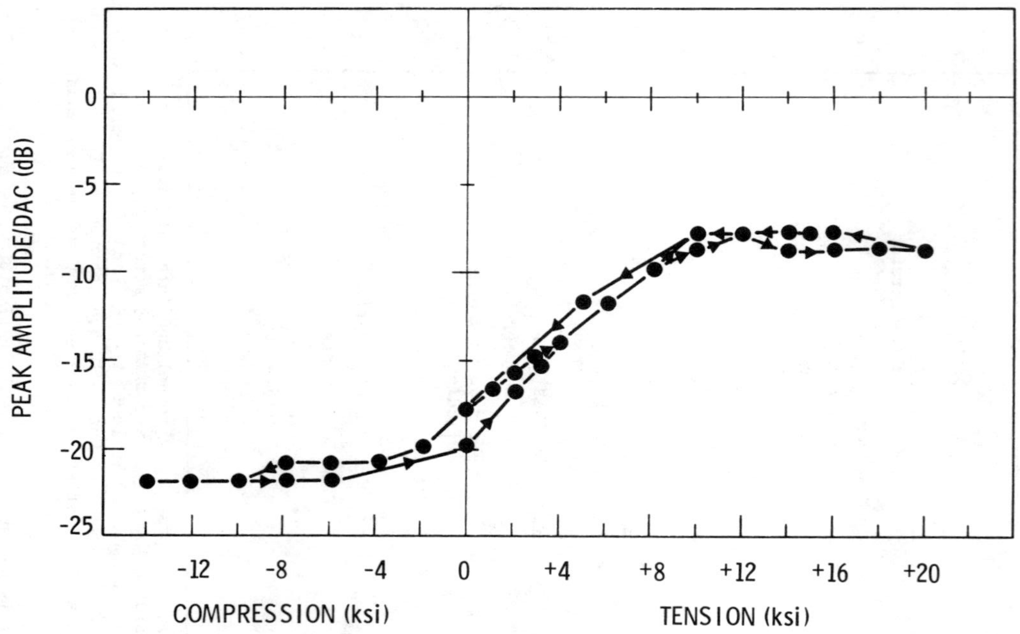

Fig. 4. - Reflected Amplitude from a Fatigue Flaw as a Function of Load in Bending. Yield Stress is Equivalent to a Load of 33 ksi. Flaw size is .193 Inch Deep (32% Through Wall).

Table II. Estimate of Standard Deviation of Inspection Variables

	Becker (5) ± dB	Silk (7) ± dB	Forli (4) ± dB
Defect Orientation	3.6	3.5	---
Defect Roughness	1.4	3.0	---
Transparency	3.7	4.0	---
σ_1 Crack-to-Crack Total	5.4	6.1	3.3
σ_2 Between Inspection Teams	2.0	---	4.4
σ_3 Within Inspector	2.0	2.0	2.6
Coupling	---	2.0	---
σ_r Total Inspection	6.1	6.7	6.0

Failure and Fracture Mechanics

A great deal of progress has been made in assessing the mechanisms of failure and the relationships between material properties, flaw characteristics and loads. Even so, we still have difficulties in handling structures of complex geometries.

Within the limitations of fracture mechanics one can develop semi-quantitative to quantitative relationships. In the context of what we need in UT reliability I will arbitrarily divide cases of interest into three categories:

(1) Situations where section size, appropriate mechanical properties (K_{Ic}), and loads indicate linear elastic-fracture mechanics (LEFM) applies and the critical flaw size is quite small;
(2) As above, LEFM applies due to thick section; however K_{Ic} is high so that the flaw size is relatively large;
(3) Situations where thinner sections are used, such as in piping. LEFM will be very conservative; valid analyses of the critical flaw size will require use of elastic-plastic fracture mechanics (EPFM).

A few simple examples will be used to indicate critical flaw sizes for various combinations of geometry, mechanical properties and loads. The values cited should be taken as indicative and not as absolute since several factors are influencing. One such is the actual versus the idealized flaw geometry. An idealized flaw is planar and of clearly defined shape whereas an actual flaw often is not planar and has a highly variable shape so that the actual flaw may have a substantially larger total area at critical size than does the idealized flaw.

The example selected for small size critical flaws under LEFM conditions is that of large steam turbine discs. The formula used for this and the following example are:

(1) $$K_I = \sigma F (\pi a)^{1/2} \quad (\sigma < \sigma_y) \tag{1}$$

(2) $$a_c = \frac{1}{\pi}\left(\frac{K_{Ic}}{\sigma F}\right)^2 \tag{2}$$

where

K_I = stress intensity factor defining the magnitude of stress distribution. This is calculated in terms of σ, a, F(a) defined as:

σ = the applied, nominally uniform stress

a = the crack depth (half depth if embedded)

F(a) = a factor depending on flaw geometry (c/a), stress gradients, and structural displacement constraints, where c is the crack length

K_{Ic} = critical value of K_I at fracture

a_c = critical flaw size at fracture.

Returning to the turbine case, values of fracture toughness for discs have been cited to be as low as 40 MPa $m^{1/2}$ (36 KSI $in^{1/2}$). The a_c for this is a 2.5 mm (0.1 in) deep crack located in a typical disc keyway. As the toughness increases the critical flaw size increases quadratically with K_{Ic}. For example a_c will be 22.5 mm (0.9 in) for K_{Ic} = 120 MPa $m^{1/2}$ (109 KSI $in^{1/2}$) and a_c will be 62.5 mm (2.46 in) for K_{Ic} = 200 MPa $m^{1/2}$ (182 KSI $in^{1/2}$).

The latter cases of 22.5 and 62.5 mm should be easier to detect plus taking much longer to grow to critical size either in terms of da/dN (fatigue) or da/dt (stress corrosion).

The first flaw case as noted previously is marginal in terms of detection, and, once detected, in terms of sizing with UT. At least one catastrophic failure has occurred due to a flaw in the general size range of 2.5-10 mm (0.1-0.4 in) (8).

In the second case, that of a thick walled pressure vessel, the same general formula for calculating the critical flaw size applies; however, the F factor differs due to differences in geometry. Also, the stresses will differ compared to the turbine case. In a worst case calculation for a loss-of-coolant accident followed by emergency core cooling system injection (of cold water) the thermal and other stresses and the assumed K_{Ic} values of 200 KSI $in^{1/2}$ (220 MPa $m^{1/2}$) yield a critical flaw depth of a_c = 6.5 in (16.5 cm) in the unirradiated zones of a pressure vessel (9). A similar accident where there is a flaw in the irradiated beltline of a vessel yields a_c = 5.9 in (15 cm). These flaws are essentially through-wall in the first instance and 50-60% through-wall in the second instance.

The third case concerns a large IGSCC crack in an austenitic stainless steel pipe. Due to the high toughness and relatively thin thickness, LEFM would yield ultra-conservative results. An alternate approach is to use some version of EPFM such as the tearing modulus model suggested by Paris, et al (10). In fact this approach was used for the Duane Arnold IGSCC case where cracks penetrated 70% of the wall. Tearing modulus (10) calculations indicated no failure for an idealized flaw at 70% penetration (a/t) and 360° around the circumference. Since cracks caused by IGSCC tend to be less severe in a fracture mechanics sense than the idealized flaw, one could assume that a flaw with 70% penetration (or a somewhat larger flaw) would not lead to failure. For purposes of UT evaluation, the detection of a 70% through-wall flaw is quite adequate.

A Sensitivity Index

If we relate flaw sizes, as discussed in the preceding examples, to some factor or factors relevant to NDE (UT), it should be possible to arrive at some suggested lower bounds for detectable flaw depths. The following is recognized as being overly simplistic; however, it can be used as a "straw man" to stimulate discussion.

An obvious criterion is that

(1) The probability of failure (P_F) approaches one, the probability of NDE detection (P_D) for $a < a_c$ should approach one

(2) As a_c increases in size, P_D for $a < a_c$ can become smaller than one.

This concept should be able to discriminate between the relatively impossible and the feasible. Table III lists parameters and suggested values for a "sensitivity index" given by the product of flaw size x geometry factor x growth factor. It should be noted also that conservative (high) growth rates should be used. The examples, given in Fig. 5 and the Appendix, represent five cases reasonably typical of actual experience. The parameters selected are quite arbitrary; their maximum value could be as high as 64 and the minimum conceivably could be as low as 0.125.

Table III. Parameters For A Sensitivity Index

Flaw Size Parameter. If:	and:	then:
$a_c/2 < 5$ mm	$a/t < 20\%$	use 0.5
$a_c/2 > 5$ mm	$a/t > 10\%$	use 10 a_c/t (max 8)
Geometry Factor. If:		
Geometry, weldment, material are simple		use 1.0
Geometry, weldment, material are complex		use 0.5
Growth Factor. If:		
da/dt or da/dN are such that the time to grow the flaw from $a_c/2$ to a_c		
is less than a year		use 0.5
> one year		nearest year (max 8)

SENSITIVITY INDEX

MAX 64

- LARGE a/t (VESSELS, PIPING, ETC.) EMBEDDED FLAWS OR LOW da/dN, FLAWS MAY BE STABLE, ONCE DETECTED; REPAIR DUE TO FLAW SIZE OR MONITOR PERIODICALLY.

10 — FLAWS (a_c) MEDIUM TO LARGE, CRACK GROWTH RATES FAIRLY HIGH TO LOW; P_D SHOULD BE GOOD; REGION OF BEST USE FOR UT

2.0 — FLAW MAY BE LARGE; P_D SHOULD BE GOOD; CRACK GROWTH RATE MAY BE SO HIGH THAT CRACK PENETRATES WALL IF NDE INTERVALS > YEAR

1.0 — DETECTION RELIABILITY MAY BE LOW; MAY REQUIRE AT LEAST TWO NDE TECHNIQUES; FLAW WILL BE SMALL (a_c); FAILURE PROBABILITY RELATIVELY HIGH

Fig. 5. – A Graphical Example of the Sensitivity Index, Indicating Relationship Between Index Values and Flaw Characteristics.

Summary

Critical inputs into any fracture mechanics model are the shape, size, orientation and location of flaws in systems being analyzed. All of these parameters will have finite error values when determined by non-destructive examination (NDE) techniques such as ultrasonics (UT). The factors influencing ultrasonic detection, sizing and location of defects are examined to permit a prediction of reliability values for the various parameters. The linear elastic fracture mechanics model used in Appendix A of ASME Section XI of the Boiler and Pressure Vessel Code is examined in the context of flaws in various locations in a pressure boundary to infer the necessary levels of reliability in detecting, locating and sizing these flaws by UT. Where the NDE reliability values were considered to be inadequate with conventional UT, based on analysis of fracture mechanics parameters, potential gains with alternate or advanced NDE systems were examined.

REFERENCES

1. Plate Inspection Steering Committee (PISC), Euratom Report, EUR 6371 en 1979, Vol. I-V:

   (a) Vol. I
   (b) Vol. II
   (c) Vol. III
   (d) Vol. IV
   (e) Vol. V
   (f) Vol. VI Analyses of the PISC Trial Results for Alternate Procedures 4/80.

2. ASME Boiler and Pressure Vessel Code Section XI, "Rules for Inservice Inspection of Nuclear Power Plant Components", 1977.

3. T. Lambert, data obtained from tests at Duane Arnold, cited in NUREG-0531, Investigation and Evaluation of Stress-Corrosion Cracking in Piping of Light Water Reactor Plants.

4. J. M. Hodge and I. L. Mogford, "UK Experience of Stress Corrosion Cracking in Steam Turbine Discs", Proc. Inst. Mech. Engrs. 193, 3/79, pp 93-109.

5. F. L. Becker, "Ultrasonic Inspection Reliability for Primary Piping Systems", 10th Nuclear Power Educational Seminar, Paper No. 29, Southwest Research Institute, San Antonio, Texas, April 21-24, 1980.

6. M. G. Silk, "Estimates of the Magnitude of Some of the Basic Sources of Error in Defect Testing", AERE-R-9023, 2/78.

7. O. Forli, "Comparison of Radiography and Ultrasonic Testing", 2nd Nordske NDT Symposium Kobenhavn, May 21-23, 1979.

8. D. Kalderon, "Steam Turbine Failure at Hinkley Point A", Proc. Inst. Mech. Engrs, 1972, 186, pp 341-377.

9. "Flaw Evaluation Procedures: ASME Section XI", EPRI NP-719-SR Special Report, August 1978, prepared by American Society of Mechanical Engineers, New York, edited by T.U. Marston, EPRI, 340 pps.

10. H. Tada, P. Paris, and R. Gamble, "Stability Analysis of Cracks in Reactor Piping Systems", NUREG/CR/0838, June 1979.

Appendix

Some Examples:

1. Small flaw, LEFM; complex geometry; avg. (2 yrs) growth
   0.5 x 0.5 x 2 = 0.5

2(a). Large flaw; EPFM; complex geometry; fast growth
   8 x 0.5 x 0.5 = 2.0

2(b). Large flaw, EPFM; complex geometry; avg. (2 yrs) growth
   5 x 0.5 x 2 = 5.0

3(a). Medium flaw, LEFM; complex geometry; slow growth
   2.5 x 0.5 x 8 = 10.0

3(b). Large flaw, LEFM; complex geometry; slow growth
   5 x 0.5 x 8 = 20.0

Example 1

Low K_{Ic} values; LEFM applies; section may or may not be thick; geometry sufficiently complex to complicate UT; crack growth rate medium.

Small Flaw	Geometry Complex	Growth to Critical Size		2 Yrs
0.5	0.5	2	$\cong$	0.5

Comments

Reliability of flaw detection probably not close to one. Specialized techniques would be necessary. Even then reliability could be inadequate. Critical flaw as low as 2.5 mm.

Example 2(a)

High toughness (high J_{Ic}); LEFM does not apply; complex geometry; relatively thin section; fast flaw growth rate.

Large Critical Flaw	Complex Geometry	Fast Growth		
8	0.5	0.5	$\cong$	2.0

Comments

Time intervals between UT examinations probably too long to assure reliability. Crack growth rates very high. Value of UT limited; a crack detection system might prove more reliable.

Example 2(b)

High toughness (high J_{Ic}); LEFM does not apply; complex geometry; growth rate is intermediate.

Large Critical Flaw	Complex Geometry	Intermediate Growth		
5	0.5	2	=	5.0

Comments

Critical flaw essentially half through-wall. a/t 50%; at 25% a/t detection relatively straight-forward providing flaw not transparent to UT and appropriate detection criteria used.

Example 3(a)

Intermediate toughness; LEFM applies; geometry complex; low crack growth rate.

Medium Flaw	Complex Geometry	Growth Rate		
2.5	0.5	8	=	10

Comments

Critical flaw at 25% a/t. Material and geometry typical of a ferritic pressure vessel. Detection by UT at 12.5% a/t considered reasonable. Very low growth rate.

Example 3(b)

High toughness ($\sim$200 MPa $m^{1/2}$); LEFM should apply due to thick ($\sim$25 cm) section. Geometry is complex; slow crack growth rate.

Large Critical Flaw	Complex Geometry	Slow Crack Growth		
5	0.5	8	=	20

Comments

Critical flaw at 50% a/t; very slow crack growth rate. A detection threshold of 25% a/t should give nearly 100% flaw detection reliability. Complex geometry or atypical flaw orientation could influence detection.

RELIABILITY STRATEGIES FOR NOMINALLY BRITTLE MATERIALS

R. Nathan Katz and Edward M. Lenoe
Army Materials and Mechanics Research Center
Watertown, Massachusetts 02172

The introduction of brittle materials as alternatives to traditional engineering materials offers many potential advantages. However, the unforgiving nature of such materials imposes the need for rigorously defined component reliability. Yet, for such non-traditional materials; data, design experience or codes of practice are often limited or lacking. Further, existent NDE and fracture mechanics based life prediction formalisms may require modification to be appropriate. The authors present several case histories of development programs where the forgoing limitations prevailed, and nevertheless reliability criteria were developed during the course of component development programs. The reliability strategies consisted of; design modification, improvements in materials and processing, and use of available NDE coupled with empirically derived proof and qualification tests.

Introduction

Many of today's high performance engineering systems, particularly in military and energy conversion technologies, require combinations of materials properties which are not available in traditional materials. In many cases the unique combinations of properties which are required are to be found only in certain ceramic or advanced fiber reinforced composite materials. All ceramic materials and many advanced composites are brittle, which tends to inhibit their utilization by structural designers. As a consequence, design experience and codes of practice or design procedures are generally lacking or ill-defined for these materials. Furthermore, data bases necessary for design are often inadequate or at best marginal. Detailed information on failure modes and failure criteria are also either ill-defined or lacking. Given this situation, a natural question is "How does one develop a reliability strategy for the successful utilization of brittle materials in engineering design?"

The point of departure in the introduction of new, brittle materials is to consciously apply the iterative design trade-off process (illustrated in Figure 1). From the preliminary design stage all the way through the "accepted" component this process must be viewed as a systems problem, and efforts made to factor in reliability into every aspect of the process. Thus, in the feasibility stage shown in the top box in Figure 1, during the materials input part of the design process, in-process quality assurance is factored in from the onset. Similarly, when performing the preliminary design and engineering calculations, estimates of failure modes and required inspection criteria are factored in as early as possible.

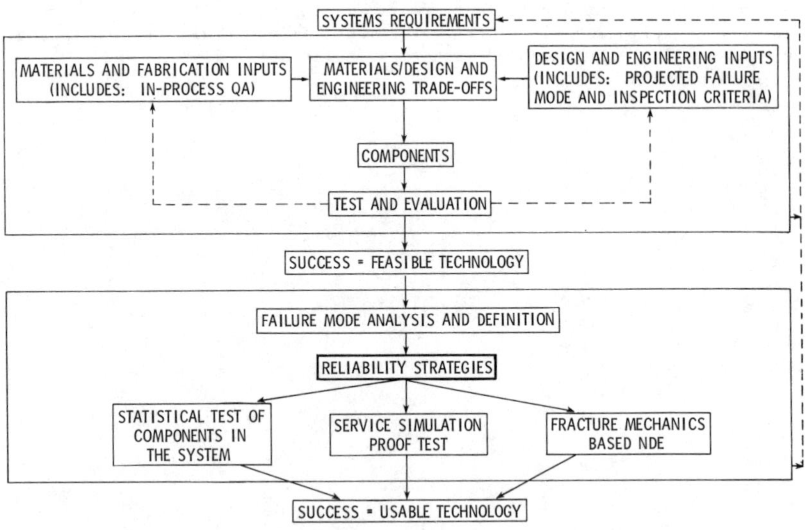

Fig. 1. Brittle materials design/reliability — systems trade-off logic.

While it is essential to obtain high levels of reliability at each sub-step, it is also eventually necessary to develop a reliability strategy which will assure acceptable performance as defined by systems requirement analyses (see Figure 1). These reliability strategies include statistically testing components in the real system. Such full scale evaluation is often too expensive and time consuming to attempt in most cases. An alternative is to use service simulation proof tests. Proof testing has the advantage of checking individual components, hopefully in a simulated service environment, and therefore providing a level of confidence in each component. However, care must be taken during the proof testing process not to induce additional or unaccounted for damage or degradation to the component. If such damage should occur the proof test becomes invalid. Another technique for reliability assurance is a fracture mechanics-based nondestructive evaluation (NDE) strategy. This is the strategy commonly used in high performance metal parts, for example, in the aircraft industry. However, it is more important to have a quantitative fracture mechanics-based NDE for ceramics and certain advanced composites than for metals because the former two are less defect tolerant.

This is a very attractive area (1) for continued development since it ultimately promises an inexpensive method of quality assurance. Another important point to note is that in both the service simulation proof test method or the fracture mechanics-based NDE method one is only investigating the existence of initial flaw populations. The very important issue of long term effects of service environment on properties, especially service generated flaw population development, is not addressed.

The remainder of this paper will focus on a brief description of the NDE-based reliability approach, and its limitations for ceramic and advanced composite materials. We will then turn to a brief description proof testing strategy and describe several areas where this has been implemented with beneficial results. Finally, description and a case history of empirically derived proof testing procedures during the early stages of component development will be described.

NDE-Based Reliability Strategies

The NDE-based approach to the reliability of brittle materials is founded on fracture mechanics estimates of critical flaw sizes for the component and service stress conditions of interest. The critical stress is dependent upon the fracture toughness, which is a materials property, and upon the critical crack length. By establishing the minimum existing flaw sizes, the NDE method seeks to insure that a safe working stress as determined from fracture mechanics calculations exists and thus failure can be precluded. Often in ceramic materials the critical crack length or critical flaw size is on the order of 25-50 microns (2). Also, there are a wide variety of possible flaws and one must determine the relative controlling influence of flaws which will be detected by NDE indications. Shown graphically in Figure 2 which is taken from the work of Evans (3), are the effects of several classes of defects on the strength of hot pressed Si_3N_4. Note that for hot pressed silicon nitride an NDE indication of a flaw which may be tungsten carbide and would show up on the radiograph as a high density flaw, is not deleterious at all, whereas an indication of an iron inclusion which would also show up as high density flaw is moderately deleterious to strength and silicon would be very deleterious to strength. Similarly a surface crack which might not show up at all on the radiograph is most deleterious to strength and voids which are very difficult to detect unless they are extremely large are also extremely deleterious. Therefore, knowing the nature of the flaw is crucial to being able to decide whether to accept or reject the part.

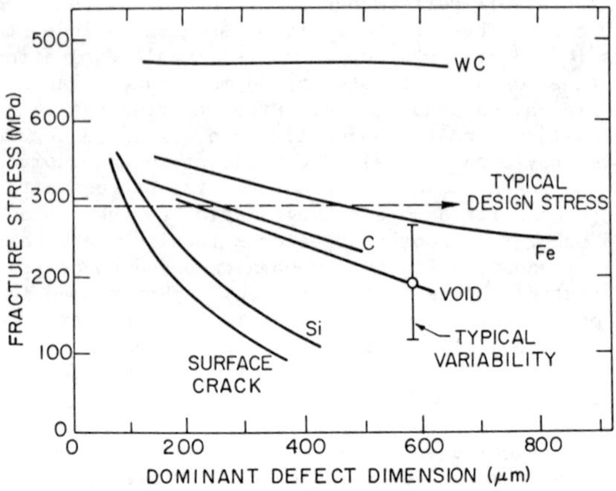

Fig. 2. Strength-size relations for various fracture initiating defects in silicon nitride [after Evans (3)].

It is well known that not only the nature of the flaw is important but specific location and orientation of a given flaw in relation to the stress field in the component is important. For example, in an area where the stress is principally compressive or slightly tensile a much larger flaw can be tolerated than in an area where the stresses rise to very high tensile levels. Thus, rejecting a piece merely due to an indication of a flaw may or may not be appropriate. In fact, much effort remains to be completed to define the influence of various types of flaws on component strength, particularly in severe, long term environmental exposures.

The current state of the art in NDE of ceramics while it has been advancing rapidly is not yet in a state where it can provide such detailed information on the nature and location of flaws in a ceramic component. However, NDE is important and is being effectively used in many ceramic component programs to screen out gross flaws and to assure that those components that go into service or into service simulation proof testing (which is an additional cost) are reasonably sound. Also NDE of green bodies or unmachined billets of ceramic gives information as to the level of quality control during the processing and as such provides feedback of possible deterioration of process controls. However, at the present time, one cannot totally rely on NDE to insure that a ceramic component which has been examined will indeed perform satisfactorily. Regarding flaws in composite materials, it must be recalled that the relative controlling influence of numerous flaws also remains to be precisely defined.

Fracture Mechanics-Based Proof Tests

Fracture mechanics-based proof testing is valuable in that it permits both screening out defective parts and also provides a methodology for estimating a minimum lifetime of a component. In using this method it is generally assumed that there is a degradation of materials properties with

ASSUME: SLOW CRACK GROWTH OR POWER LAW MATERIAL DEGRADATION WITH TIME

$$V = AK_I^{\,n} = dc/dt$$

AND

$$\sigma = K_{Ic}/Y\sqrt{c}$$

COMBINING AND INTEGRATING OVER t

$$t_f = BS^{n-2}\sigma_a^{-n}$$

$$B = 2/AY^2(n-2)K_{Ic}^{\,n-2}$$

IF APPROPRIATELY PROOF TESTED

$$t_{f\,min} = BS_{min}^{\,n-2}\sigma_a^{-n}$$

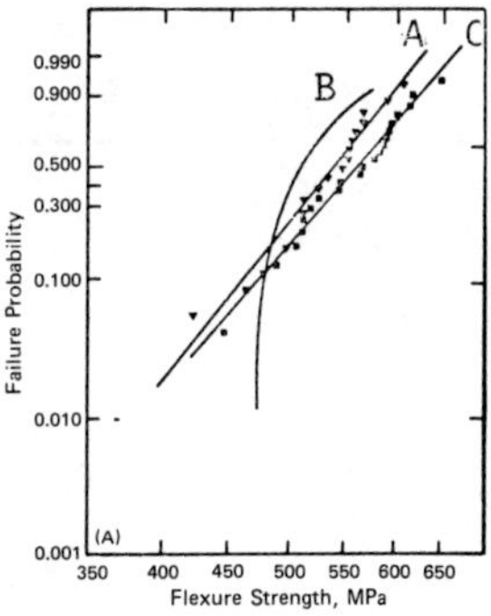

Fig. 3. Fracture mechanics-based proof test to minimum time to failure.

time which follows a given power law relationship.* The power law relationship is usually associated with a slow crack growth phenomena. It is important to realize that various mechanisms of material degradation can lead to failure. Once the nature of material failure is understood then, in principle, an appropriate proof test may be developed. Regarding slow crack growth as idealized by the power law, materials parameters are determined by slow crack growth experiments on either flexure bars (4) or double torsion specimens (5); and the fracture toughness and critical crack length are measured from a separate set of strength tests. The proof test sets a minimum materials or component strength under the conditions in which the test was run. The problem with using this fracture mechanics proof test is that it is possible that at different temperatures or after a given period of service, a flaw population other than the flaw population which was initially tested will control strength. This is shown in Figure 3 which was taken from the work of Wiederhorn and Tighe (6). Illustrated are probability of failure versus strength in a Weibull type plot. Curve A is for a hot pressed Si_3N_4 as-machined. The material was proof tested at room temperature and a truncation in failure stress was subsequently observed on retesting the population (curve B). However, when these samples were tested at high temperature (1200°C) instead of the samples following the truncation curve (curve B) the test results indicated that there was no truncation and the high temperature data followed curve C. Thus, it is evident that proof testing must be very carefully conducted to assure that the simulation to service is in fact attained.

Notwithstanding the above potential problem in the use of proof tests, such procedures have been utilized in several ceramic and advanced composite systems. These uses are listed in Table I. The applications range from high volume commercial products such as vitreous bonded grinding wheels and specialty military products such as filament wound motor cases and fused silica radomes. In these three cases the applicability of proof testing has been verified. The use of proof testing in the space shuttle window has yet to be verified by flight testing; however, previous use of this technique in the Skylab window indicates high levels of confidence in the method.

---

* This relationship $V = AK^n$ and the other equations used in life prediction are shown in Figure 3. Definitions and a more detailed discussion can be found in Chapter 4 of Reference 1.

Table I: Analytically Derived Proof Tests Used in Advanced Systems

SYSTEM	CLASS OF MANUFACTURING	TYPICAL PROOF STRESS RATIO
VITREOUS BONDED GRINDING WHEELS	MASS PRODUCTION	1.8 - 1.5
SMALL FILAMENT-WOUND ROCKET MOTOR CASES	HIGH VOLUME PRODUCTION	~2.5
SLIP-CAST FUSED SiO_2 RADOMES	LIMITED VOLUME PRODUCTION	<2
SPACE SHUTTLE WINDOWS	INDIVIDUALLY CRAFTED	3 - 2.5

Table II: Proof Testing of Vitreous Bonded Grinding Wheels (VBG) [After Ritter, et al. (7)]

MANUFACTURER/ MATERIAL		FRACTURE MECHANICS-BASED PROOF STRESS (psi)	SPR	EMPIRICALLY DERIVED SPR
VBG-1	STATIC FATIGUE TESTS	2715	1.57	1.80
	DYNAMIC FATIGUE TESTS	2825	1.60	
VBG-2	STATIC FATIGUE TESTS	2285	1.45	1.50
	DYNAMIC FATIGUE TESTS	2280	1.45	

• NO KNOWN FATIGUE FAILURES HAVE OCCURRED IN SERVICE

Ritter and his co-workers (7) have recently done extensive work on proof testing of vitreous bonded grinding wheels. This work is summarized in Table II. Using two test techniques, both static fatigue tests (stress rupture testing) and dynamic fatigue tests (variable strain rate tests), Ritter and his co-workers determined the fracture mechanics based-proof stresses for the various grinding wheels. Furthermore, they calculated a spin proof ratio required to guarantee survival of the grinding wheels. What is interesting to note is that the manufacturers of grinding wheels have been using empirically derived spin proof ratios which in one case were very close and in another case were in significant excess of the analytically derived spin proof ratio. This information has allowed one manufacturer to have the option to reduce the spin proof ratio which would result in rejecting fewer grinding wheels, thus lowering cost. It is important to realize that since spin proof testing has been used in the vitreous bonded grinding wheel industry in the United States no known fatigue failures have occurred in service. This clearly demonstrates the power of this technique in providing a reliability strategy for the use of a brittle material.

Service simulation spin proof tests are also being factored in to several advanced gas turbine programs where the use of ceramic turbine blades is being demonstrated. The advantage of being able to spin test individual ceramic blades is that one can address the problem of service simulation at high temperatures fairly conveniently. In most hybrid metal hub-ceramic bladed rotor designs where a ceramic dovetail is inserted into a metal rotor disk, the maximum stresses and the usual modes of failure are in the dovetail region. During operation of the rotor, even at temperatures of 1200-1370°C on the blade airfoil, the dovetail will function at a maximum temperature of the order of 760-870°C. The maximum stresses in the dovetail are therefore at a temperature of only 760-870°C. Thus, by placing the entire rotor assembly into a hot vacuum spin pit, operating at 870°C one can conduct overspeed proof testing at the service temperature. This does not take into account any possible effects of long term deterioration or generation of new populations of flaws which might occur in an oxidizing environment.

Case Study of the Development of Empirical Proof Tests

Let us now turn our attention to the development of empirical proof tests in a specific development program. One of the best documented case histories of such an effort is provided by a stator development program* (8). The goal was to demonstrate an uncooled reaction bonded Si_3N_4 stator over a representative engine duty cycle comprising 175 hours at 1050°C and 25 hours at 1370°C in a test rig (an engine minus rotors). Short duration engine tests (10 to 100 hr) were to be conducted with metallic and, later, ceramic rotors to validate the overall engine feasibility.

Table III lists some of the systems considerations and requirements which were subjected to the trade-off analysis illustrated in Figure 1. Injection-molded reaction-bonded silicon nitride (RBSN) and reaction-sintered silicon carbide were chosen to cycle through the iterative design/materials/ process refinement procedure Figure 1). During the development of the RBSN stator, many empirically derived qualification tests were developed. The iterative development of this stator proceeded over a period of seven years during which five major iterations (plus many more subiterations) were carried out. These included four iterations of design and three iterations of materials, summarized in Figure 4.

The original design, A in Figure 4, included an integral first-stage rotor shroud but did not include an inner shroud. Individual vane segments were injection molded and bonded together with a silicon metal slip in the green state to form the full first-stage stator. The vane segments attained a nitrided density of 2.2 gcm^{-3} while the bonding material had a somewhat lower density. During tests of this stator in the static engine rig the bonding material in the outer shroud often failed. More significantly, vane root thermal cracks were initiated during thermal transients. These led to

* Carried out jointly between Defense Advanced Research Project Agency (DARPA), Department of Energy (DOE) and Ford Motor Company.

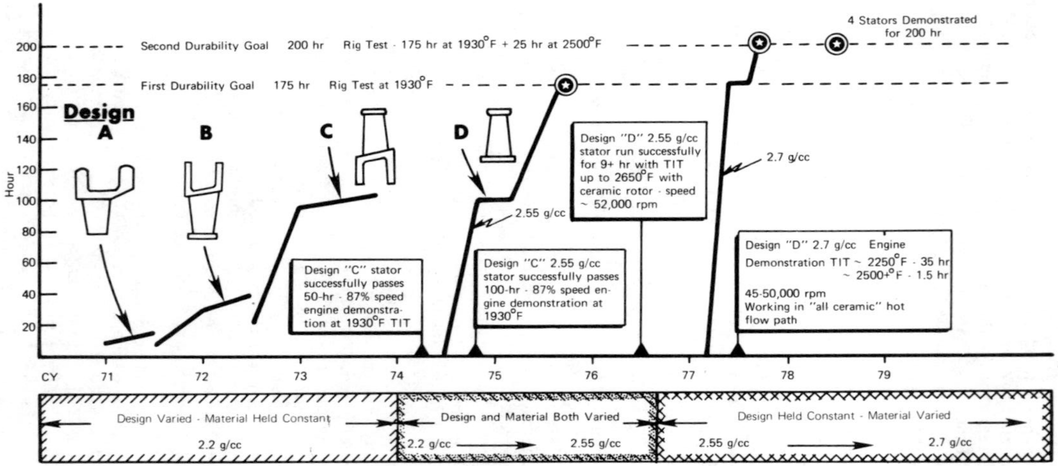

Fig. 4. Engineering development history - first stage RBSN stator ARPA/Ford engine - a cast study of iterative "Brittle Materials Design".

Table III: Materials/Design Considerations

STATOR FOR TYPICAL SMALL AXIAL AUTOMOTIVE G.T. (150-250 HP, ~100 KW)

SELECTION CRITERIA:
- HIGH TEMPERATURE CAPABILITY
- THERMAL SHOCK AND THERMAL FATIGUE RESISTANCE
- CORROSION/EROSION/OXIDATION RESISTANCE
- FABRICABILITY
- STRENGTH

MOST IMPORTANT SELECTION CRITERIA:
- COST

Duty Cycle	LIFE OR TBO	STEADY STATE, σ_{MAX}	TRANSIENT, σ_{MAX}
VERY SEVERE, MANY TRANSIENTS, 2500 F (TIT)	~3500 HR	~1-3,000 PSI	12,000 PSI

POSSIBLE ENGINEERING CERAMICS: INJ MOLD RBSN, SLIP CAST RBSN, R.S. SiC

FAVORED CERAMIC/FABRICATION ROUTE: INJ MOLD RBSN 2.7 g/cc or R.S. SiC

loss of vanes on light-off or shut-down. Two dimensional finite element stress analysis was used to attempt to understand the thermal shock problems. This resulted in elimination of the integral rotor shroud and a thinning down of the outer stator shroud channel. Aerodynamic considerations led to the introduction of inner vane shrouds. This first iteration, Design B, is shown in Figure 4. Again, vanes were injection molded as individual segments, bonded together, and nitrided to 2.2 gcm^{-3}. Design B led to significantly longer test rig life than Design A. An important factor in the improved performance was the better shroud-to-vane mass balance in Design B. The vane cracking, which was not predicted by the 2-D analysis, moved to the mid-span of the vane, and outer shroud cracking continued as before. The use of a laboratory thermal shock rig, which could be considered a kind of proof test, (Figure 5) allowed empirical evaluation of several modified vane profiles.

Based on this empirical design input a second iteration, Design C, was defined in which the vanes had their trailing edges cut back. Stators were still assembled from individual segments and nitrided to 2.2 gcm^{-3}. These design changes with material being held constant increased maximum stator life time from approximately 10 hours to approximately 100 hours. It is significant to note that the Design C stator was used in the first major engine demonstration of 50 hours at up to 87% design speed and 1050°C. (Full speed occurs at 1370°C and therefore required a ceramic rotor, which was unavailable.)

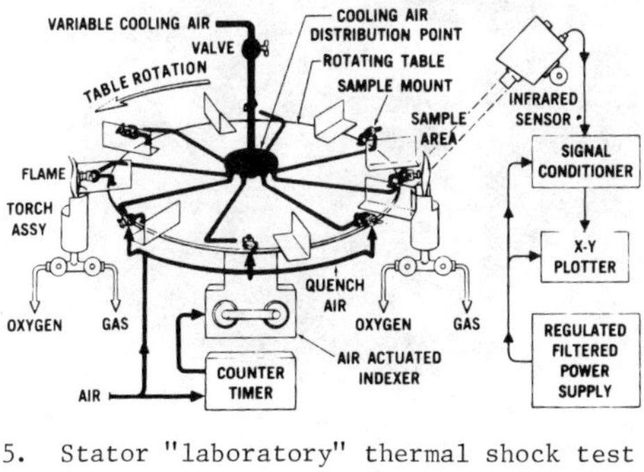

Fig. 5. Stator "laboratory" thermal shock test rig.

At this stage of development failures in the bond joint between the individual vane segments were attributed to oxidation. Factoring in the difference in thermal expansion of oxidized bond material and oxidized vane material (the two materials had different densities and thus oxidation rates) into a detailed three dimensional finite element analysis confirmed that in order to meet the program goals a one-piece stator would be required. Tooling for a one-piece stator was procured in the third design iteration D. This design had the same vane geometry as Design C but the outer shroud channels were removed and the gaps in the inner shroud were precisely defined. Simultaneously with the introduction of Design D, the first materials iteration, 2.55 gcm^{-3} RBSN, was incorporated. The combined effect of this improvement in design configuration and material enabled the program to achieve two important program goals: a 175-hour rig durability demonstration, and a 100-hour engine test demonstration. These are also noted in Figure 4.

Outer shroud cracking was still a problem with the Design D 2.55 gcm^{-3} materials, possibly related to oxidation of the RBSN. At that time it was considered that the elimination of the outer shroud cracking could be most directly addressed via materials improvement, thus design of the stator was frozen. The final iteration was the introduction of 2.7 gcm^{-3} material. This last iteration permitted the stator to achieve a 200-hour program demonstration as well as several engine tests with ceramic rotors at temperatures of up to 1400°C. Thus, the demonstrated durability of the first-stage stator increased from approximately 10 hours in 1971 to 200 hours, reproducibly demonstrated in 1977.

Between 1976 and 1978 several empirically derived proof tests were developed which allowed some degree of prediction as to which stators would survive the demonstration test; these included stator vane bend tests, Figure 6, stator outer shroud pressure tests, Figure 7, and an oxidative weight gain proof test, Figure 8.

Fig. 6. Stator vane bend test.

It is important to note that these predictive proof "tests", especially that utilizing oxidative weight gain during relatively short exposure, resulted from experience gained working with the materials and components and could not have been anticipated in the advance of that experience. In the course of the iterative stator development, significant engine demonstrations were made which showed the stators would perform for 35 to 100 hours in a real engine environment, Figure 4. The U.S. Department of Energy is currently supporting further testing on both RBSN and reaction-sintered SiC stators in a more rigorous automotive duty cycle. Lives of approximately 500 hours incorporating over 30,000 severe thermal transient cycles have been demonstrated on RBSN stators in the Department of Energy Program (9).

The DARPA/DOE/Ford stator is a case study in successful design of a ceramic structure, which was critically dependent upon the development of appropriate empirical proof tests.

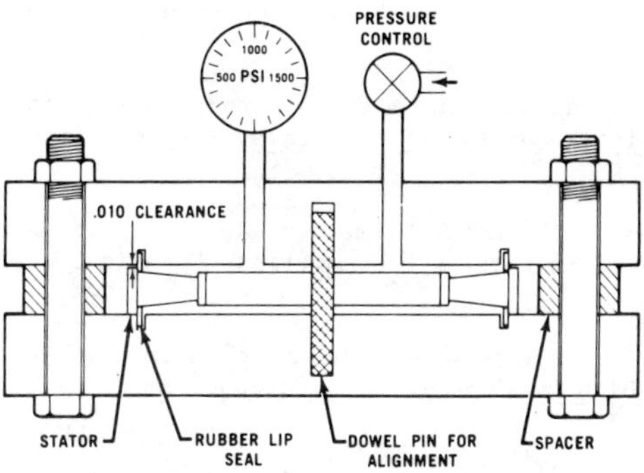

Fig. 7. Stator outer shroud pressure test.

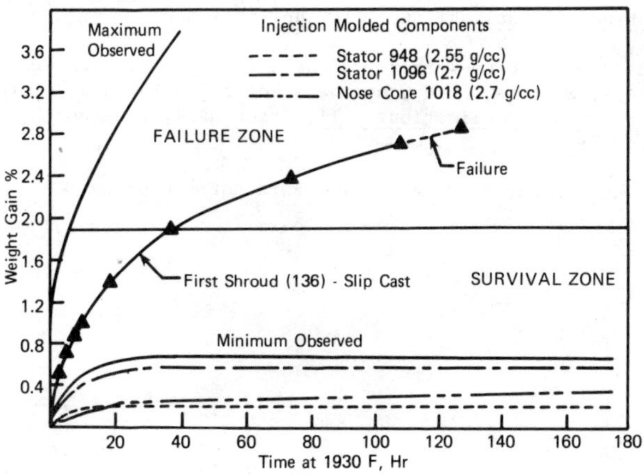

Fig. 8. Oxidative weight gain used to predict stator quality and survival.

Summary

The relatively recent emergence of ceramics and composites as structural materials has tended to limit experience in assuring reliability of these materials when compared to more traditional structural materials such as metals. Nevertheless, it has been shown above that the application of reliability strategies which combine in-process quality assurance (QA), fracture mechanics-based or empirically derived proof tests, and NDE to the extent possible, has resulted in successful utilization of such materials in many areas. It is these successful applications, themselves, which will provide the experience base required for the confident use of brittle materials in future applications.

References

1. Reliability of Ceramics for Heat Engine Applications, NMAB Report NMAB-357, 1980.

2. F. I. Baratta, G. W. Driscoll and R. N. Katz, Ceramics for High Performance Applications, Brook Hill, Chestnut Hill, MA, 1974, pp. 445-476.

3. A. G. Evans, "Structural and Microstructural Design in Brittle Materials," Mechanical Behavior of Materials, Pergamon Press, Oxford 1980, pp. 279ff.

4. G. D. Quinn and R. N. Katz, "Time Dependent High Temperature Strength of Sintered Silicon Carbide," J. American Ceramic Society, 63, 117-119, 1980.

5. A. G. Evans, "High Temperature Slow Crack Growth in Ceramic Materials," Ceramics for High Performance Applications, Brook Hill, Chestnut Hill, MA, 1974, pp. 373-396.

6. S. M. Wiederhorn and N. J. Tighe, Reliability of Ceramics for Heat Engine Applications, NMAB Report NMAB-357, pp. 96-97.

7. J. Ritter and S. Wiederhorn, "Applications of Fracture Mechanics to Assure Reliability of Ceramic Components," Ceramics for High Performance Applications III-Reliability, (In publication).

8. D. L. Hartsock and co-workers, "Test and Development of Ceramic Combustors, Stators, Nose Cones, and Rotor Tip Shrouds," Ceramics for High Performance Applications - II, Brook Hill, Chestnut Hill, MA, 1978, pp. 293-315.

9. Unpublished research, W. A. Trela, Ford Motor Company.

NON-DESTRUCTIVE FAILURE PREDICTION FOR BRITTLE SOLIDS

A.G. Evans
Lawrence Berkeley Laboratory
Materials Science and Mineral Engineering
University of California, Berkeley, CA 94720

Structural design with brittle materials requires that the stress level in the component correspond to a material survival probability that exceeds the minimum survival probability permitted in that application. This can be achieved by developing failure models that fully account for the probability of fracture from defects within the material (including considerations of fracture statistics, fracture mechanics and stress analysis), coupled with non-destructive techniques that determine the size of the large extreme of critical defects. Approaches for obtaining the requisite information are described in this paper.

1. Introduction

The design of structural components from brittle solids is, in concept, quite straightforward. It simply requires that the stress level in the component should not exceed the strength of the material, at the permissible level of survival probability. The implementation of this concept is, however, rather involved. It requires the combination of information derived from the disciplines of fracture statistics, fracture mechanics and flaw detection (or non-destructive evaluation). The description of the general scientific framework for structural design, utilizing these disciplines, is the primary intent of the present paper.

Ultimately, design might take the form of a computer simulation of crack growth in real microstructures, coupled with microstructural characterization techniques (such as acoustic scattering). Presently, however, useful progress is being achieved using a partially decoupled approach. The evolution of failure from defects and the crack extension mechanisms are studied separately, and merged where possible. This approach has influenced the structure of the paper, which includes separate considerations of fracture initiating flaws and defect characterization.

The character of the design problem is illustrated in Fig. 1a, which plots the probability of fracture of a ceramic (measured, say, in flexure) as a function of stress level. It might be construed that for design purposes, it is simply necessary to superimpose the permissible level of failure onto this figure, to obtain a maximum allowable stress in the component: and then to design the component accordingly. The limitations of this approach are exposed when it is appreciated that the fracture probability curve can be substantially perturbed by a wide variety of phenomena. These include: the incidence of slow crack growth (Fig. 1b), the occurrence of undetected flaw populations in the inevitable region of extrapolation (Fig. 1c), and effects of stress state (Fig. 1d). Because of the problems associated with the direct use of statistical design procedures, alternate approaches have been sought, which attempt to effectively truncate the strength distribution at a level above the design stress (Fig. 1e). One such approach, involving the characterization of fracture initiating defects and of the evolution of failure, is emphasized in the present paper.

Fracture in brittle solids usually occurs either by the direct extension of a single pre-existent flaw (from the large extreme of the flaw population) or by the coalescence of small flaws. The level of stress needed to activate these flaws relates to the size of the flaw in a manner that depends upon the interaction of the flaw with the surrounding microstructure. The

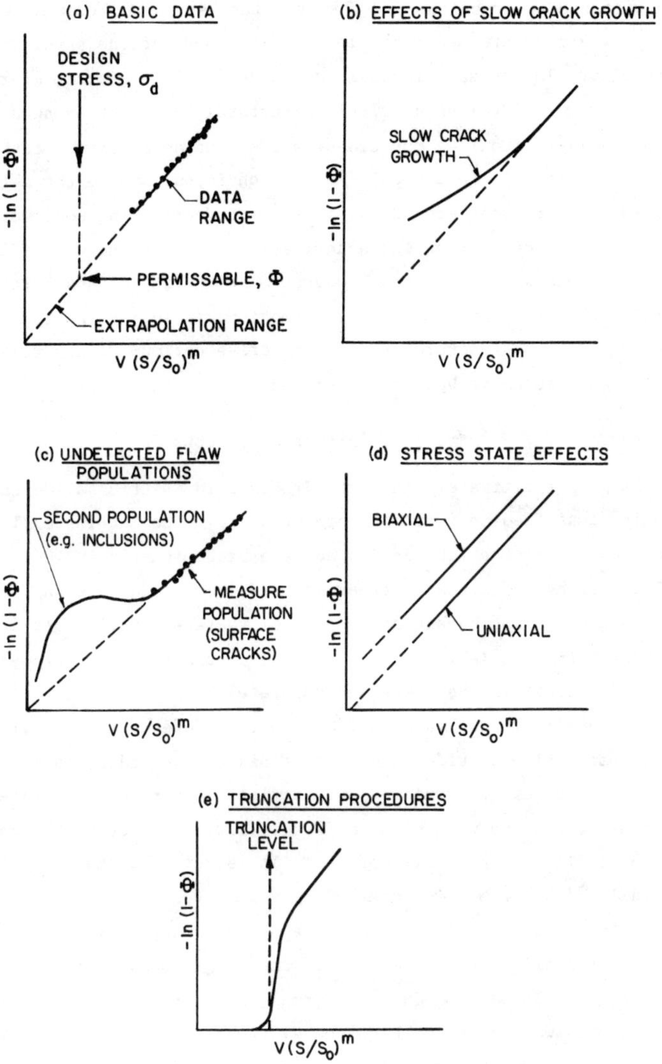

Fig. 1. A schematic illustrating some of the issues that limit the use of a direct statistical approach for structural design. The diagrams relate the fracture probability Φ to the strength level S and the sample volume V.

character of these interactions and the resultant strength, flaw size, probability relations are described in the first part of the paper, for each of the prevalent flaw types: inclusions, voids, surface cracks, microcracks.

The ultimate survival of a brittle structural component at an acceptable survival probability requires the use of a flaw characterization technique in conjunction with a failure model. Such techniques involve the detection and analysis of waves scattered or absorbed by defects. The most versatile and sophisticated mode of flaw characterization involves the use of acoustic waves: either bulk waves of surface waves. The utility of acoustic waves for providing the requisite survival information, including the combination of the measurement and fracture results to derive optimum accept/reject decision schemes, is described in the second part of the paper.

2. Fracture Initiating Defects

The flaws that ultimately initiate fracture in brittle solids can be conveniently classified as intrinsic or extrinsic. The intrinsic flaws are introduced during the fabrication and are predominantly inclusions or voids. The extrinsic flaws are stress induced cracks, such as the surface cracks introduced during machining and the microcracks that result from large residual stresses (e.g., due to thermal contraction anisotropy). Each class of defect will be discussed separately.

The only available analyses of fracture from defects that provide a consistent description of effects of defect size, type and shape invoke the existence of preexistent microflaws, activated by the concentrated stress fields around and within the defects. (1,2,3) However, the character of these small flaws is not well-defined. It is supposed that the flaws are the small voids (or precipitates) that typically occur at grain boundaries (e.g., at triple points). These flaws are prone to activation at relatively small levels of applied stress because of the large residual stresses that can exist at grain boundaries due to thermal contraction anisotropy (4) (Fig. 2); such flaws located on high energy boundaries would be particularly susceptible to microcrack formation. Direct evidence of this mode of microcracking has not been obtained, however, and the concept must be treated as phenomenological at this juncture.

The statistical character of the microcracking process can be conveniently posed by commencing with the premise that the microflaws exhibit an extreme value size distribution that leads to the probabilistic relation (5,6);

$$\Phi(a) = 1 - \exp[-(A/A_o)(a_o/a)^k] \quad (1)$$

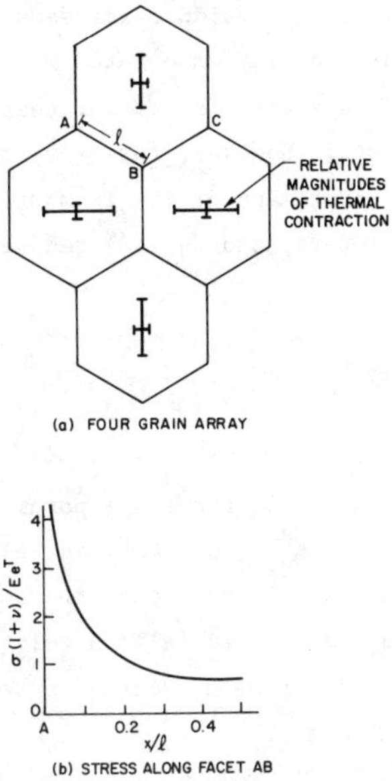

Fig. 2. Stresses caused by thermal expansion anisotropy.

where $\Phi(a)$ is the probability of finding a microflaw larger than a on a grain boundary of area A, a_o is a scale parameter, k is a shape parameter and A_o is a normalizing constant. Noting that a flaw will extend under the condition that the stress intensity factor K reaches the critical value for grain boundary fracture, $K^c_{g.b}$, then gives the approximate result (7,8);

$$\frac{\pi\left(K^c_{g.b}\right)^2}{4a} \approx \sigma^2 + \frac{4\tau^2}{(2-\nu)^2} \tag{2}$$

where σ is the total stress (applied plus residual) normal to the boundary and τ is the total in-plane shear stress needed to induce crack extension. Substituting a from Eq. (2) into Eq. (1) gives the probability of microcracking as a function of applied stress $(\sigma_\infty, \tau_\infty)$ as:

$$\Phi(\sigma_\infty, \tau_\infty) = 1-\exp\left[-\left(\frac{1}{A_o}\right)\left(\frac{4a_o}{\pi}\right)^k \int_A \left(\frac{(2-\nu)^2(\sigma_\infty+\sigma_R)^2 + 4(\tau_\infty+\tau_R)^2}{K^c_{g.b}(2-\nu)^2}\right)^k dA\right]. \tag{3}$$

It should be noted that, since the residual stresses and the toughness are variables, the microcrack probability associated with a specific boundary is not uniquely related to the applied stress; rather, a distribution of probabilities generally exists. However, for many problems, the probability of crack formation averaged over many grains is most pertinent. The average residual stress must then be zero, and Eq. (3) reduces to the simple form;

$$\Phi(\sigma_A) = 1 - \exp\left[-\frac{1}{V_o}\int_V (\sigma_A/S_o)^m \, dV\right] \tag{4}$$

where σ_A is the applied stress, m is the shape parameter (=2k), S_o is the scale parameter (which includes $K^c_{g.b}$, a_o and ν as well as a coefficient that reflects an averaging of the normal and shear stresses over the grain boundaries) and V is the volume of material. A relation similar to Eq. (4) is also generally assumed to describe microcrack extension, except that S_o will have a different significance.

2.1 Intrinsic Defects

(a) Voids. The probability of fracture from a void can be ascertained by combining the void stress field with the appropriate statistical relation for extension of microcracks existing in the vicinity of the void. If the microcracks are very much smaller than the void radius, a direct statistical analysis using Eq. (4) suffices. (1,2) For example, when the microcracks predominate at the void surface (Fig. 3), the surface stress field (9);

$$\begin{aligned} \sigma_\theta/\sigma_A &= [3/2(7-5\nu)]\,[(4-5\nu) + 5\cos 2\theta] \\ \sigma_\alpha/\sigma_A &= [3/2(7-5\nu)]\,[5\nu\cos 2\theta - 1] \end{aligned} \tag{5}$$

can be combined with Eq. (4), and integrated over the tensile portion of the void surface, to yield (2);

$$\xi \equiv -\ln[1-\Phi] \approx 8r^2(\sigma_A/S_o)^m \exp[0.52m-1.4] \tag{6}$$

where r is the void radius. A stronger dependence on r emerges for volume distributed microflaws, viz., $\xi \propto r^3$, as deduced by Vardar, et al. (1)

When the microcracks are not small, vis-a-vis the void radius, a stress gradient correction derived from fracture mechanics solutions must be applied. (2) This correction arises because the stress intensity factor for

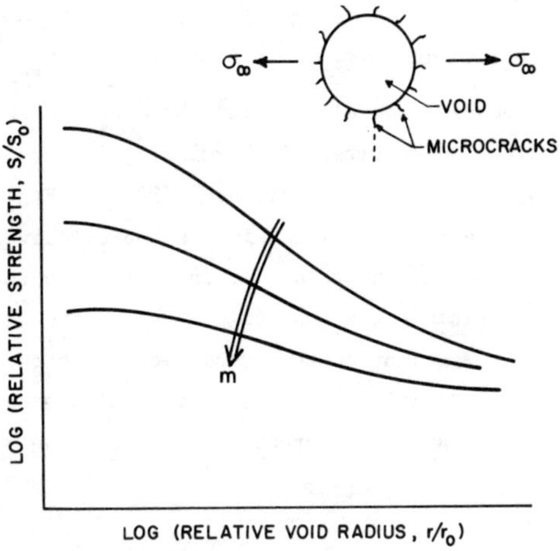

Fig. 3. A schematic illustrating flaws distributed at the void surface, and the fracture strength, void size relation derived at constant probability.

a flaw located in a rapidly varying stress field depends sensitivity on the exact flaw location and on its size relative to the gradient. The effect is especially manifest for surface located microcracks, which are subject to the following approximate peak stress intensity factor (2);

$$\hat{K} = \frac{2\sigma_A \sqrt{a}}{\sqrt{\pi}\,[1 + 0.3\ (0.2 + a/r)]} . \tag{7}$$

This relation for $\hat{K}$ can be used to obtain an effective stress σ_{eff}, that replaces the applied stress in Eq. (16), given by;

$$\sigma_{eff} = \sigma_A \left\{ 0.3 + 0.7[1 + \alpha/2)^2]^{-1} \right\}$$

where $\alpha = (1/r)(K_c/\sigma_A)^2$. Typical void size, strength relations at constant probability, predicted by this analysis, are plotted in Fig. 3.

The basic pertinence of the statistical approach for describing strength in the presence of voids, has been substantiated for voids in silicon nitride (3) and in PZT (1). A detailed statistical analysis was conducted for the experiments performed on silicon nitride. This analysis revealed a maximum likelihood estimate of the void radius dependence of 2.1 and demonstrated that the coefficients m and S_o were independent of the void radius, with maximum likelihood estimates of $\hat{m} = 4.6$ and $\hat{S}_o = 106$ MPa.

(b) Inclusions. Several modes of failure have been associated with the presence of inclusions. The first distinguishing feature is the tendency for cracking due to thermal contraction mismatch (10,11) (Fig. 4). If the thermal expansion coefficient of the inclusion is appreciably lower than that for the matrix, radial matrix cracks can initiate when the defect exceeds a critical size. This situation can produce severe strength degradation. This is, however, an unusual condition for structural brittle materials, which must have an intrinsically low thermal expansion coefficient in order to resist thermal shock. Alternately, if the expansion coefficient

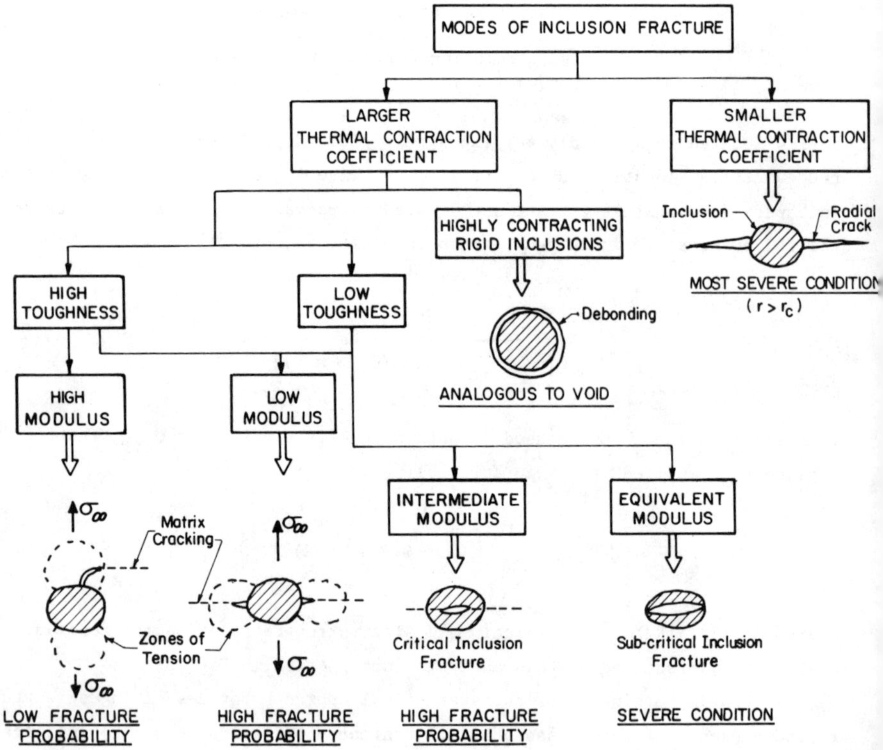

Fig. 4. A schematic indicating the various cracking responses that can occur in the presence of inclusions.

of the inclusion exceeds that of the matrix, several possibilities can result. Highly contracting, high modulus inclusions will tend to detach from the matrix, tending to produce a defect comparable in character to a void. Inclusions that are more compliant or exhibit smaller relative contractions, remain attached to the matrix. Thereupon, several modes of failure are possible, as exemplified by the results for several types of inclusion in silicon nitride (3) (Fig. 5). The expected failure mode depends upon the elastic modulus and fracture toughness of the inclusion, vis-a-vis the matrix. When the inclusion has a larger toughness than the matrix (an unusual occurrence) fracture initiates within the matrix, usually from microflaws located within (or adjacent to) the interface. The process then resembles the void fracture problem. However, one additional distinction must be introduced. When the bulk modulus of the inclusion exceeds that of the matrix, the tensile stresses (in a direction suitable for continued extension of the crack due to the applied stress) are confined to a relatively small zone near the poles of the inclusion (Fig. 4). The fracture probability can then be anticipated to be relatively small, as exemplified by the high survival probability for WC inclusions in silicon nitride. Alternatively, when the modulus of the inclusion is smaller than that for the matrix, the maximum tensile stresses occur near the equatorial plane. The fracture condition is then comparable to that for a void, modified by a stress coefficient λ that depends on the modulus ratio;

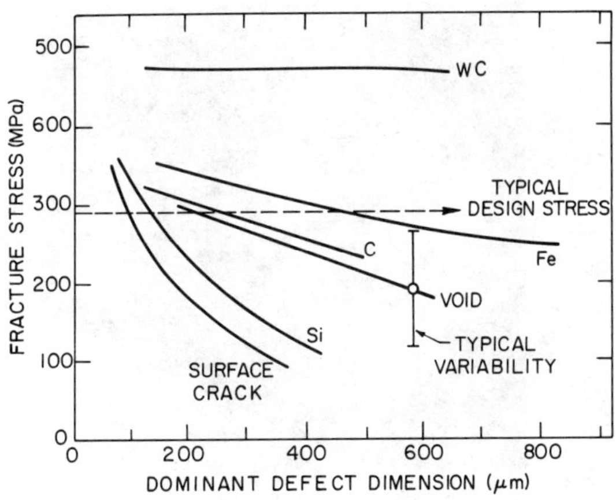

Fig. 5. Strength, size relations for various fracture initiating defects in silicon nitride.

$$\lambda = 1 + \frac{2[(\kappa_i/\kappa_m)-1](1-2\nu)}{3(1-\nu)} \left[\frac{4\mu_m + 3\kappa_m}{4\mu_m + 3\kappa_i}\right] \tag{9}$$

where κ is the bulk modulus and μ is the shear modulus. This case is expected to be an important one in ceramics, because the inclusions are often porous (3) (following high temperature mass transport driven by thermal contraction anisotropy) and hence, of low effective modulus.

Most inclusions typically encountered in brittle matrices are of low toughness, because they are usually the friable product of chemical reaction with the matrix (12) (Fig. 6). If such an inclusion also has a relatively high modulus (approaching that of the matrix), the inclusion can fracture sub-critically to create a crack of dimensions comparable to the cross-section of the inclusion. The ultimate fracture strength is then dictated by the usual fracture mechanics relation for an internal crack (13)

$$\sigma = z(a/c)\, K_c^M a^{\beta} F(\kappa_m/\kappa_i) \tag{10}$$

where a and c are the dimensions of the crack, K_c^M is the effective toughness of the matrix phase neighboring the defect and β is an exponent (≈ 0.5) that depends on the modulus ratio. This type of defect is the most deleterious

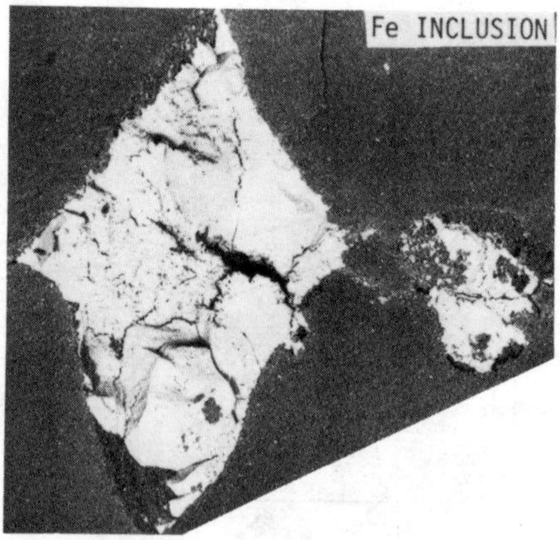

Fig. 6. A scanning electron micrograph of an iron silicide inclusion in silicon nitride.

of the high expansion defects (Fig. 4). Defects in this category are exemplified by Si inclusions in Si_3N_4 (Fig. 5). When the modulus of the defect becomes very small, because of extensive porosity (Fig. 6), the stresses do not attain a sufficient level to induce defect fracture (despite their friability); the situation is then identical to that of low modulus, high toughness inclusions. However, an intermediate condition is also possible; wherein fracture can initiate within the defect and then propagate directly into the matrix to cause complete failure. In this situation, fracture is dictated by the probability of activating microflaws within the inclusion, and the fracture probability becomes (3);

$$\Phi = 1 - \exp\left[-V_i\left(\frac{\lambda\sigma_\infty + \sigma_\alpha}{\rho_o}\right)^{\gamma}\right] \quad (11)$$

where ρ_o is the scale parameter, γ is the shape parameter, and V_i is the inclusion volume. Fracture results obtained for iron silicide inclusions in silicon nitride satisfy a joint fracture relation involving a combination of the critical defect fracture model (Eq. 11) and the matrix fracture model pertinent to low modulus inclusions. (3) This is a plausible situation considering the potential for a transition, with decrease in size, from inclusion initiated fracture (volume dependent) to matrix fracture (area dependent).

2.2 Extrinsic Defects

Extrinsic defects are usually cracks produced by large transient or localized stress states. The most common sources of extrinsic defects are surface cracks produced by machining (14), impact (15) or thermal shock. The machining induced cracks are the most prevalent (and comparable in character to the cracks introduced by projectile impact). (15) The evolution of the cracks, and their resultant influence on strength, is analogous to the cracking that occurs during indentation (16) (Fig. 7). The ultimate dimensions of the cracks are dictated by the residual indentation field, as controlled by the hardness, H, toughness, K_c, and modulus, E, of the material. A specific relation recently derived for the strength controlling radial cracks is (16);

$$a^{3/2} = 2.10^{-2}(\cot^{2/3}\psi)(E/H)^{2/3}K_c^{-1}P_N \quad (12)$$

where ψ is the included angle of the grinding particle and P_N is the normal

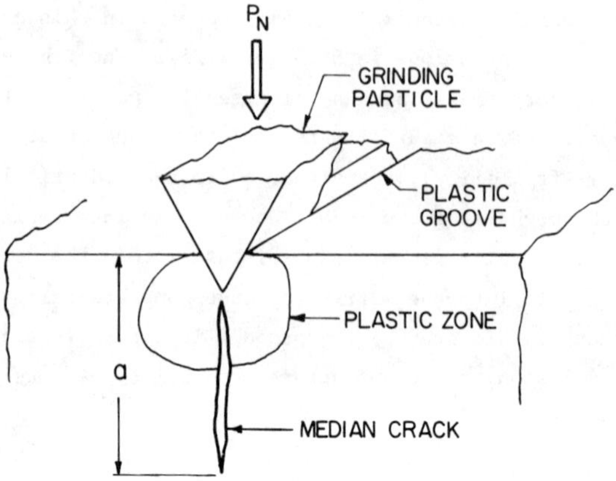

Fig. 7. The median cracking that accompanies the grinding of ceramic surfaces.

force applied to the particle. The term E/H arises because fracture is a residual stress dominated process. (17) The radial cracks are usually semi-circular, because of the symmetry of the residual field.

The extension of the surface cracks introduced by grinding is explicable using standard fracture mechanics relations for mode I[13],

$$K = F(\theta)\,\frac{2}{\sqrt{\pi}}\,\sigma\sqrt{a} \tag{13}$$

where F(θ) is a function that describes the variation in K_I around the crack periphery. Extension to the mixed mode fracture of inclined cracks appears to be adequately described, over an appreciable angular range, by the simple coplanar strain energy release rate criterion. (18) However, the effective stress that produces crack extension can include a significant residual component particularly in coarse grinding situations, where the plastic zone is not removed by subsequent fine grinding. Consequently, the applied stress at fracture can exhibit both systematic and random variations from that anticipated by direct application of Eq. (13) (with the peak stress intensity factor $\hat{K}$ equated to K_c). Available test results suggest a probability of fracture given by the normal distribution (19)

$$\Phi(\sigma_\infty|\sigma_\rho) = \frac{1}{v\sqrt{2\pi}}\int_{-\infty}^{\sigma_\infty} \exp\left[\frac{\sigma_\infty - a + \beta\sigma_\rho}{\sqrt{2}\,v}\right]^2 d\sigma_\infty \tag{14}$$

where σ_ρ is the predicted strength obtained from Eq. (13);

$$\sigma_\rho = \frac{K_c}{\hat{F}(\theta)} \frac{\sqrt{\pi}}{2\sqrt{a}} \tag{15}$$

$\hat{F}(\theta)$ corresponds to the value at $\hat{K}$, v is the variance in σ_ρ, β is a systematic error coefficient and α is a parameter related to the mean strength.

3. Non-Destructive Defect Characterization

3.1 Accept/Reject Criteria

Accept/reject decisions based on a non-destructive measurements of scattering from a defect must recognize the probabilistic character of the problem. (20) At least three probabilities enter the analysis: the failure probability, given the defect dimensions (discussed above) $\phi(\sigma_\infty^c|a)d\sigma$; the joint probability of identifying the defect type and of estimating its size, $\phi(a_{es}|a)da_{es}$; the a priori distribution of defect size, $\phi(a)da$. These probabilities are combined and integrated to various inspection levels, a_{es}^*, to obtain two interrelated probabilities: the false-accept probability Φ_A and the false-reject probability, Φ_R (Fig. 8a):

$$\Phi_A = \int_0^{\sigma_A} \int_0^{a_{es}^*} \int_0^{\infty} \left[\phi(\sigma_\infty|a)d\sigma_\infty\right]\left[\phi(a_{es}|a)da_{es}\right]\left[\phi(a)da\right] \tag{32}$$

$$\Phi_R = \int_{\sigma_A}^{\infty} \int_{a_{es}^*}^{\infty} \int_0^{\infty} \left[\phi(\sigma_\infty^c|a)d\sigma_\infty\right]\left[\phi(a_{es}|a)da_{es}\right]\left[\phi(a)da\right]$$

where σ_A is the level of the applied tension in the volume element containing the defect. The inspection level a_{es}^* refers to the estimated defect dimension(s) selected for the rejection or acceptance of the component, e.g., all components with an estimated maximum dimension less than a_{es}^* are accepted and all components with an estimated dimension greater than a_{es}^* are rejected. The false-accept probability Φ_A is thus the probability that components accepted in accord with the specified inspection level will contain defects more severe than indicated by the estimate, and will actually fail in service (i.e., related to the failure probability). This probability decreases, of course, as the inspection level decreases (Fig. 8b). The false-reject probability Φ_R is the (related) probability that rejected components would, in fact, have performed satisfactorily in service, because the defect severity has been overestimated by the selected inspection level.

This probability increases as a^*_{es} decreases (Fig. 8b). However, it is crucial to recognize that these probabilities are interrelated, i.e., they merely represent different ranges of integration of the same combination of probability functions (Eq. 16). This interdependence is exemplified in Fig. 8a, which is a typical plot relating the false-accept and false-reject probabilities. Once one of these probabilities has been selected, the other probability, as well as the associated inspection level, are necessarily defined. It is now apparent from Fig. 8b that the inspection technique, or combination of techniques, that would be preferred is that which yields a curve as close as possible to the probability axes. For example, technique B is preferred over technique A, because the rejection of satisfactory components required to satisfy the failure probability requirements is much lower. Such curves thus represent a quantitative method for characterizing the failure prediction capabilities of various inspection techniques, for a given material and service condition.

Scrutiny of the available inspection methods pertinent to ceramics indicates that acoustic methods are preferred, because acoustic waves are appreciably scattered by all of the critical defect types encountered in structural ceramics. The most promising measurement algorithms and their future potential are thus briefly reviewed.

3.2 Acoustic Measurement Algorithms

Surface Waves. The most directly applicable acoustic method is the use of surface acoustic waves to predict failure from surface cracks: in particular, the use of long wavelength, $\lambda \gg a$, surface waves. (21,22) In the long wave length limit, the scattering of an acoustic wave (stress wave) by a crack is closely analogous to the interaction of the crack with an applied stress field. In particular, both the scattering coefficient S_1 and the strain energy release rate $\mathcal{G}$ are related to the crack surface integral (21),

$$\int_{A_s} \sigma_{ij} \Delta u'_j n_i dA_s$$

where σ_{ij} is the stress across the crack plane in the absence of the crack, $\Delta u'_j$ is the displacement of the crack surfaces and A_s is the crack surface area. Hence, it is straightforward to demonstrate that the scattering coefficient when the acoustic beam is aligned with the tensile axis in the compound is directly related to the crack extension stress σ_c by (22),

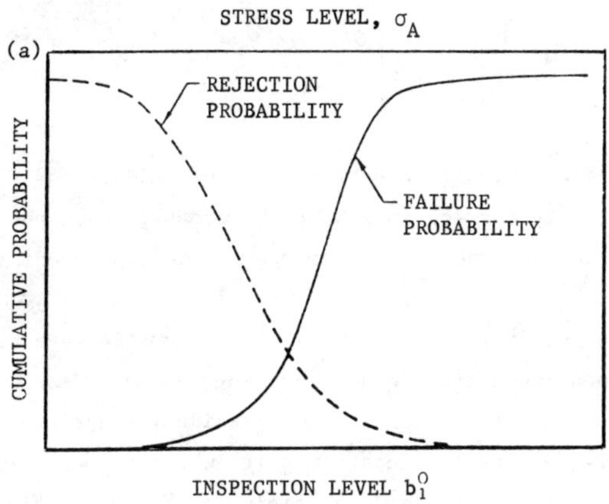

Fig. 8a. False-accept, false-reject curves for two hypothetical measurement techniques, A and B.

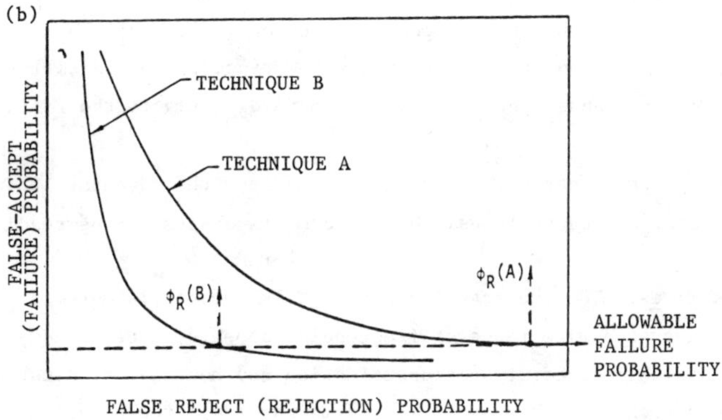

Fig. 8b. The variations of failure probability and rejection probability with inspection level.

$$\frac{K_c}{\sigma_c} = 2\left[\frac{6(1-\nu)\lambda^2 S_i w}{\pi^5 f_z \nu}\right]^{1/6} \tag{33}$$

where w is the beam width, η is the transducer efficienty and $f_z \sim 0.4$. This result is strictly correct when both the acoustic wave and the applied stress are normal to the crack plane, as exemplified by recent results for glass (22).

The prediction of failure based on surface acoustic wave measurements is more complex when the surface cracks are introduced by machining. (19) The crack formation occurs in response to a residual elastic/plastic field (16), that normally remains (at least in part) when the machining operation has been completed. (23) This residual field causes sub-critical crack extension when a remote tension is applied to the system. This crack extension is manifest in an irreversible change in the acoustic scattering coefficient as an external stress is applied (19). The crack length at fracture is thus appreciably larger than the initial length (by ~ 2.5) and hence, a failure estimate based on the initial length would provide a substantial underestimate of the failure stress.

The residual stresses can be largely removed by annealing at ~ 1200°C. The amplitude of the acoustic signal is thus only influenced by the load through the separation of contacting crack surfaces. The initial acoustic measurement after annealing thus relates more directly to the failure prediction.

Finally, it is noted that surface waves have the advantage that they propagate over curved surfaces, so that complex shapes can be readily probed. (22)

Bulk Waves. The characterization of bulk defects is more complex. Information over a wide range of frequencies appears to be needed to obtain a highly probable defect type classification and hence, a size estimation. Appropriate techniques are available including (24): the scanning laser acoustic microscope, 200 MHz ZnO transducers and conventional (2-50 MHz) transducers. Rapid scanning methods for defect location have also been developed. The most critical issue, therefore, concerns the appropriate choice of the algorithms to obtain the most reliable defect characterization. A typical set of algorithms and their interaction are illustrated in Table I, using low and high frequency information as well as acoustic microscopy. This set has not yet been fully evaluated, so redundancies may exist. Four algorighms are employed in this scheme: (i) long wave length scattering, (ii) intermediate wavelength Born approximation, (iii) high frequency

Table I. A Typical Set of Algorithms and Their Interaction

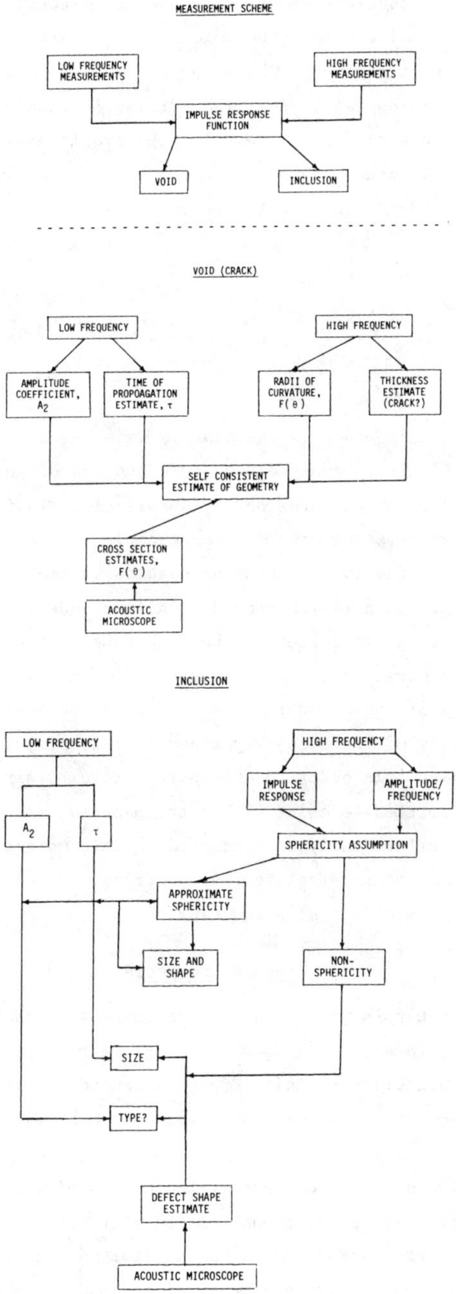

spectroscopy and (iv) cross sectional information from acoustic microscopy. The impulse response functions are firstly used to determine whether the defect is a void or an inclusion; the void has an impulse response function characteristic of the transducer, while inclusions have more complex functions. Thereafter, voids can be analyzed straightforwardly, using a variety of algorithms. For example, a long wavelength algorithm similar to that described for surface cracks may be employed. In the long wavelength limit the scattered amplitude is related to the void volume V by (20);

$$A = \frac{V\omega^2}{(4\pi c^2)^2}\left[1 + \frac{1+\nu}{7-2\nu} + \frac{10(1-2\nu)}{7-5\nu}\right]^2 \tag{34}$$

where ω is the frequency and c is the elastic wave speed in the host. Inclusions are more difficult to analyze; the combined use of several algorithms is almost certainly required. For nearly spherical inclusions, the interpretation is relatively straightforward. For example, a combination of the long wavelength algorithm (which contains coupled volume and type information) and the Born approximation (which provides an independent estimate of the distance from the geometric center to the back face of the inclusion) can yield the requisite size and type information. A typical result, obtained for a 100 μm radius Si inclusion in Si_3N_4, is illustrated in Fig. 9; wherein the joint probability of the defect type and size is plotted as a function of the estimated size. In order to obtain this result, six possible inclusion types were permitted to exist within the material (selected on the basis of detailed failure analyses conducted on this material). Alternatively, high frequency measurements displayed in the frequency domain would provide close estimates of the defect size and type.

3. Future Prospects

It is hoped that this paper conveys the impression that a positive start has been made in establishing the scientific framework for microstructural design with brittle materials. Certain rewarding research directions have emerged and several exciting near term, and more remote, prospects seem viable.

Further studies aimed at characterizing models of fracture from defects are very pertinent. The incisive combination of inputs from mechanics, materials and statistics demonstrated on the limited set of problems addressed thus far should provide some direction and scope for continued

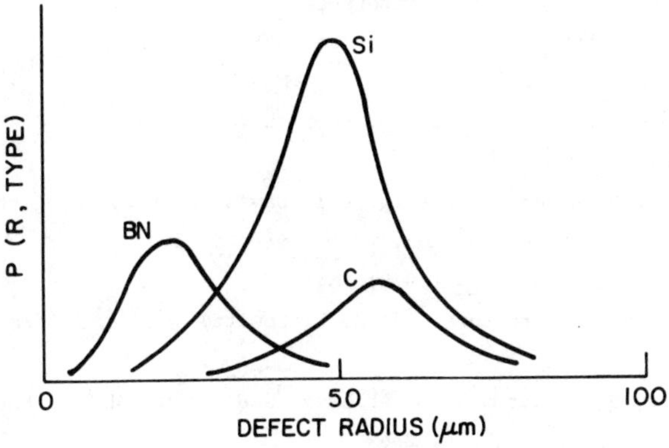

Fig. 9. The joint probability of defect type and size deduced from a coupled long wavelength, Born approximation algorithm.

activity. Important defects not yet considered include: void clusters, sub-surface inclusions, surface crack arrays. Progress toward the comprehension of fracture from these defects could utilize existing (or marginally extended) stress analyses coupled with advanced statistical methods and fracture mechanics solutions.

Imminent advances can be anticipated in ultrasonic flaw characterization. A comprehensive set of inversion algorithms already exist, and initial results imply that good estimates of defect size and type are possible, using combinations of these algorithms. Future prospects for devising effective accept/reject schemes pertinent to ceramics are thus very exciting.

Acknowledgments

The author wishes to thank the Advanced Research Projects Agency for funding the work on fracture models and accept/reject criteria under Contract F 33615-74-5180, and the Office of Naval Research (A.G.E.) for supporting the work on surface crack detection, under Contract No. N0014-79-C-0159, and the U.S. Department of Energy under Contract W-7405-ENG-48.

References

1. O. Vardar, I. Finnie, D.R. Biswas and R.M. Fulrath, Int'l. Jnl. Frac., 13 (1977) 218.
2. A.G. Evans, D.R. Biswas and R.M. Fulrath, Jnl. Amer. Ceram. Soc., 62 (1979) 101.
3. A.G. Evans, G. Meyer, K. Fertig, B.I. Davis and H.R. Baumgartner, Jnl. of Non-Destructive Evaluation, (Nov. 1980) in press.
4. A.G. Evans, Acta Met., 26 (1978) 1845.
5. A.M. Freudenthal, Fracture (Ed. H. Liebowitz) Vol. 2, Academic Press, N.Y. (1969) p. 592.
6. A.G. Evans, B.R. Tittmann, L. Ahlberg, G.S. Kino and B.T. Khuri-Yakub, Jnl. Appl. Phys., 49 (1978) 2669.
7. S.B. Batdorf and H.L. Heinish, Jnl. Amer. Ceram. Soc., 61 (1978) 355.
8. A.G. Evans, Jnl. Amer. Ceram. Soc., 61 (1978) 302.
9. S. Timoshenko and J.M. Goodier, Theory of Elasticity, McGraw Hill, N.Y. (1951).
10. A.G. Evans, Jnl. Mater. Sci., 9 (1974) 1145.
11. R.W. Davidge and T.J. Green, Jnl. Mater. Sci., 3 (1968) 629.
12. H.R. Baumgartner and D. Richerson, Fracture Mechanics of Ceramics (Ed. R.C. Bradt, D.P.H. Hasselman and F.F. Lange) Vol. 1 (1974) p. 367.
13. G.C. Sih, Handbook of Stress Intensity Factors, Lehigh Univ. Press (1973).
14. M.V. Swain, Fracture Mechanics of Ceramics, ibid. Vol. 3 (1978) p. 257.
15. A.G. Evans, M.E. Gulden and M. Rosenblatt, Proc. Roy. Soc., A361 (1978) 343.
16. B.R. Lawn, A.G. Evans and D.B. Marshall, Jnl. Amer. Ceram. Soc., in 63 (1980) 574.
17. D.B. Marshall and B.R. Lawn, Jnl. Mater. Sci., in press.
18. J.J. Petrovic and M.G. Mendiratta, Jnl. Amer. Ceram. Soc., 59 (1976) 163.
19. J. Tien, B.T. Khuri-Yakub, G.S. Kino, D.B. Marshall and A.G. Evans, Jnl. Amer. Ceram. Soc. to be published.
20. J.M. Richardson and A.G. Evans, Jnl. of Non-Destructive Evaluation, 1 (1980) 368.
21. B. Budiansky and J.R. Rice, Jnl. Appl. Mech., 45 (1978) 2.
22. B.T. Khuri-Yakub, G.S. Kino and A.G. Evans, Jnl. Amer. Ceram. 63 (1980) 65.
23. D.B. Marshall, R.R. Lawn, and J. Mecholsky, Jnl. Amer. Ceram. Soc., 63 (1980) 358.
24. R.B. Thompson, this issue.

II. NDE Characterization of Microstructure

EFFECT OF METALLIC MICROSTRUCTURE ON ULTRASONIC ATTENUATION

Robert E. Green, Jr.

Materials Science and Engineering Department
The Johns Hopkins University
Baltimore, Maryland 21218

Ultrasonic techniques offer very useful and versatile nondestructive methods for investigating the microstructure and associated mechanical properties of materials. There are various mechanisms by which energy can be lost from ultrasonic waves propagating through real materials, and measurement of this ultrasonic attenuation can complement ultrasonic velocity measurements in yielding valuable information about the mechanical properties of the material. Among the inhomogeneities which can cause ultrasonic attenuation either by absorption or scattering are precipitates, inclusions, voids, cracks, grain boundaries, interphase boundaries, twin boundaries, magnetic domain walls, dislocations, substitutional impurities, vacancies, and interstitial impurities. The present paper will present a condensed overview of the current state-of-the-art of the use of ultrasonic attenuation to determine metallic microstructure.

Introduction

Ultrasonic techniques offer very useful and versatile nondestructive methods for investigating the microstructure and associated mechanical properties of materials. An ultrasonic wave propagating through a solid body can be used as a probe to give indication of internal material microstructure and microstructural alterations throughout the volume of the body. The use of such an internal probe offers definite advantages over other techniques which rely on surface measurements, since it is well known that many properties exhibited by the surface layer of a solid are not identical with the behavior of the bulk material. Surface ultrasonic waves may also be used to advantage, however, to probe this surface layer since the penetration depth of such waves increases with decreasing frequency.

Although most industrially practical uses of ultrasonics are applied to solid materials which are polycrystalline aggregates and therefore assumed to be isotropic, with real crystalline solids the condition of ideal isotropy is extremely difficult if not impossible to attain. The prime reason for this is that the solidification or forming and heating processes associated with material processing does not permit a random distribution of crystallographic orientations among the aggregation of grains and often does not even permit a uniformity of grain size. Therefore most real polycrystalline aggregates possess a "texture" and this texture strongly influences the mechanical properties of the material including ultrasonic wave propagation.

The use of ultrasonic waves as nondestructive internal probes of metallic microstructure has as a prerequisite the careful documentation of the propagational characteristics of the ultrasonic waves themselves. Since in nondestructive evaluation applications it is not desirable for the ultrasonic waves to alter the material through which they pass, it is necessary to work with very low amplitude waves, which normally are regarded to obey linear elasticity theory. However, the actual existence of ideal linear elastic waves is extremely doubtful despite the wide use of such terminology in the literature. Nonlinear effects in ultrasonic wave propagation may arise from several different causes. First, the amplitude may be sufficiently large so that finite strains arise. Second, a material may behave in a nonlinear fashion, even when very low amplitude ultrasonic waves are propagated, provided that a sufficient amount of external stress is superimposed or provided that a residual stress is present in the material. This residual stress may be the result of a previously applied external stress or the result of various internal stress concentrators such as precipitates or dislocations. These internal stress concentrators also cause loss of energy from the ultrasonic wave either by absorption or scattering. There are various mechanisms by which energy can be lost from ultrasonic waves propagating through real materials, and measurement of this ultrasonic attenuation can complement ultrasonic velocity measurements in yielding valuable information about the mechanical properties of the material.

Energy losses from the ultrasonic waves are due to both geometrical and intrinsic effects. The physical size of the test specimen, the parallelism of opposite faces, the diameter of the transducer relative to the wavelength of the ultrasound beam, the curvature of the ultrasonic wavefront, and other geometrical effects exert a strong influence on the propagational characteristics of ultrasonic waves and on the energy loss from such waves. When an ultrasonic wave propagating inside a solid material encounters a free surface, an interface separating two different phases, or a grain boundary separating two crystals of the same material of different crystallographic orientation, then reflection and refraction accompanied by mode conversion take place. For a boundary between different anisotropic media, a longitudinal wave or a shear wave incident at an angle less than the critical angle will result in one reflected quasilongitudinal wave and two different reflected quasitransverse waves and one refracted quasi-longitudinal wave and two different refracted quasitransverse waves in the second medium. Thus, the presence of a single bounding surface complicates the propagational characteristics of ultrasonic waves in solid materials and can lead to erroneous interpretation of attenuation measurements. The presence of many bounding surfaces, in general, complicates the propagational characteristics even more, and except in a few special cases, the problem has not been solved analytically. Finally, in all real cases the elastic energy emanating from an ultrasonic transducer does not remain confined in a parallel beam, but diverges as it progresses through a material. This energy should be confined to a beam that diverges as little as possible or portions of the beam striking the side walls of the specimen will undergo mode conversion and energy will be lost from the main beam.

Although geometrical effects can cause energy to be lost from the primary ultrasonic beam, such losses are not indicative of intrinsic loss mechanisms associated with the microstructure. A number of true intrinsic energy loss mechanisms are of extreme importance to specific areas of solid state physics due to their electric or magnetic origins, but except in a few special cases such mechanisms do not influence the mechanical properties of a material. Among the primary mechanisms of energy loss from ultrasonic waves propagating through real polycrystalline materials, which also play a primary role in determining mechanical properties, are those mechanisms causing scattering of the waves. Such scattering may be caused by any inhomogeneity in the material which is of sufficient size and orientation, individually or in aggregation, to interact with the ultrasonic beam. An important parameter in this regard is the wavelength of the ultrasonic wave relative to the size of the scattering center. Among the inhomogeneities which can cause ultrasonic attenuation either by absorption or scattering are precipitates, inclusions, voids, cracks, grain boundaries, interphase boundaries, twin boundaries, magnetic domain walls, dislocations, substitutional impurities, interstitial impurities, and vacancies.

The present paper will present a condensed overview of the current state-of-the-art of the use of ultrasonic attenuation to determine metallic microstructure. In this regard the term microstructure will be taken in a broad sense to refer to any internal structural feature which influences the mechanical properties of metals.

Historical Background

Although space does not permit an exhaustive literature survey of past work on the effect of metallic microstructure on ultrasonic attenuation, reference to some key publications and review articles is deemed necessary.

As early as 1947, Mason and McSkimin (1,2) investigated the frequency and grain size dependence of ultrasonic attenuation in polycrystalline aluminum. Subsequently, Roth (3), Firestone (4), and Sokolov (5) performed additional experiments which were in qualitative agreement in that the attenuation was observed to increase with increasing frequency and grain size. In 1950, Huntington (6) published a theoretical treatment of ultrasonic wave scattering and, with consideration for the earlier theory of Mason and McSkimin, concluded that three theories existed, namely:

Rayleigh scattering for low frequencies, random-phase theory for intermediate frequencies, and at very high frequencies losses from specular reflection and refraction termed diffusion losses. The same year Lifshitz and Parkhomovskii (7), in Russia, presented a theory which was quantitatively in sharp disagreement with that of Mason and McSkimin. For several metals the ultrasonic attenuation due to scattering as calculated by the two theories differed by a factor of almost thirty.

Subsequent work in this field was reported by Kerversau, Bleton, and Bastien (8), Roderick and Truell (9), Hirone and Kamigaki (10), Krainer and Krainer (11), Seemann and Bentz (12), Hikata (13), and Ying and Truell (14). In a series of papers beginning in 1956, Merkulov (15,16,17) presented some new experimental results and summarized the Russian view of the theory of ultrasonic scattering. Additional experimental work was conducted by Hirone and Kamigaki (18), Kamigaki (19), Hikata and Truell (20), and Bayer (21). In 1959, Bhatia (22) and Bhatia and Moore (23) again considered the theoretical aspects of ultrasonic wave scattering in polycrystalline materials endeavoring to establish a more realistic model by proper accounting for the elastic properties of the individual grains.

Since these early beginnings research efforts in this field have progressed in ever increasing numbers over the past twenty years. During this time major survey articles were published by Papadakis (24), in 1968; Truell, Elbaum, and Chick (25), 1969; Gericke (26), 1970; Papadakis (27), 1976; Krautkramer (28), 1977; Goebbels (29), 1980; and Vary (30), 1980.

The generally accepted formulae for ultrasonic wave scattering in polycrystatline materials in the different frequency regimes are shown in Table I.

Table I.

SCATTERING OF ULTRASONIC WAVES

RAYLEIGH	$\lambda >> D$	$\alpha = A_1 D^3 f^4$
STOCHASTIC	$\lambda \approx D$	$\alpha = A_2 D f^2$
DIFFUSIVE	$\lambda << D$	$\alpha = \frac{A_3}{D}$

λ = WAVELENGTH OF ULTRASONIC WAVE

D = AVERAGE GRAIN DIAMETER

α = ULTRASONIC ATTENUATION

A_1, A_2, A_3 = COEFFICIENTS DEPENDENT ON ELASTIC MODULI

f = ULTRASONIC FREQUENCY

Selected Experimental Results

In this section selected experimental results are presented which illustrate the variety of microstructural features studied by ultrasonic attenuation measurements. Figure 1, taken from the work of Papadakis (24), shows a comparison of the experimental results obtained for equiaxed polycrystalline iron-30% nickel specimens with theoretical calculations. As can be seen, the longitudinal wave ultrasonic attenuation values agree well with theory over the entire frequency range. The theoretical point with the dashed line labeled slope 2 drawn through it was computed from stochastic scattering theory, while the theoretical point with the dashed line labeled slope 4 drawn through it was computed from Rayleigh scattering theory. f_R corresponds to the frequency for Rayleigh scattering, f_I corresponds to the frequency for the intermediate or stochastic scattering, and f_B corresponds to the frequency bounding the two regimes based on the Lifshitz-Parkhomovskii theory. Note that the data points lie in this boundary region.

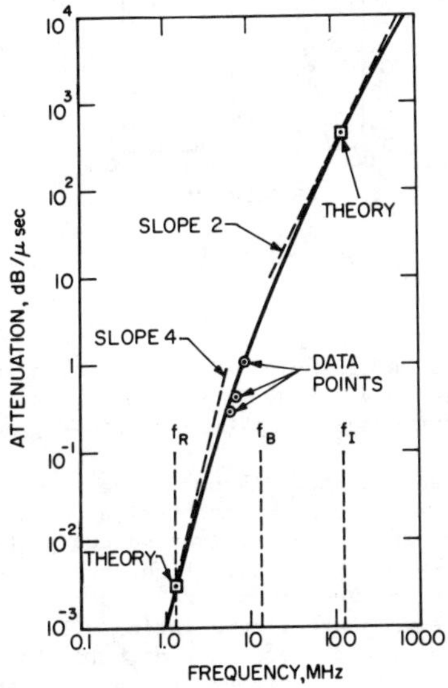

Fig. 1. Longitudinal ultrasonic wave attenuation versus frequency for grain scattering in an equiaxed polycrystalline Fe-30% Ni specimen [Papadakis (24)]

A similar plot of shear wave ultrasonic attenuation versus frequency for an equiaxed polycrystalline nickel specimen is reproduced in Fig. 2. This figure, also from Papadakis (24), shows that for this case the data points in the boundary region lie above the curve drawn through the theoretically computed points in the Rayleigh and stochastic regions. Papadakis attributed this discrepancy to the fact that the measured attenuation was excessive because no diffraction correction was available for the shear wave propagation.

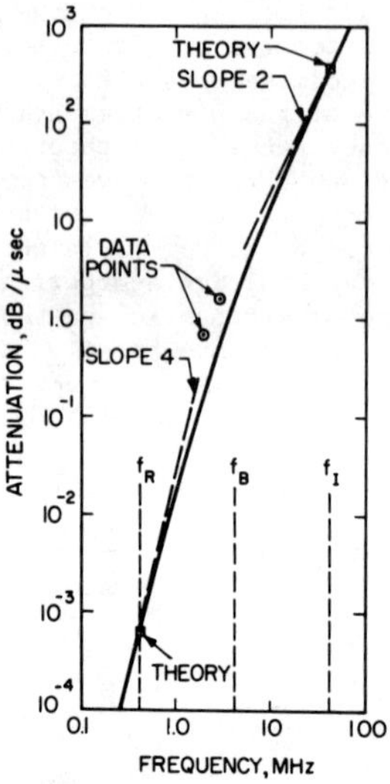

Fig. 2. Shear ultrasonic wave attenuation versus frequency for grain scattering in an equiaxed polycrystalline nickel specimen [Papadakis (24)]

Most recently Klinman and Stephenson (31) have conducted an experimental study of longitudinal ultrasonic wave attenuation versus grain size measurements for both laboratory and commercially produced plates of plain carbon-manganese ferrite-pearlite steel. Figures 3 and 4 show their experimental results. The curves drawn through the data points are least squares fits to the Rayleigh scattering relation. In order to obtain the data shown in Fig. 3 measurements were made at both 5 and 10 MHz. It is evident that for this material the higher frequency established a much more definitive relationship between attenuation and metallographically measured grain size. However, comparison between the 10 MHz data in Fig. 3 with the data in Fig. 4, also measured at 10 MHz, shows the much larger scatter in data obtained with the factory produced sheet. The usefulness of such measurements for industrial quality control is self-evident.

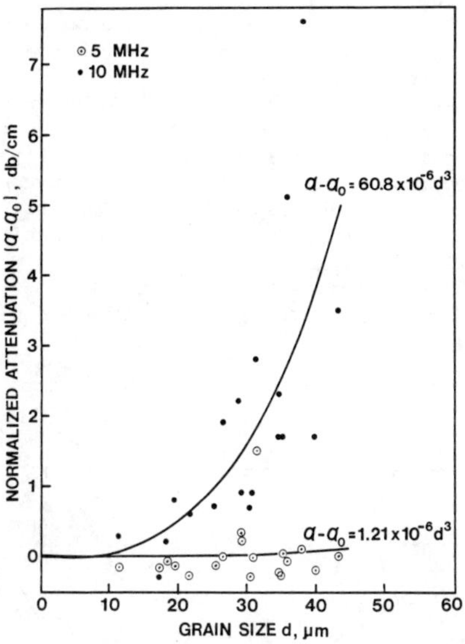

Fig. 3. Longitudinal ultrasonic wave attenuation versus grain size for laboratory produced steel plate [Klinman and Stephenson (31)].

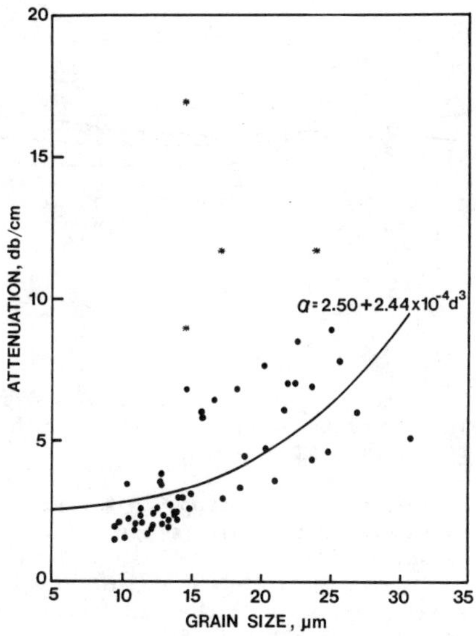

Fig. 4. Longitudinal 10 MHz ultrasonic wave attenuation versus grain size for factory produced steel plate [Klinman and Stephenson (31)].

Figure 5, taken from the work of Latiff and Fiore (32), illustrates the fact that meaningful ultrasonic attenuation measurements can be made on two-phase systems. Longitudinal ultrasonic wave attenuation measurements were made over a frequency range from 10 to 45 MHz on spheroidized Fe-0.8C steel. The test samples were austenitized in dried argon at 843°C for 45 min and quenched into a room-temperature agitated 10% brine solution. Then they were tempered in argon at 700°C for various times in order to develop microstructures consisting of a ferrite matrix containing controlled number densities of roughly spherical Fe_3C particles. They found that the attenuation passed through a minimum as a function of tempering time, because for short tempering times the attenuation was dominated by particle scattering which decreased with increasing time, while for longer tempering times grain scattering which increased with increasing time dominated.

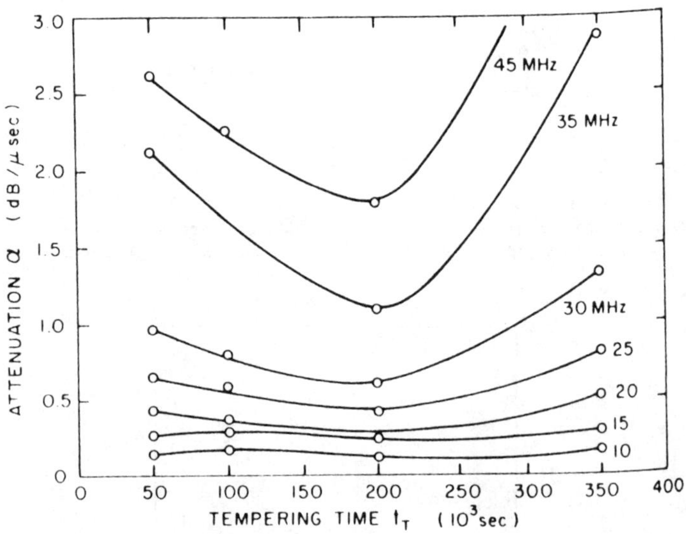

Fig. 5. Longitudinal ultrasonic wave attenuation versus tempering time for a two-phase steel. Minimum in curves is due to transition from particle scattering at short times to grain scattering at long times [Latiff and Fiore (32)].

An example of the role which anisotropy in the grain structure can play in influencing ultrasonic attenuation measurements is illustrated by Figs. 6 and 7, from Papadakis (24). Figure 6 shows schematic drawings of specimens cut from zinc bar stock having a axes of grains aligned strongly with the bar axis, while along the bar diameter crystalline a and c axes occur almost equally. As shown in Fig. 7, the attenuation of 3 MHz longitudinal waves in the radial direction was much greater than in the axial direction. Papadakis reasoned that, since the dislocation damping and diffraction corrections were probably small, the major portion of the attenuation must come from scattering. For propagation along the bar axis, because of the alignment of the a axes, the boundary conditions were matched and no scattered wave resulted. On the other hand, in the radial direction the individual grains were rotated various degrees about the common axial a axis and hence the propagating wave encountered both a and c axes causing boundary condition mismatches and scattering. Therefore, the radial propagation direction should be one of maximum scattering as found experimentally.

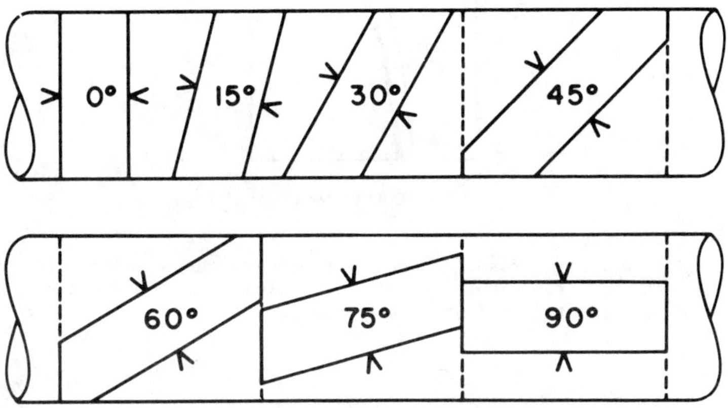

Fig. 6. Schematic illustration of location of test specimens cut from zinc bar stock having a axis of grains aligned with bar axis [Papadakis (24)].

The fact that ultrasonic surface wave attenuation measurements may be used to determine average grain size in polycrystalline sheets is shown in Fig. 8, from the original work of Kettler, Latiff, and Fiore (33). Kettler et al. measured the attenuation of surface waves propagating in Fe-3% Si sheet as a function of grain size for the frequencies indicated. Frequencies below 6.75 MHz were not employed because their long wavelength became comparable to the sheet thickness and propagation of a surface wave under such conditions was not possible. The measurements were made parallel to the rolling direction in order to minimize the effects of texture. The dashed curve was calculated from diffusive scattering theory and gives a good fit to the experimental data.

Figure 9 shows the experimental results obtained by West (34) who measured the temperature dependence of the ultrasonic attenuation of a 10 MHz longitudinal wave propagating in the [110] direction of a nickel

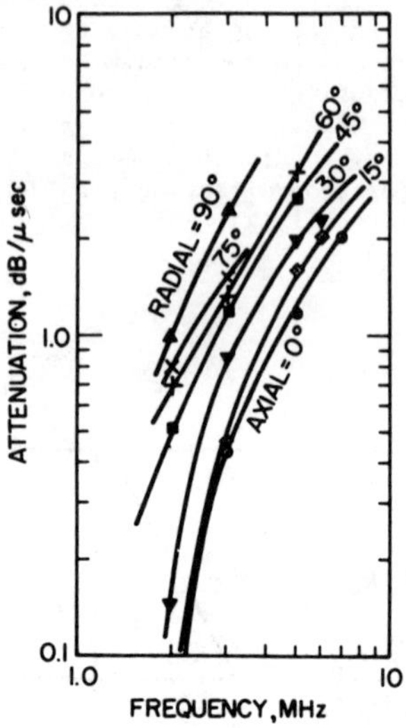

Fig. 7. Longitudinal ultrasonic wave attenuation versus frequency for scattering from highly anisotropic grains in zinc bar stock cut as illustrated in Fig. 6 [Papadakis (24)].

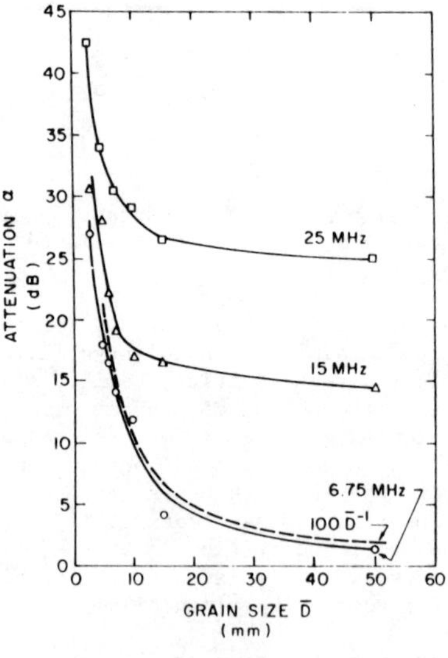

Fig. 8. Surface ultrasonic wave attenuation versus average grain size in Fe-3% Si sheet [Kettler, Latiff, and Fiore (33)].

single crystal. These results show that ultrasonic attenuation measurements may be used to study magnetic domain wall motion in ferromagnetic materials.

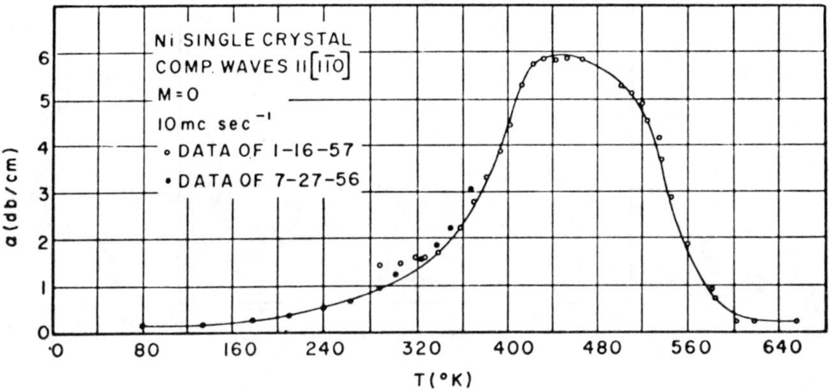

Fig. 9. Temperature dependence of ultrasonic attenuation in a nickel single crystal attributed to magnetic domain motion [West (34)].

Chiao (35) used ultrasonic attenuation to measure the mixed microstructures in steel. Figure 10 shows the relationship he obtained between 5 MHz longitudinal wave attenuation and temperature of transformation to 100% bainitic structure in various alloys. The reason for the dashed curve on the right side of the figure is that Chiao initially questioned whether the point for Alloy 113-A (A.C.) should be considered valid because it fell far from the curve containing the other data points. However, when he recognized that dislocation damping theory predicted a minimum in a plot of ultrasonic attenuation versus dislocation density it was decided that the curve should be drawn as shown in Fig. 10 even though Alloy 113-A had a quite different composition (higher boron and cobalt content) than the other alloys. Chiao used two aspects of dislocation damping theory to explain the maximum observed in Fig. 10. From the point for Alloy 168-A to that for Alloy 168-D, he proposed that the attenuation decreased because of dislocation intersections and entanglements which reduce the dislocation loop length free to vibrate and therefore reduces the attenuation. For Alloy 113-A, he proposed that the low attenuation is due to pinning of dislocations by dissolved boron and cobalt atoms. These results are of interest because it shows that substructural changes smaller than the grain size can easily be detected by ultrasonic attenuation measurements.

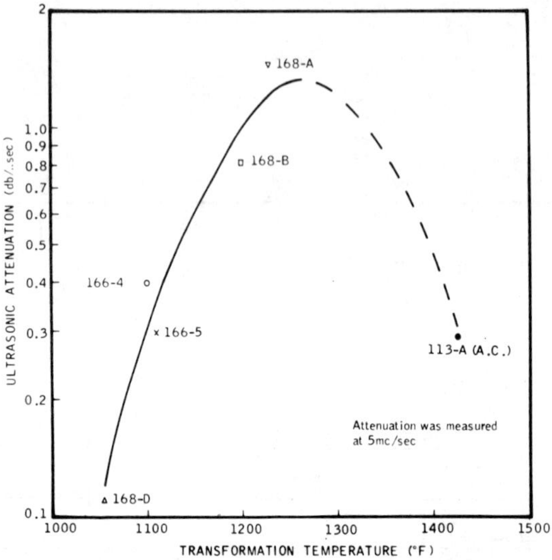

Fig. 10. Relationship between ultrasonic attenuation and temperature of transformation to 100% bainitic structure in various steel alloys [Chiao (35)].

Figure 11 demonstrates the ability of ultrasonic attenuation measurements to detect extremely small microstructural alterations during fatigue testing and hence give early warning of fatigue induced failure. This figure was taken from the work of Joshi and Green (36) who used ultrasonic attenuation measurements as a continuous monitor of fatigue damage in polycrystalline steel specimens which were cycled in reverse bending as cantilever beams to fracture at 30 Hz. Note that the ultrasonic attenuation change indicated microstructural alterations, probably microcrack formation, prior to detection of an additional ultrasonic pulse on the A-scan display as is used in conventional nondestructive ultrasonic evaluation to detect crack formation.

Included in the measurements of Merkulov (17) on the use of ultrasonic waves to investigate the structure of steels are the experimental results reproduced in Fig. 12. This figure shows plots of ultrasonic attenuation versus frequency for chrome-nickel steel specimens with both a) tempered and b) untempered microstructures. From the curves shown it is evident that the losses for both groups of specimens are due to elastic hysteresis up to 8 MHz, while above 15 MHz the attenuation is basically due to scattering. The most interesting fact about these results is that the ultrasonic attenuation is appreciably greater in specimens with an untempered structure than in specimens with a tempered structure, although the average austenite grain size in both groups of specimens was completely identical. As pointed out by Merkulov, this demonstrates the

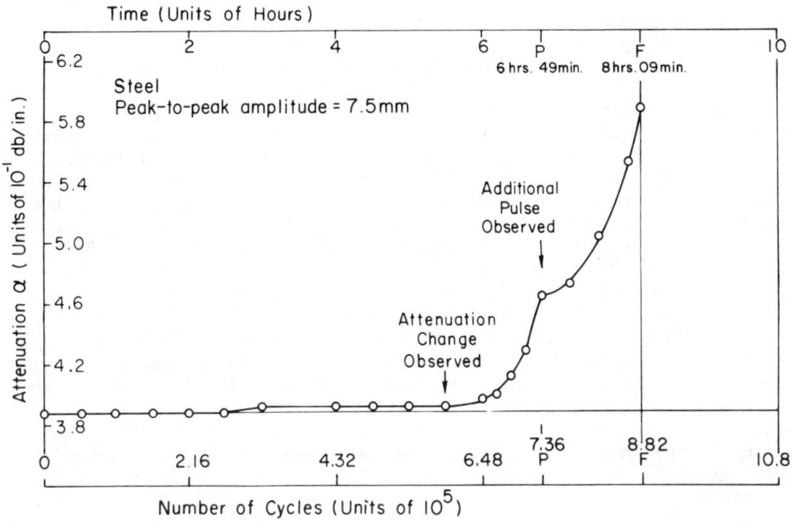

Fig. 11. Longitudinal ultrasonic wave attenuation versus number of fatigue cycles for polycrystalline steel specimen [Joshi and Green (36)].

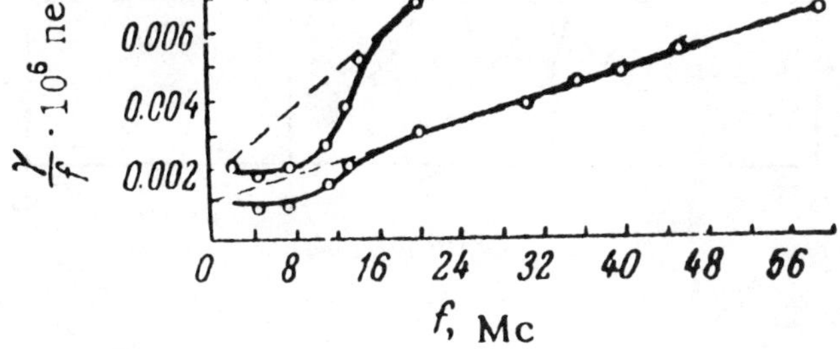

Fig. 12. Ultrasonic attenuation of longitudinal waves in chrome-nickel steel specimens [Merkulov (17)].
a) untempered microstructure; b) tempered microstructure

extremely high resolving power of the ultrasonic attenuation method. This resolving power makes it possible to detect extremely small microstructural inhomogeneities in the structure of a metal and shows that ultrasonic attenuation monitoring can be successfully applied to practical problems of determining untempered microstructures in steel parts.

Gefen and Rosen (37) performed innovative studies of the relationship between the attenuation of ultrasonic waves and changes in the precipitate structures occurring during the aging process in age hardening alloys. Figure 13 from their work shows the variation in ultrasonic attenuation of 10 MHz longitudinal waves in Duraluminum 2024 specimens during isothermal aging at the temperatures indicated. Prior to the aging process, solution treatment was carried out for 2 hrs at 495°C, followed by rapid cooling to 0°C at a rate of 2000°C/sec. Gefen and Rosen attributed the observed ultrasonic attenuation maxima to relaxation processes caused by the growth of precipitates and the accompanying structural and coherency changes. They pointed out that Duraluminum 2024 should exhibit two maxima in the ultrasonic attenuation due to the formation of GPB and GP zones, respectively. The maxima in attenuation occurs at aging times when the interaction of the ultrasonic waves with the configuration of stress around the precipitates is also a maximum. This was confirmed by additional experiments which showed that at higher ultrasonic frequencies the maximum interaction occurred for smaller precipitates at shorter aging times.

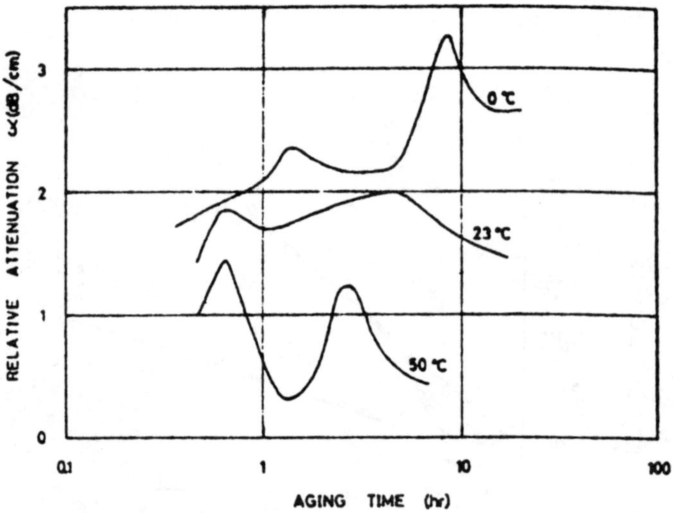

Fig. 13. Variation of the ultrasonic attenuation during isothermal aging of Duraluminum 2024 at the temperature indicated [Gefen and Rosen (37)].

As a final illustration of the extreme sensitivity of ultrasonic attenuation measurements to microstructural alterations which are difficult if not impossible to detect by other nondestructive testing techniques, Fig. 14, from the work of Sachse and Green (38) is reproduced. This figure shows the results of an experiment in which an aluminum crystal which had previously been loaded to a plastic load of 180 kg was loaded to 60 kg and maintained at this load for one minute, unloaded to 45 kg and maintained for one minute, unloaded to 30 kg and maintained for one minute, unloaded to 15 kg and maintained for one minute, and then unloaded completely. Upon subsequent reloading to 70 kg and unloading, the ultrasonic attenuation displayed dips at those load values which had been maintained for a short time during the previous cycle. Note that even though the attenuation "remembered" the discontinuous nature of the first load-unload cycle, the load-time curve (or equivalently the stress-strain curve) gave no indication of it. The observed results were attributed to the pinning of dislocation loops by point defects which were preferentially located at the position they occupied when the load was maintained constant for the one minute intervals. Upon reloading, these point defects pinned the dislocation loops again when the dislocations arrived at the location of the defects thus causing the dips in attenuation. These results show that ultrasonic attenuation measurements are sensitive to migration of point defects and their interaction with dislocations.

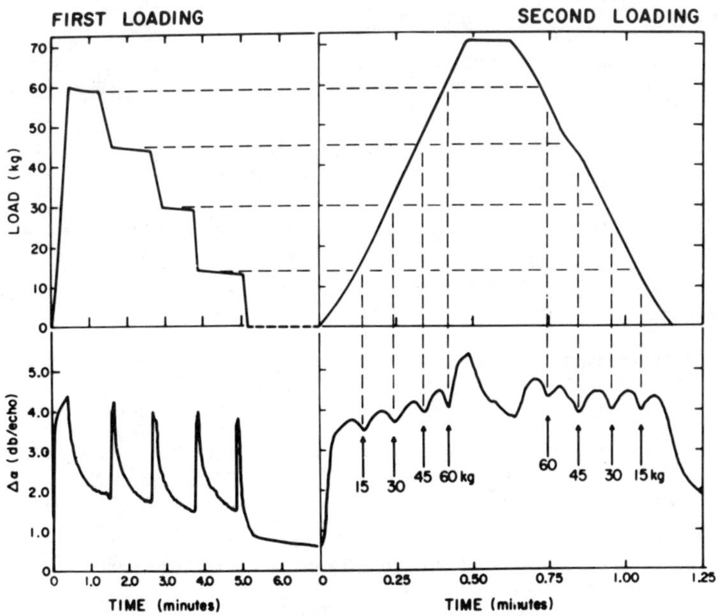

Fig. 14. Load versus time and change of ultrasonic attenuation versus time plots obtained during a load-unload-hold sequence followed by a reload-unload sequence showing the load-memory dips in ultrasonic attenuation [Sachse and Green (38)].

Conclusions

Ultrasonic attenuation measurements serve as a very sensitive indicator of microstructures and microstructural alterations in metallic materials. This sensitivity derives from the ability of ultrasonic waves to interact with volume defects such as cracks, microcracks, included particles, and precipitates; surface defects such as grain boundaries, interphase boundaries, and magnetic domain walls; and dislocation line defects. Moreover, ultrasonic waves are sensitive to the interaction of point defects such as impurity atoms and vacancies with dislocations. Although numerous investigations have been conducted in this area, much more research needs to be performed before ultrasonic attenuation measurements can be fully exploited in practical applications.

Acknowledgments

The author would like to thank his wife, Sydney T. Green, for her invaluable assistance with the literature search and compilation of the camera-ready manuscript. A special note of appreciation is due Corinne Harness for her care and professionalism in typing the paper.

References

1. W. P. Mason and H. J. McSkimin, "Attenuation and Scattering of High Frequency Sound Waves in Metals and Glasses," The Journal of the Acoustical Society of America, 19 (3) (1947), pp. 464-473.

2. W. P. Mason and H. J. McSkimin, "Energy Losses of Sound Waves in Metals due to Scattering and Diffusion," Journal of Applied Physics, 19 (1948) pp. 940-946.

3. W. Roth, "Scattering of Ultrasonic Radiation in Polycrystalline Metals," Journal of Applied Physics, 19 (1948) pp. 901-910.

4. Floyd A. Firestone, "Tricks with the Supersonic Reflectoscope," Non-Destructive Testing, 7 (1948) pp. 5-19.

5. S. Ya. Sokolov, "Attenuation of Ultrasonic Vibrations in Solid Bodies," Doklady Akademii Nauk SSSR, 59 (1948) pp. 883-886.

6. H. B. Huntington, "On Ultrasonic Scattering by Polycrystals," The Journal of the Acoustical Society of America, 22 (3) (1950) pp. 362-364.

7. I. M. Lifshitz and G. D. Parkhomovskii, "Theory of Propagation of Ultrasonic Waves in Polycrystals," Zhurnal Eksperimental'noi i Teoreticheskoi Fiziki, 20 (1950) pp. 175-182.

8. E. de Kerversau, J. Bleton, and P. Bastien, "Anomalies de Propagation des Ultra-sons dans les Metaux en Liaison avec la Structure," Revue de Metallurgie, 47 (1950) pp. 421-446.

9. R. L. Roderick and Rohn Truell, "The Measurement of Ultrasonic Attenuation in Solids by the Pulse Technique and Some Results in Steel," Journal of Applied Physics, 23 (2) (1952) pp. 267-279.

10. Tokutaro Hirone and Kazuo Kamigaki, "The Attenuation of Ultrasonic Waves in Metal (I). Aluminum," Japan Institute of Metals, 17 (1953) pp. 412-416.

11. Helmut Krainer und Ekkehart Krainer, "Überschallprüfung nach dem Durchschallungs-und Impulsecho-Verfahren," Archiv für das Eisenhüttenwesen, 24 (1953) pp. 229-236.

12. Hugo Josef Seemann und Werner Bentz, "Untersuchungen über den Einfluss des Gefüges auf die Extinktion von Ultraschallwellen in Metallischen Stoffen," Zeitschrift fuer Metallkunde, 45 (1954) pp. 663-669.

13. Akira Hikata, "Ultrasonic Attenuation in Polycrystalline Steel," Government Mechanical Laboratory Bulletin Japan, 1 (1954) pp. 1-7.

14. C. F. Ying and Rohn Truell, "Scattering of a Plane Longitudinal Wave by a Spherical Obstacle in an Isotropically Elastic Solid," Journal of Applied Physics, 27 (9) (1956) pp. 1086-1097.

15. L. G. Merkulov, "Investigation of Ultrasonic Scattering in Metals," Soviet Physics - Technical Physics, 1 (1956) pp. 59-69.

16. L. G. Merkulov, "Absorption and Diffusive Scattering of Ultrasonic Waves in Metals," Soviet Physics - Technical Physics, 2 (1957) pp. 953-957.

17. L. G. Merkulov, "The Use of Ultrasonic Waves in Investigating the Structure of Steels," Soviet Physics - Technical Physics, 2 (1957) pp. 1282-1286.

18. Tokutaro Hirone and Kazuo Kamigaki, "Attenuation of Ultrasonic Waves in Metals (7th Report) Stainless Steel," Japan Institute of Metals, 21 (1957) pp. 245-248.

19. Kazuo Kamigaki, "Ultrasonic Attenuation in Steel and Cast Iron," Sendai Tohoku University Science Reports, A9 (1957) pp. 48-77.

20. Akira Hikata and Rohn Truell, "Attenuation and Velocity Measurements in 1013 Steel as a Function of Deformation and Heat Treatment," Metals Research Laboratory Report WAL 143/14-51, Brown University, Jan., 1958.

21. Raymond G. Bayer, "Ultrasonic Attenuation in Mild Steel under Deformation," Metals Research Laboratory Report AFOSR TN 58-913 AD 204567, Brown University, Sept., 1958.

22. A. B. Bhatia, "Scattering of High-Frequency Sound Waves in Polycrystalline Materials," The Journal of the Acoustical Society of America, 31 (1) (1959) pp. 16-23.

23. A. B. Bhatia and R. A. Moore, "Scattering of High Frequency Sound Waves in Polycrystalline Materials.II," The Journal of the Acoustical Society of America, 31 (1959) pp. 1140-1141.

24. Emmanuel P. Papadakis, "Ultrasonic Attenuation Caused by Scattering in Polycrystalline Media," pp. 269-328 in Physical Acoustics Principles and Methods Vol. IV Part B, Warren P. Mason, ed.; Academic Press, New York, 1968.

25. Rohn Truell, Charles Elbaum, and Bruce B. Chick, Ultrasonic Methods in Solid State Physics, pp. 159-350; Academic Press, New York, 1969.

26. O. R. Gericke, "Ultrasonic Spectroscopy," pp. 31-61 in Research Techniques in Nondestructive Testing, R. S. Sharpe, ed.; Academic Press, London, 1970.

27. Emmanuel P. Papadakis, "Ultrasonic Velocity and Attenuation: Measurement Methods with Scientific and Industrial Applications," pp. 277-374 in Physical Acoustics Principles and Methods Vol. XII, Warren P. Mason and R. N. Thurston, ed.; Academic Press, New York, 1976.

28. Josef Krautkrämer and Herbert Krautkrämer, Ultrasonic Testing of Materials, 2nd ed.; Springer-Verlag, Berlin, 1977.

29. K. Goebbels, "Structure Analysis by Scattered Ultrasonic Radiation," pp. 87-157 in Research Techniques in Nondestructive Testing Vol. IV, R. S. Sharpe, ed.; Academic Press, London, 1980.

30. A. Vary, "Ultrasonic Measurement of Material Properties," pp. 159-204 in Research Techniques in Nondestructive Testing Vol. IV, R. S. Sharpe, ed.; Academic Press, London, 1980.

31. R. Klinman and E. T. Stephenson, "Relation Between Mechanical Properties, Grain Size, and Ultrasonic Attenuation in Plain Carbon Steel," Research Department, Bethlehem Steel Corporation, Bethlehem, Pa., 18016, paper submitted for publication in Metallurgical Transactions, 1980.

32. R. H. Latiff and N. F. Fiore, "Ultrasonic Attenuation in Spheroidized Steel," Journal of Applied Physics, 45 (12) (1974) pp. 5182-5186.

33. R. Kettler, R. Latiff, and N. F. Fiore, "Surface Ultrasonic Measurement of Grain Size of Silicon Iron Sheet," Metallurgical Transactions, 5 (1974) pp. 952-953.

34. Forrest G. West, "Temperature Dependence of the Absorption of Ultrasound in a Nickel Single Crystal from 77° to 650°K," Journal of Applied Physics, 29 (3) (1958) pp. 480-482.

35. W. F. Chiao, "Ultrasonic Attenuation Studies of Mixed Microstructures in Steel," Transactions of the Metallurgical Society of AIME, 242 (1968) pp. 1543-1551.

36. Narayan R. Joshi and Robert E. Green, Jr., "Ultrasonic Detection of Fatigue Damage," Engineering Fracture Mechanics, 4 (1972) pp. 577-583.

37. Y. Gefen and M. Rosen, "Behavior of Ultrasonic Attenuation During Aging of Duraluminum 2024," Materials Science Engineering, 8 (1971) pp. 246-247.

38. Wolfgang Sachse and Robert E. Green, Jr., "Deformation Rate Effects on the Attenuation During Loading, Unloading, and Reloading of Aluminum Crystals," Journal of Physics and Chemistry of Solids, 31 (1970) pp. 1955-1961.

ACOUSTIC PROPERTIES AS MICROSTRUCTURE DEPENDENT MATERIALS PROPERTIES*

J. C. Shyne, N. Grayeli, G. S. Kino

Stanford University
Stanford, California 94305

Acoustic properties, usually well understood for single crystals, often exhibit a complex dependence upon microstructure in polycrystalline, multi-phase solids. Grain size and shape, preferred texture, the distribution of phases, and other microstructural details can strongly influence acoustic behavior of materials. This paper discusses acoustic velocity, acoustic attenuation, and the acoustoelastic effect as they are affected by variations in microstructure. The understanding and utilization of acoustic property/microstructure interrelationships have important implications for practical acoustic NDE methods.

* This work was supported by the U. S. Air Force Office of Scientific Research.

Most researchers and practitioners of acoustic nondestructive evaluation (NDE) have been experts in instrumentation or physical acoustics. Relatively few researchers have been concerned about the structure sensitivity of acoustic properties. As advances in the art and science of acoustic NDE have occurred, it has become ever more apparent that it is not adequate to treat solids simply as homogeneous and isotropic. The dependence of acoustic properties on their microstructure can complicate the practice of NDE, but it can also suggest new NDE applications. For these reasons, more metallurgists and ceramists are now contributing to advances in acoustic NDE, and the acoustic properties of materials are being increasingly considered from the structure/property viewpoint of materials science.

In this paper, examples of acoustic properties that exhibit sensitivity to microstructure are described. Acoustic velocity, attenuation, and the acoustoelastic effect can be influenced by such microstructural details as grain size, preferred texture, and the amounts and distribution of phases.

Acoustic Velocity

The propagation velocity of acoustic waves in a solid is relatively simple to measure with good precision. Velocity values accurate to within one or two parts in 10^4 are routinely obtainable; the precision is limited by the accuracy of the acoustic path length measurement, i.e., the specimen thickness. Since accurate acoustic velocities are readily measured, velocity measurements offer some interesting possibilities for microstructural characterization despite the rather weak influence of microstructure variations on acoustic velocity.

In ideally elastically isotropic polycrystalline materials acoustic velocity is simply related to the elastic constants and density. Longitudinal compression wave velocity, v_p , and shear wave velocity, v_s, are given by:

$$v_p = \sqrt{\frac{\lambda + 2\mu}{\rho}} \qquad \text{and} \qquad v_s = \sqrt{\frac{\mu}{\rho}} \tag{1}$$

where λ and μ are the Lamé elastic constants and ρ is density. The elastic constants and density are weakly dependent on microstructure, therefore acoustic velocity must be somewhat influenced by microstructure.

Velocity Comparison for Carbon Steels

The longitudinal acoustic wave velocity was measured in a set of plain carbon steels; the carbon content (Table I) varied from 0.10 to 0.93 weight pct. In the normallized condition, air cooled from 900°C, the microstructure of these steels consisted of two phases, ferrite (BCC α-Fe) and cementite (Fe_3C). The proportions of the two phases varied with carbon content, and the cementite was distributed in quite normal mixtures of proeutectoid ferrite and pearlite (lamellar eutectoid of α-Fe and Fe_3C); the volume percentage of pearlite varied with carbon content from about 12 to 100 pct. It is easy to predict from the known densities of α-Fe and Fe_3C (1), and from the reported elastic modulus for Fe-C alloys (2), how the longitudinal acoustic wave velocity, v_p, should vary in Fe-C alloys as a function of carbon content. This is shown in Figure 1; v_p should decrease by 0.98 pct. for each weight pct. C. Assuming an absolute accuracy of acoustic velocity measurement of 1 in 10^4 implies that acoustic velocity measurements could discriminate between steels differing by as little as 0.01 weight pct. C. Also shown in Figure 1 are the measured velocities. The measured values are all higher than predicted for pure Fe plus C, and they exhibit a random scatter that completely obscures any systematic variation with carbon content. Measurements made perpendicular to and parallel to the rolling direction of the rolled bar stock differed only slightly and identically for all compositions, indicating that preferred texture was very slight and practically identical in all the steels. The observed scatter was caused by random variations in composition, especially manganese content. As shown in Table I, the Mn content varied from 0.45 to 0.80 weight pct. Small variations in silicon and trace impurities also occurred. Using the atomic weights of Fe and Mn and the known influence of Mn on the lattice parameter of α-Fe, the influence of Mn content on density can be calculated (3). Unfortunately that effect will cause random variations in acoustic velocity larger than the systematic influence of carbon content. This ignores other possible velocity changes resulting from the influence of Mn on the elastic moduli of steel. Random variations in Si content and trace impurities must also contribute to the variation in acoustic velocity.

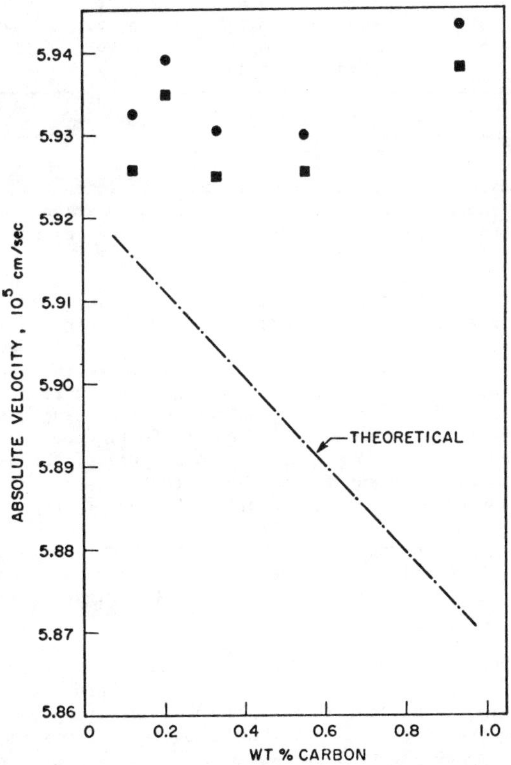

Fig. 1 Longitudinal acoustic wave velocity in plain carbon steel, ● parallel to rolling direction, ■ perpendicular. The theoretical curve was calculated for pure Fe-C alloys.

The conclusion to be drawn is that acoustic velocity measurements on samples of different commercial steels cannot be usefully correlated to variations in microstructure, because the influence of microstructural variation will be outweighed by the random effect of normal composition variance.

TABLE I

Composition of Plain Carbon Steels used for Acoustic Velocity Measurements

Steel Type (AISI)	C	Mn	Si	P	S
1010	0.10	0.45	0.03	0.007	0.032
1020	0.20	0.52	0.21	0.009	0.028
1035	0.33	0.72	0.17	0.018	0.025
1060	0.55	0.80	0.19	0.012	0.023
1095	0.93	0.50	0.22	0.009	0.030

Relative Velocity Acoustic End Quench Test

By relative acoustic velocity we mean velocities obtained on specimens of identical composition, e. g. from different locations in the same piece of material. Such measurements are not subject to uncertainties related to random composition variations.

The end quench test for measuring the hardening response of steel provides examples of microstructural variation within a piece of steel of constant composition. End quench test bars are circular cylinders of hardenable alloy steel that have been heated and cooled in a standard way so that a reproducible gradient in the cooling rate occurs from one end of the bar to the other. This causes a gradient in microstructure ranging from all martensite in the rapidly cooled end to ferrite/pearlite at the other end. The gradient in microstructure is a measure of the hardening response. The microstructure gradient in end quench test bars is conventionally measured by surveying the hardness along the bar length.

We have surveyed different composition, alloy steel end quench test bars both by the usual hardness method and by measuring the longitudinal acoustic velocity change along the bar length. Figure 2 is a typical example.

The conventional Rockwell C hardness scan and the acoustic velocity scan reflect the microstructural gradient in the test bar with comparable resolution. The acoustic velocity method of surveying end quench test bars may have some practical utility. The acoustic scan was made in three minutes with a computer controlled test rig; hardness scans are performed manually and require considerably more time and labor.

Preferred Texture and Acoustic Velocity

Polycrystalline metals and alloys invariably have some degree of elastic anisotropy resulting from preferred crystal orientation or texture. Preferred texture can have important influences on material properties. X-ray pole figures showing diffracted line intensity vs. spatial orientation

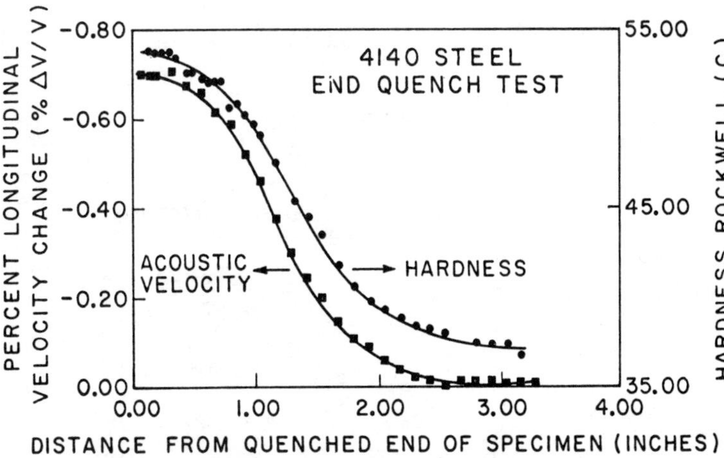

Fig. 2. End quench test bar of 4140 steel. Acoustic velocity scan along bar compared to conventional hardness scan. Velocity variation, % $\Delta v/v$, is normalized relative to slow cooled end.

are the usual way of displaying rolling textures. However, X-ray texture determination is a tedious process not readily adaptable to quality assurance testing. Several authors have demonstrated that acoustic velocity measurements offer a convenient means for quantifying the degree of texture in wrought materials (4). The elastic anisotropy resulting from the preferred texture causes a corresponding anisotropy of acoustic velocity.

The accuracy of acoustic velocity measurements can provide a more sensitive measure of preferred orientation in weakly textured materials than X-ray methods do. This is demonstrated by some acoustic measurements made on a 3 inch diameter round bar of hot rolled type 304 stainless steel. The fiber texture expected in a wrought round bar of this FCC alloy is [111] and/or [100] (5). That is, the individual crystallites (grains) are expected to have a higher than random average tendency to be oriented with a [111] or a [100] crystal direction parallel to the rolling direction.

Figure 3 shows isovelocity contours resulting from a scan of the longitudinal acoustic velocity in the rolling direction over a cross section of the 304 stainless steel bar; the circular symmetry of the velocity pattern is apparent. Figure 4 compares compression wave acoustic velocity values and X-ray diffraction line intensity data at various radial positions in the bar measured on specimens cut transverse to and parallel with the rolling direction. The X-ray intensity ratio of the (111) and (200) diffraction peaks are shown for each location tested. These X-ray data are consistent with weak [111] and [100] fiber texture components with the ratio of [111] to [100] components lowest at the center and increasing toward the surface of the bar. When measured perpendicular to the rod axis the acoustic velocity was almost independent of radial position. When the acoustic path was parallel to the bar axis, the velocity was lowest at the center and increased monotonically at positions closer to the surface. This is exactly as to be expected for the observed increasing ratio of [111] to [100] fiber texture components from center to surface. In FCC metal crystals the longitudinal elastic modulus is always lowest in the [100] direction and highest in the [111] direction. Thus the [100] fiber texture component tends to decrease the acoustic velocity in the rolling direction, while the [111] component tends to increase it.

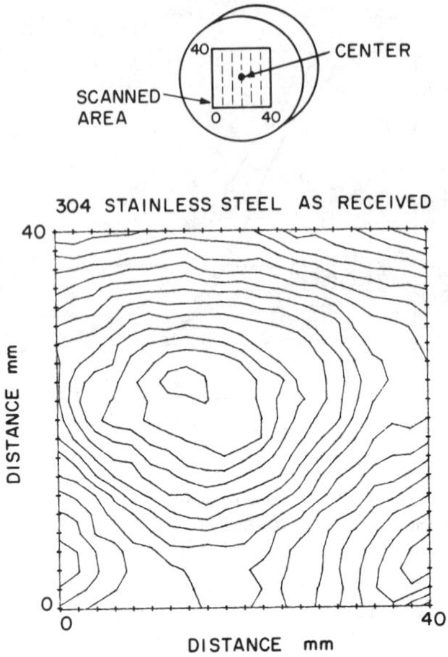

Fig. 3. Isovelocity contours on cross section of rolled round bar of 304 stainless steel.

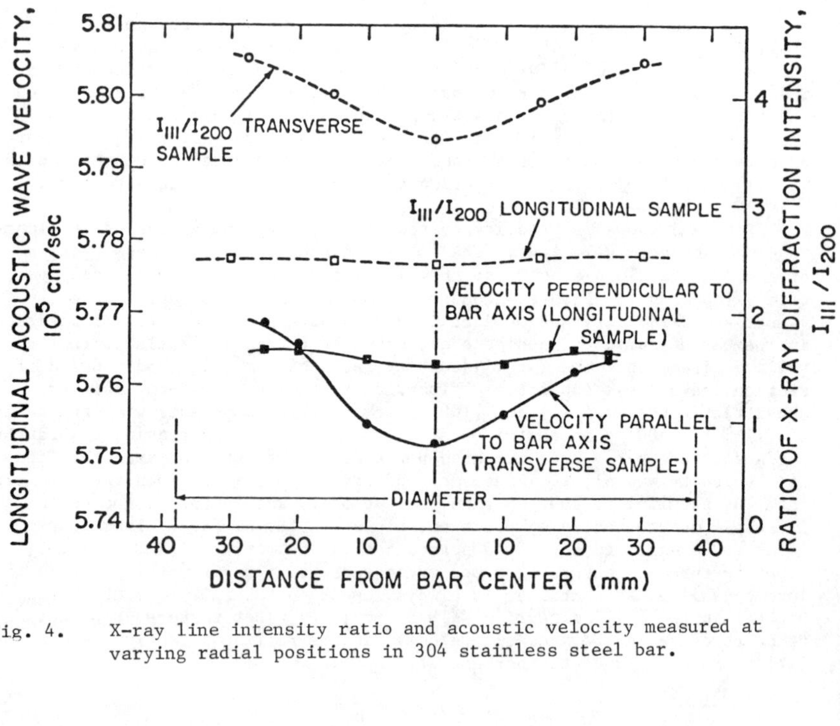

Fig. 4. X-ray line intensity ratio and acoustic velocity measured at varying radial positions in 304 stainless steel bar.

Acoustic Attenuation

In contrast to acoustic velocity, acoustic attenuation is strongly influenced by microstructure. However, acoustic attenuation measurements are more difficult and much less precise than acoustic velocity measurement, and while microstructure strongly influences attenuation, the interpretation of attenuation data can be difficult.

As an acoustic plane wave passes through a medium it dissipates energy by a variety of processes, and the wave amplitude decreases accordingly. Letting A be the wave's displacement amplitude and x be the acoustic path length:

$$A = A_o \exp(-\alpha x) \qquad (2)$$

where A_o is the amplitude at x = 0 and α is the attenuation coefficient, α is usually expressed as dB per unit distance.

The acoustic attenuation coefficient consists of two parts, $\alpha = \alpha_a + \alpha_s$. α_a represents mechanical energy converted to heat by any internal friction process; α_s is the result of energy scattered from the acoustic wave as the wave encounters discontinuities in acoustic impedance at grain boundaries and other microstructural features. At the 1 to 20 MHz range of acoustic frequencies most convenient for attenuation measurements, scattering losses are the dominant cause of acoustic wave attenuation.

The only well developed microstructure acoustic scattering theories are for grain boundary scattering from equiaxed grains in single phase, polycrystalline microstructures. For grain diameters, D, small compared to the acoustic wave length, Λ, i.e., for $\Lambda > 2\pi D$, each grain is treated as a Rayleigh scatterer suspended in a medium consisting of all the others (6). Because each randomly oriented crystallite is elastically anisotropic, there will be some average difference in elastic constants across the grain boundaries providing an average acoustic impedance mismatch. This model yields:

$$\alpha = S\beta^2 D^3 f^4 \qquad (3)$$

where S includes the acoustic velocities, β is the single crystal anisotropy factor ($\beta = [c_{11} - c_{12} - 2c_{44}]$, the c's being the usual crystalline elastic constants for cubic metals), D is the average grain diameter, and f is acoustic frequency. With higher frequencies or with coarser grain sizes so that $\Lambda \approx D$, the acoustic wave is assumed to be scattered stochastically by grain boundaries encountered along the wave path (7). In that case

$$\alpha = S'\beta^2 D f^2 \qquad (4)$$

An example of grain size variations causing attenuation differences is provided by the same 3 inch diameter rolled bar of type 304 stainless steel used to illustrate the influence of preferred texture on acoustic velocity. The average grain size in the hot rolled bar varied from center to surface as shown in the photomicrographs of Figure 5 and as plotted in Figure 6. The attenuation was measured for acoustic compression wave pulses passing through a 1 cm. thick slice of the bar cut perpendicular to the rolling direction. The measuring technique (8) involved comparison of the acoustic amplitude of reflections from the front and rear faces of the planar specimen, the second echo having been attenuated by passing through the thickness twice. The acoustic pulses were generated by a broad band transducer centered about 10MHz. A computer was used for on-line data processing. Taking the Fourier transform of the amplitude-time acoustic pulses converted the data into the amplitude-frequency domain thus allowing computer comparison of the acoustic pulse echoes and yielding attenuation as a function of frequency.

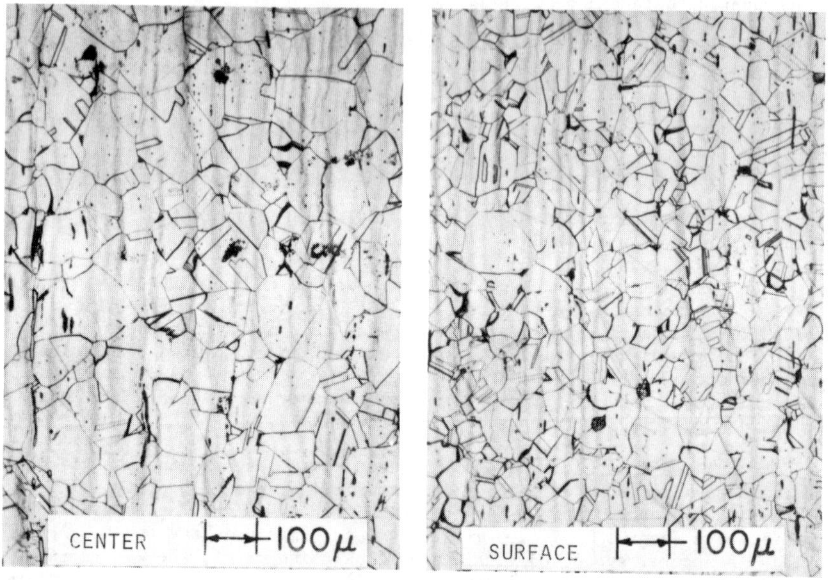

Fig. 5. Photomicrographs of the 304 stainless steel bar grain structure at center and surface radial positions.

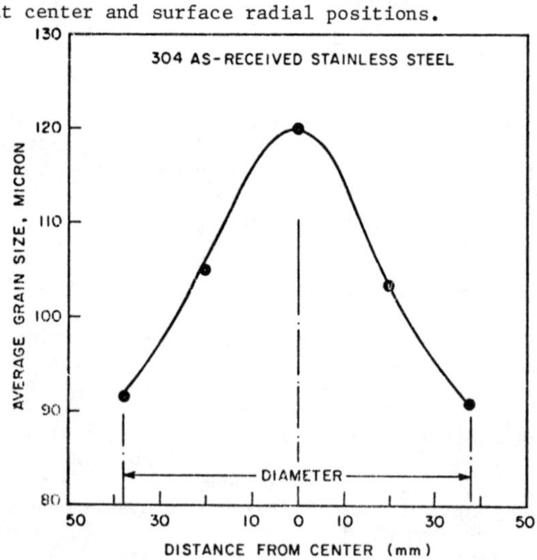

Fig. 6. Average grain diameter vs. radial position in 304 stainless steel bar.

Figure 7 shows attenuation vs. acoustic frequency measured at different radial locations within the 304 stainless steel bar. The nearly straight lines on the log-log coordinates have slopes ranging from 3.43 to 3.96. Since that is nearly 4, it can be concluded that the Rayleigh grain scattering model adequately describes the attenuation (as suggested by the f^4 in Equation (3)). This is to be expected since, at 10MHz, Λ is 570μm, while the largest average grain size (at the center) is 120μm; thus $\Lambda \approx 2\pi D$.

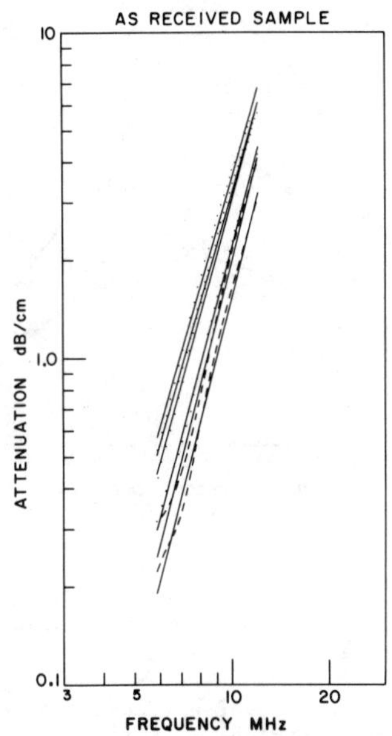

Fig. 7. Acoustic attenuation vs. frequency measured at various radial positions in the 304 stainless steel bar. Attenuation proportional to the fourth power of frequency.

Figure 8 shows the measured attenuation coefficient at different radial positions for three acoustic frequencies. Comparison of Figures 6 and 8 indicates that for a given acoustic frequency the attenuation coefficient varies as the grain size cubed, consistent with Rayleigh grain scattering.

Acoustoelastic Coefficients

The measurement of stress by acoustic means is currently the focus of considerable research. Acoustic stress measurement, a technique with important technological potential, makes use of the acoustoelastic effect.

The velocity of sound in solids is dependent upon the state of stress in the acoustic medium. This phenomenon, the acoustoelastic effect, is a manifestation of nonlinear elasticity. For most purposes Hooke's law, which assumes a linear elastic stress-strain relationship, is adequate, however for large elastic strains or when small deviations from linear elasticity are significant, as with the acoustoelastic effect, nonlinear elastic theory must be applied. Nonlinear elasticity requires the use of higher order elastic constants in addition to the familiar Hookean elastic moduli.

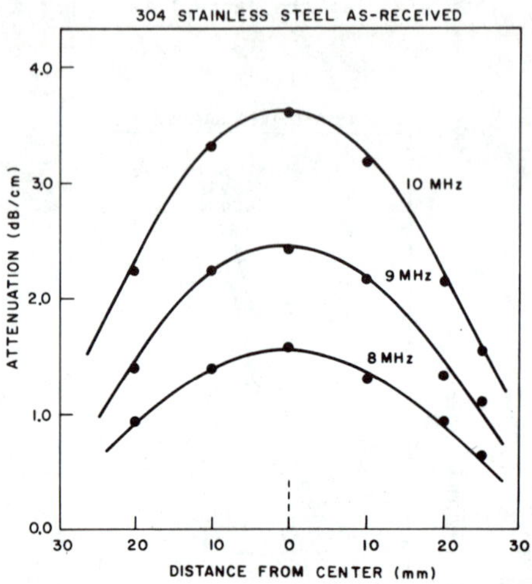

Fig. 8. Acoustic attenuation coefficient vs. radial position in the 304 stainless steel bar.

The elastic strain energy, Φ , of a volume element of a solid can be expressed as an infinite power series of Lagrangian strain components, ε;

$$\Phi(\varepsilon) = \Phi_o + \sum_{n=1}^{\infty}{}' \frac{1}{n!} \frac{\partial^n \Phi}{\partial \varepsilon^n} \varepsilon^n \quad (5)$$

The first two terms in the series are equated to zero by defining the unstrained state as zero energy and the initial state as unstrained.

Thus:

$$\Phi(\varepsilon) = \tfrac{1}{2} c_{ij}\varepsilon_i\varepsilon_j + 1/6\ C_{ijk}\varepsilon_i\varepsilon_j\varepsilon_k + \text{- - -} \quad (6)$$

where the c_{ij}'s are the usual Hookean second order elastic constants, and the C_{ijk}'s are third order elastic constants. For finite elasticity and for acoustoelasticity, second and third order elastic constants suffice; fourth and higher order constants can be neglected. Crystal symmetry reduces the number of independent third order elastic constants to 6 for the cubic metals For elastically isotropic solids there are only 3 independent third order elastic constants to be considered along with the familiar 2 second order constants.

For the case of a longitudinal acoustic wave passing through the thickness of a plate made from elastically isotropic material in a static state of plane stress, the acoustic wave velocity, v_p^σ, is

$$v_p^\sigma = v_p^o \, [1+B(\sigma_x+\sigma_y)] \qquad (7)$$

where v_p^σ is the acoustic velocity unstressed, σ_x and σ_y are the principal values of the static plane stress tensor, and B is an elastic property of the material, an <u>acoustoelastic coefficient</u>. The acoustoelastic coefficient B, depends upon the second and third order elastic constants (9).

$$B = \frac{\mu l - \lambda(m+\lambda+2\mu)}{\mu(\lambda+2\mu)(3\lambda+2\mu)} \qquad (8)$$

Here λ and μ are the second order Lamé elastic constants, while l and m are 2 of Murnaghan's 3 third order elastic constants (10). For shear waves, equations similar to Equation (7) but with other acoustoelastic coefficients describe how shear wave velocity depends on the static stress field. The acoustoelastic coefficients can be calculated if the second and third order elastic constants are known, or they can be obtained empirically.

If the appropriate acoustoelastic coefficients are known from calculation or measurement, and if the acoustic velocity can be measured with sufficient precision, then acoustoelastic stress measurements are possible. There have been many laboratory demonstrations of the feasibility of acoustoelastic stress measurement, and much current effort is directed toward the development of its practical applications.

One difficulty complicating the development of practical acoustoelastic stress measurement is the variance and anisotropy of the acoustoelastic coefficients, so that the acoustoelastic coefficients must be viewed as microstructure sensitive material properties. Polycrystalline materials invariably have some degree of preferred crystal texture; this must affect the acoustoelastic coefficients.

Figure 9 shows the change in longitudinal acoustic wave velocity measured in 6061-T6 aluminum plate specimens stressed in uniaxial tension; the stress axis was oriented parallel to the rolling direction (11). The slope of the lines is equal to B, the acoustoelastic coefficient in Equation (7). These data were obtained with 6061-T6 aluminum plate stock from three different suppliers. Obviously variations in the processing histories caused considerable variation in B, the largest slope being more than 50% greater than the smallest. Variations in the rolling texture must be assumed responsible for the variation in the measured acoustoelastic coefficient.

Another example of variations in the acoustoelastic coefficient B was observed with the 304 stainless steel bar described above. A piece was removed from the 3 inch diameter round bar as depicted in Figure 10. Pin grip end pieces were welded onto the stainless steel sample so that a tensile stress could be applied; the stress axis coinsided with a diameter of the original bar stock. The acoustic velocity of longitudinal acoustic waves was measured along the gauge section of the composite tension test specimen at different values of applied tensile stress. From those data the acoustoelastic coefficient, B, was obtained as a function of radial position in the 304 stainless steel bar. This is plotted in Figure 11.

The pronounced variation in the acoustoelastic coefficient plotted in Figure 11 can be related to the radially varying fiber texture discussed in connection with acoustic velocity measurements and to the welds used to construct the specimen. The large fluctuations in B values measured 30 mm and farther from the original center line of the round bar occur in the welds (weld material is also 304 stainless steel). Between the welds the measured B values are highest at the center line and are lower toward the outer surface of the bar. This variation is consistent with the radially varying fiber texture pattern observed in the bar. The fine scale variation in B near the center of the bar is not understood. Possibly it may reflect fine scale variations in the texture. The acoustic measurements were made using an acoustic beam 1.5 mm in diameter capable of spatially resolving the plotted fine scale variations in B.

While explanations for the variability encountered in measuring acoustoelastic coefficients are incomplete and necessarily tentative, it is clear that large variations do occur even from point to point in the same piece of material. Such variations must be better understood if the full potential of acoustoelastic stress measurements is to be achieved.

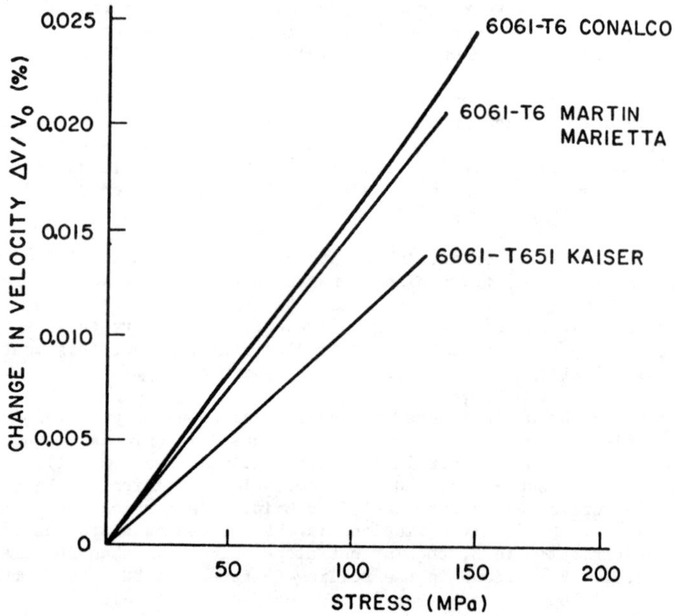

Fig. 9. Acoustoelastic effect in 6061-T6 aluminum alloy plate samples obtained from three different suppliers (11).

Summary

This paper describes several examples of microstructure dependent acoustic properties. Some of these microstructural effects on the acoustic behavior of solids can suggest novel NDE techniques for materials characterization, while others can complicate the NDE by acoustic methods. For these reasons, the study of acoustic properties as microstructurally sensitive materials properties will advance acoustic NDE. Acoustic methods provide the most versatile means for examining the interior of solid objects nondestructively. To achieve the full potential of acoustic NDE, however, the acoustic properties of the materials must be better understood.

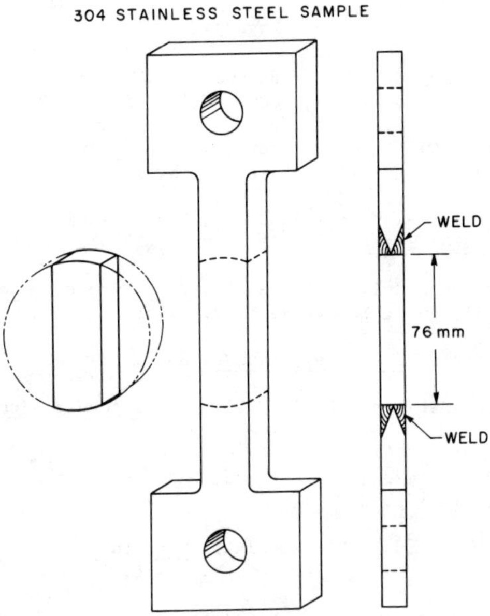

Fig. 10. Tension test specimen constructed from 304 stainless steel bar to measure acoustoelastic coefficient.

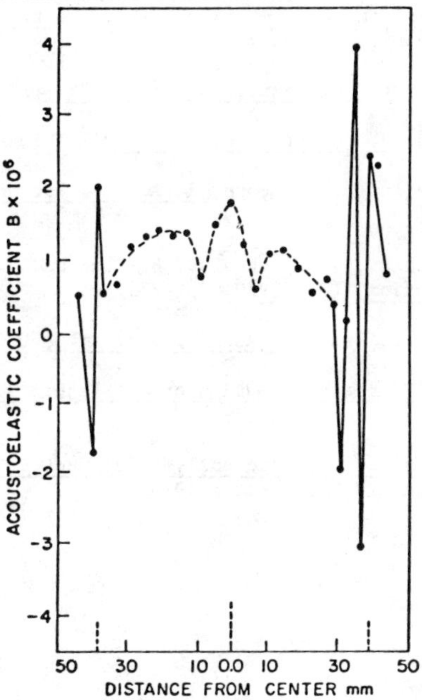

Fig. 11. Acoustoelastic coefficient vs. radial position in 304 stainless steel bar.

REFERENCES

1. (a) (for α-Fe) C. S. Barrett and T. B. Massalski, Structure of Metals 3rd edition, p. 628, McGraw Hill Inc., New York, N. Y., 1966.

   (b) (for Fe_3C) W. Hume-Rothery, The Structure of Alloys of Iron, p. 140, Pergamon Press Inc., Elmsford, N. Y. 1966.

2. L. A. Glikman, A. M. Kartashov, Z. M. Ru-ashkina, and A. F. Lobov, Met. Sci. Heat Treat., 17 (May-June), (1975), p. 398.

3. G. S. Barrett and T. B. Massalski, Structure of Metals, 3rd ed., p. 371; McGraw-Hill, Inc., New York, N. Y., 1966.

4. (a) J. Weerts, Z. f. Metallkunde, 25 (1933), p. 101.

   (b) E. Papadakis and P. F. Sullivan, J. Acoust. Soc. Am., 33 (1961), p. 1622.

   (c) B. R. Tittmann, G. A. Alers, and L. J. Graham, Metallurgical Trans., 7A (1976), p. 229.

5. C. S. Barrett and T. B. Massalski, Structure of Metals, 3rd ed., p. 543; McGraw-Hill, Inc., New York, N. Y., 1966.

6 (a) W. P. Mason and H. J. McSkimin, J. Acoust. Soc. Am., 19 (1947), p. 464.

   (b) I. M. Lifshits and G. D. Parkhomovskii, Zhur. Eksp. i Theoret. Fiz., 20 (1950), p. 175.

7. (a) H. B. Huntington, J. Acoust. Soc. Am., 22 (1950), p. 362.

   (b) A. B. Bhatia, J. Acoust. Soc. Am., 31 (1959), p. 16.

   (c) A. B. Bhatia and R. A. Moore, J. Acoust. Soc. Am., 31 (1959), p. 1140.

8. N. Grayeli, F. Stanke, G. H. Chou, D. B. Ilic and J. C. Shyne, IEEE 1979 Ultrasonics Symposium, (May 1980), p. 273.

9. D. M. Egle and D. E. Bray, J. Acoust. Soc. Am., 60 (1976), p. 741

10. F. D. Murnaghan, Finite Deformation of an Elastic Solid; Wiley, New York, 1951.

11. J. B. C. Hunter, Acoustoelastic Analysis of Center-Cracked Panels, Ph.D. Dissertation, Stanford University, 1979.

IDENTIFICATION BY SMALL ANGLE NEUTRON SCATTERING OF MICROSTRUCTURAL CHANGES IN METALS AND ALLOYS*

J. R. Weertman

Department of Materials Science and Engineering
and
Materials Research Center
Northwestern University
Evanston, IL 60201

Small angle neutron scattering (SANS) is a promising technique for the nondestructive evaluation of metals and alloys. The size range of microstructural features which are most easily studied by small angle scattering coincides with that which is important in the degradation processes leading to failure: less than 10 nm to about a micrometer. Neutrons have a large penetrating power, so that observations derived from neutron scattering pertain to bulk effects.

In this paper we first outline briefly a number of methods whereby scattering curves are analyzed to yield information on the scatterers. These methods then are illustrated with scattering data from a model scattering system, fatigue-induced grain boundary cavities in a pure metal. Finally we describe a number of examples, taken from the literature, of the use of SANS in NDE in technologically important situations.

* This work was supported by NSF-MRL Grant DMR7923573 and NSF Grant DMR-7826026.

Introduction

The phenomenon of small angle neutron scattering (SANS) has several inherent features which make it an attractive technique for the nondestructive examination of metals and alloys. Although a large initial investment must be made in providing a neutron source and building a camera which is capable of making measurements over the wide range of conditions required for an adequate analysis, SANS has numerous offsetting advantages. The size range of microstructural features which are most easily studied by small angle scattering techniques is precisely that which is important in monitoring the degradation processes leading to failure: less than 10 nm to about a micrometer. (Larger inhomogeneities cause a broadening of the beam through multiple refraction.) Unlike X-rays, which detect irregularities only in a thin surface region, neutrons have a considerable power of penetration. SANS metal samples typically are 5-10 mm thick. The information which can be extracted from scattering curves can be varied and extensive: average size and size distribution of the scattering entities, their concentration, shape, occasionally even their nature.

In this paper we first outline briefly a number of methods whereby scattering curves are analyzed to yield information on the scatterers. These methods then are illustrated with scattering data from a model scattering system, fatigue-induced grain boundary cavities in a pure metal. Finally we describe a number of examples of the use of SANS in nondestructive evaluation (NDE) in practical situations.

The reader is referred to books by Bacon (1) and by Kostorz (2) for in depth studies of neutron diffraction and small angle neutron scattering. An extensive compilation of NDE experiments involving SANS has been made by H. Herman (3).

Analysis of Scattering Data

A schematic drawing of an arrangement for carrying out SANS measurements is shown in Fig. 1. Slow neutrons are led well away from the neighborhood of the reactor by means of a neutron guide tube. A collimated beam of

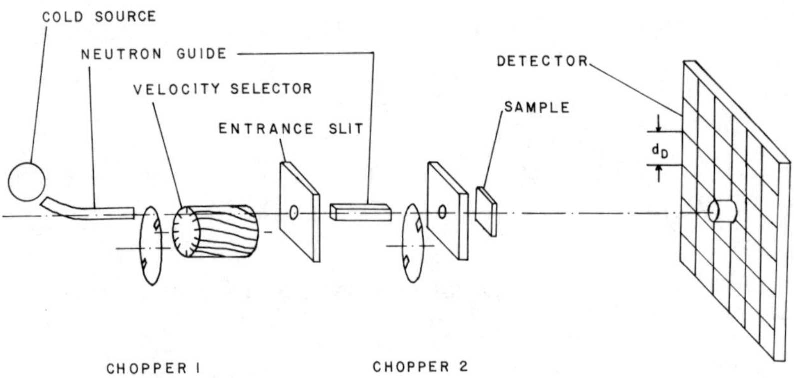

Fig. 1 - Experimental arrangement for SANS measurements.

neutrons which has been made nearly monochromatic by a velocity selector impinges on the sample. The scattered neutrons are counted by a two-dimensional position-sensitive detector. Neutron wave lengths λ typically are 0.5 to 1 nm, so that the complication of multiple Bragg scattering is eliminated. Measurements of coherent scattering are obtained over $q = |\vec{q}|$ ranges of about 2×10^{-2} to 2 nm^{-1}. Here the scattering vector, $\vec{q}$, is defined as $\vec{K}' - \vec{K}$, where $\vec{K}$ and $\vec{K}'$ are the wave vectors of the incident and scattered neutrons, respectively. $|\vec{K}| = |\vec{K}'| = 2\pi/\lambda$, so $q = 4\pi \sin\theta/\lambda$, where θ is the Bragg angle, i.e., one half the scattering angle. At the end of a test the total number of scattered neutrons recorded at the detector as a function of $\vec{q}$ is corrected for instrumentation effects and absorption by the sample (2). The corrected number of neutrons $I(\vec{q})$ scattered into the detector cell corresponding to the vector $\vec{q}$ and subtending a solid angle $\Delta\Omega$ at the sample is given by the equation:

$$I(\vec{q}) = I_o\, t\, s\, e\, \frac{d\Sigma(\vec{q})}{d\Omega}\, \Delta\Omega + I_\infty, \tag{1}$$

where I_o is the (number) flux of neutrons incident on the sample, t is time of test, s is area of the sample exposed to the beam, e is sample thickness, $d\Sigma(\vec{q})/d\Omega$ is the coherent differential scattering cross section per unit sample volume, and I_∞ is the number of neutrons incoherently scattered into the detector cell during the test. (The asymptotic value of $I(\vec{q})$ at large q's is I_∞). Absolute values for $d\Sigma/d\Omega$ can be obtained by comparing neutron scattering from the sample with that from vanadium, an incoherent scatterer whose scattering cross section is known.

$$\frac{I(\vec{q}) - I_\infty}{I_{\infty_v}} = \frac{I_o\, t\, s\, e\, \frac{d\Sigma(\vec{q})}{d\Omega}\, \Delta\Omega}{I_o\, t_v\, s_v\, e_v\, \frac{\Sigma_{i_v}}{4\pi}\, \Delta\Omega} \tag{2}$$

Here the subscript v denotes quantities associated with vanadium and Σ_i is total incoherent cross section. All quantities appearing in Eqn. (2) other than $d\Sigma/d\Omega$ are known or cancel. Thus may be obtained curves of $d\Sigma/d\Omega$ vs $\vec{q}$, or, if the scattering is isotropic about the axis of the incident beam, vs q.

Typically, the area of the sample normal to the beam, s, is of the order of 10^2 mm^2 and the optimum sample thickness ranges from several to some tens of millimeters.

Let us now turn to the problem of determining what information can be extracted from a curve of $d\Sigma/d\Omega$ measured as a function of the scattering vector $\vec{q}$. Basic diffraction theory (4) gives the following equation for the coherent differential scattering cross section $d\Sigma/d\Omega$ per unit volume of sample:

$$\frac{d\Sigma}{d\Omega} = \frac{1}{V} \left| \sum_{\vec{r}} b(\vec{r}) e^{-i\vec{q}\cdot\vec{r}} \right|^2 \tag{3}$$

where $b(\vec{r})$ is the coherent scattering length of the particular atom occupying a site with position vector $\vec{r}$, and V is the volume of the sample exposed to the neutron beam. Since the smallest size inhomogeneity that can be distinguished by SANS is of the order of a few nanometers, it is permissible to replace the coherent scattering length $b(\vec{r})$ of an atom by a locally averaged scattering length density, $\rho(\vec{r}) = b(\vec{r})/\omega(\vec{r})$, where ω is the atomic volume of the atom at $\vec{r}$. Thus the coherent differential scattering cross section can be written as:

$$\frac{d\Sigma}{d\Omega} = \frac{1}{V}\left|\int_V \rho(\vec{r})\, e^{-i\vec{q}\cdot\vec{r}}\, d^3\vec{r}\right|^2 \qquad (4)$$

Typically the value of b is of the order of 1 to 10×10^{-15} m, and ρ ranges from 0.05 to $1 \times 10^{15}/m^2$. Unlike the case of x-ray scattering lengths, which increase monotonically with atomic number, the neutron scattering lengths appear more randomly dependent on nuclear charge. A number of light elements are excellent scatterers. Therefore their presence may be detected by neutron scattering whereas their contribution to x-ray scattering is negligible. A compilation of neutron scattering lengths and cross sections is given in reference (2).

We shall assume that the material we are investigating by SANS is a metal or alloy which contains a number of inhomogeneities (e.g., precipitates, voids, second phase particles, Guinier-Preston zones, dislocations) in a uniform crystalline matrix. Small angle scattering by such a sample is caused by the spatial variation in the scattering length density of the material.

$$\frac{d\Sigma}{d\Omega} = \frac{1}{V}\left|\int_{V_m} \rho_m\, e^{-i\vec{q}\cdot\vec{r}}\, d^3\vec{r} + \sum_i \int_{V_{p_i}} \rho_{p_i}\, e^{-i\vec{q}\cdot\vec{r}}\, d^3\vec{r}\right|^2 \qquad (5)$$

$$\frac{d\Sigma}{d\Omega} = \frac{1}{V}\left|\int_V \rho_m\, e^{-i\vec{q}\cdot\vec{r}}\, d^3\vec{r} + \sum_i \int_{V_{p_i}} (\rho_{p_i} - \rho_m)\, e^{-i\vec{q}\cdot\vec{r}}\, d^3\vec{r}\right|^2 \qquad (6)$$

Here ρ_m is the scattering length density of the matrix (assumed to be independent of $\vec{r}$), ρ_{p_i} is the scattering length density of particle i (also assumed to be independent of $\vec{r}$ but may vary from particle to particle if different types of inhomogeneities are involved), and V_{p_i} and V_m are the volume of the ith particle and of the matrix, respectively.[1] So long as the volume V of the sample exposed to the beam is macroscopic, the first integral in Eqn. (6) is vanishingly small. If the inhomogeneities are well separated and randomly distributed, the cross terms which result from the multiplication of the sum in Eqn. (6) by its complex conjugate tend to cancel. The expression for the coherent differential scattering cross section in a sample containing a dilute dispersion of scattering entities in a uniform matrix then becomes:

$$\frac{d\Sigma(\vec{q})}{d\Omega} = \frac{1}{V}\sum_i \left|(\Delta\rho_i)\int_{V_{p_i}} e^{-i\vec{q}\cdot\vec{r}}\, d^3\vec{r}\right|^2$$

$$= \frac{1}{V}\sum_i N_{p_i}(\Delta\rho_i)^2\, V_{p_i}^{\,2}\left|F_{p_i}(\vec{q})\right|^2 \qquad (7)$$

where N_{p_i} is the number of particles in the sample with size, shape, chemical composition, and orientation with respect to the beam characterized by the subscript "i", $F_{p_i}(\vec{q})$ is the single particle form factor for the ith type of particle,

$$F_{p_i}(\vec{q}) = \frac{1}{V_{p_i}} \int_{V_{p_i}} e^{-i\vec{q}\cdot\vec{r}} \, d^3\vec{r} \tag{8}$$

and $\Delta\rho_i = \rho_{p_i} - \rho_m$. If the dispersion of scattering entities is not dilute, cross terms appear in the equation for $d\Sigma/d\Omega$ which have the form

$$e^{-i\vec{q}\cdot(\vec{R}_i - \vec{R}_j)} \int_{V_{p_i}} e^{-i\vec{q}\cdot\vec{r}_i} \, d^3\vec{r}_i \int_{V_{p_j}} e^{i\vec{q}\cdot\vec{r}_j} \, d^3\vec{r}_j$$

Here $\vec{R}_i$ is a position vector from the common origin to a reference origin in the ith particle and $\vec{r}_i$ is measured from this reference origin. Such cross terms are appreciable when the exponents $\vec{q}\cdot(\vec{R}_i - \vec{R}_j)$ are so small that terms of the type $\cos\vec{q}\cdot(\vec{R}_i - \vec{R}_j)$ do not cancel. Evidently a contribution to the scattering curve from interparticle interference first becomes apparent, upon increasing concentration of scatterers, at small q's. A peak in the scattering curve at small angles provides information on interparticle spacing (5).

We assume in the following paragraphs that the scattering entities causing SANS in our material are sufficiently dilute that Eqn. (7) is valid. What information concerning the nature of the scatterers can be obtained from curves of $d\Sigma/d\Omega$ vs $\vec{q}$?

Guinier Approximation

Consider an ensemble of N_p scattering particles (voids, precipitates, carbides, G-P zones, etc.) embedded in a matrix. The particles have similar shape, size, and chemical composition, but are randomly oriented. Their size is characterized by the linear dimension a. Guinier (6) has shown that, if the product qa is sufficiently small, the differential scattering cross section can be approximated by the following equation:

$$\frac{d\Sigma}{d\Omega}(q) \approx \frac{N_p}{V} V_p^2 (\Delta\rho)^2 e^{-q^2 R_G^2/3} \tag{9}$$

where V_p is the volume of a scattering particle, $\Delta\rho$ is the difference in scattering length density between matrix and particle, V is volume of sample in the neutron beam, and R_G, the Guinier radius, is defined as:

$$R_G^2 = \frac{1}{V_p} \int_{V_p} \vec{r}\cdot\vec{r} \, d^3\vec{r} \tag{10}$$

In evaluating R_G the origin of the position vector $\vec{r}$ is placed at the center of gravity of the particle. What constitutes a sufficiently small value of the product qa for the Guinier approximation to be valid depends somewhat on the shape of the particles responsible for the scattering. In the case of spheres qR_G should be less than about 1.2 (2).

If the scattering particles consist of spherical inhomogeneities with a distribution of sizes,

$$\frac{d\Sigma}{d\Omega} \approx \frac{N_p}{V} (\Delta\rho)^2 \langle V_p^2 \rangle e^{-\frac{q^2}{5}\frac{\langle R^8 \rangle}{\langle R^6 \rangle}} \tag{11}$$

in the Guinier approximation. Here the symbol $\langle\rangle$ denotes "average". For Eqn. (11) to be valid, it is now necessary that the Guinier radius of the largest spheres in the ensemble satisfy the condition $qR_G < 1.2$. A plot of $\ln(d\Sigma/d\Omega)$ against q^2 yields a straight line in the Guinier region with slope $-\langle R^8 \rangle / 5\langle R^6 \rangle$. Thus from such a plot it is possible to obtain one measure of a characteristic particle size, $[\langle R^8 \rangle / \langle R^6 \rangle]^{\frac{1}{2}}$. Obviously this characteristic radius is heavily weighted in favor of the larger spheres in the distribution.

Equation (11) can be generalized to the case of scattering particles of various sizes but similar shape if they are randomly oriented in the matrix. The R in Eqn. (11) is replaced by an effective radius, the radius of a sphere which has the same Guinier radius as that of the particle.

Porod Approximation

At the other end of the $d\Sigma/d\Omega$ vs q curve, the region of large q's, the coherent differential scattering cross section approaches another asymptotic value. The particular limiting value depends on the shape and orientation of the particles. Porod (7) has shown that, in the case of a distribution of spherical scattering particles of various sizes in a matrix, the cross section drops off as q^{-4}:

$$\lim_{q \to \infty} \frac{d\Sigma(q)}{d\Omega} \approx \frac{2\pi (\Delta\rho)^2}{q^4} \frac{S}{V} \tag{12}$$

Here S is the total surface area of the spherical scatterers in the volume V of the sample exposed to the beam. It is assumed, in deriving Eqn. (12), that the transition from matrix to particle at the interface is sharp. In order that the Porod approximation be valid, the value of q must be sufficiently large that the product of q with the radius of the smallest sphere in the distribution be greater than about 3.

Integrated Intensity

If SANS measurements have been carried out properly, the data will extend into the Guinier region at one end and the Porod range at the other. Therefore, although it is impossible to obtain direct measurements of $d\Sigma/d\Omega$ over all q values, the curves can be extrapolated so that the entire q range from zero to infinity is covered. If $d\Sigma/d\Omega$ depends only on $|\vec{q}|$, the integral of $d\Sigma/d\Omega$ over all reciprocal space can be written as:

$$\text{Integrated intensity} = \int_0^\infty \frac{d\Sigma}{d\Omega} 4\pi q^2 \, dq \tag{13}$$

It can be shown (8) that, in the case of a dilute distribution of scatterers in a matrix, the integrated intensity is directly proportional to the total particle volume ΔV in the sample:

$$\int_0^\infty \frac{d\Sigma}{d\Omega} 4\pi q^2 \, dq = 8\pi^3 (\Delta\rho)^2 \frac{\Delta V}{V} \tag{14}$$

Porod Radius and Number of Scatterers

For samples containing a dilute dispersion of spherical scatterers, another measure of "average" size of scatterer can be obtained by combining measurements in the Porod region and a calculation of the integrated intensity:

$$R_P = \frac{3\Delta V}{S} = \frac{\langle R^3 \rangle}{\langle R^2 \rangle} \tag{15}$$

The Porod radius, R_P, is less heavily weighted in favor of the large spheres in the dispersion than is the radius obtained from Guinier plots, $\{\langle R^8 \rangle / \langle R^6 \rangle\}^{\frac{1}{2}}$. However R_P still lies toward the high end of the distribution. An estimate of the number of scattering particles N_P in the sample can be obtained from the requirement:

$$N_P \approx \frac{S}{4\pi R_P^2} = \frac{3\Delta V}{4\pi R_P^3} \tag{16}$$

Size Distribution of Scattering Entities

Assume that the differential scattering cross section is measured over a range of q's broad enough that the Guinier criterion is satisfied at one end of the curve for even the largest scatterers in the sample and, at the other end, the smallest particles are well into the Porod region. The curve for $d\Sigma/d\Omega$ then can be extrapolated with confidence over the complete range of q from zero to infinity. In such cases it is possible to work backwards from the given $d\Sigma/d\Omega$ vs q curve to deduce what the size distribution of scatterers must have been. A number of methods have been devised to carry out this inversion (e.g., 9,10). It is essential to have some idea of the shape and nature of the scattering entities if the inversion is to succeed. Usually such information comes from TEM or SEM measurements. Unfortunately small errors or uncertainties in the $d\Sigma/d\Omega$ curve can produce large changes in the size distribution curve. For example, very minor changes in the extrapolation of the $d\Sigma/d\Omega$ curve at the high q end can produce violent fluctuations in the distribution curve at the small particle size end (11).

Problems and Complications in Interpreting SANS Data

It is evident from the preceding sections that some idea must be at hand of the nature, shape, orientation, etc., of the scattering entities before results of SANS measurements can be interpreted with confidence. Information retrieval from scattering and diffraction data is, by its nature, an indirect process. Quite different arrangements of scattering inhomogeneities can give rise to similar curves of $d\Sigma/d\Omega$. Therefore it is essential that as much information as possible be obtained by other methods, e.g., TEM, SEM, extraction replicas, to assist in the interpretation of the SANS data.

Samples are seldom so obligingly simple as those discussed in the previous paragraphs. Close spacing or clustering of scattering particles introduces complicated interference effects into the $d\Sigma/d\Omega$ curve. Frequently several different scattering particles may be present at once, for example, carbides and voids in an alloy fatigued at elevated temperatures, or a variety of carbides and gamma prime particles in a superalloy. It may be difficult to sort out the scattering attributable to each species. Surfaces which are scratched or are covered by oxides (12) are strong scatterers. Multiple scattering becomes a problem in thick samples which contain a high density of scatterers (5,13). Compositional fluctuations in the vicinity of the interface between matrix and precipitate or elsewhere in the matrix introduce additional scattering. Multiple refraction leads to appreciable line broadening when the inhomogeneities become large (of the order of several micrometers) (1). Magnetic domains in unmagnetized ferromagnetic material cause line broadening (2). (Such broadening can be eliminated by magnetizing the sample during the SANS measurements.) Multiple refraction effects can be separated from diffraction scattering by repeating measurements with neutrons of different wave length. Curves of $d\Sigma/d\Omega$ vs q remain unaltered when the wave length of the neutrons is changed if scattering is caused by diffraction, whereas scattering cross sections associated with multiple refraction are dependent on λ.

Analysis of SANS Data from a Model System

In order to demonstrate the wealth of information which can be extracted from SANS measurements we apply the methods developed in the preceding section to scattering data from samples of 99.999% pure copper which have been fatigued at elevated temperature (14,15,16). The small voids which nucleated and grew on the grain boundaries of the samples in the course of fatiguing act as scattering entities. By comparing scattering curves from a series of samples, each fatigued under identical conditions with only the time of cycling varying, it is possible to follow the kinetics of cavitation under that particular set of loading parameters. Analysis of the data is simplified by the observation, made in the course of extensive HVEM studies of the voids,* that the voids are roughly spherical at the stage of fatigue life covered in the SANS experiment (17). All samples were fatigued in fully reversed loading at a stress amplitude of 34.0 MPa at a frequency of 17 Hz in an environment of purified argon. Tests were conducted at 405°C (½ the absolute melting temperature), 486° , and 567°C. The SANS measurements were carried out on the D11A instrument at the Institut Laue-Langevin. Experimental details are given in (14) and (16). Coherent differential scattering cross sections were measured over a range of q values sufficiently broad that the scattering extended into the Guinier region at one end and the Porod range at the other. Thus the data can be extrapolated over all values of q and the integrated intensity calculated to give values of total void volume per unit sample volume, $\Delta V/V$ (Eqn. (14)). Figure 2 shows the development of void volume as a function of fatigue time for the three test temperatures.

*HVEM measurements were carried out at Argonne National Laboratory.

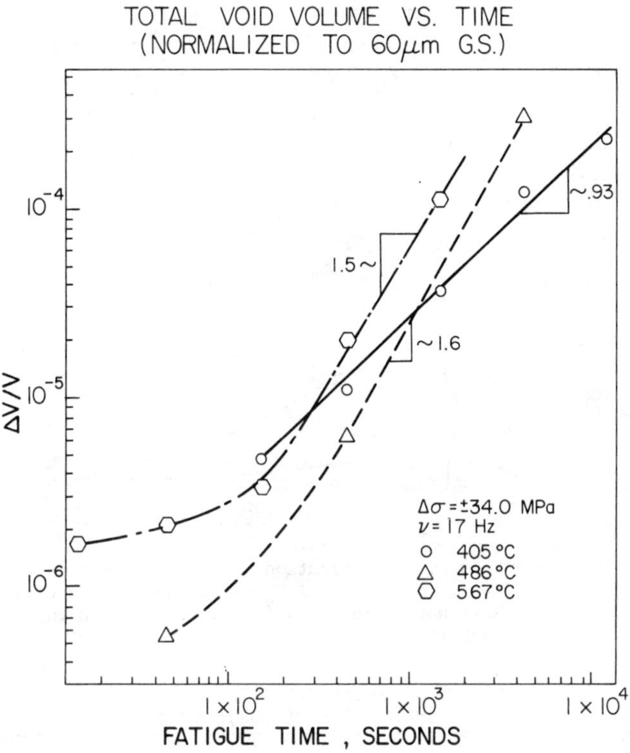

Fig. 2 - Total void volume fraction, ΔV/V, as function of fatigue time for three test temperatures, 405°C, 486°C, 567°C. (From Page, Weertman and Roth (14).)

Note the extreme sensitivity of the method. Void volume fractions of 10^{-6} or less can be determined, two orders of magnitude below that detectable by density measurements (18). Scattering from voids is evident after only a few seconds of fatiguing. Extrapolation to time zero provides information on incubation time of void nucleation.

With knowledge just of the rate of growth of the total void volume it is impossible to determine what portion of the total growth is caused by individual void growth and what results from continued void nucleation. SANS provides an almost unique method for untangling the two effects. Since $d\Sigma/d\Omega$ is known over the entire range of q's, it is possible to carry out an inversion of the scattering data to obtain a distribution curve of void sizes, dN/dR vs R. The development of size distribution in the course of fatiguing at 405°C is shown in Fig. 3. The total number of voids in each of the samples (Fig. 4) can be determined by integrating the area under the size distribution curve. Finally, from Figs. 2 and 4 it is possible to calculate void nucleation rates and individual void growth rates (14). By varying the fatigue conditions it is possible, through SANS measurements,

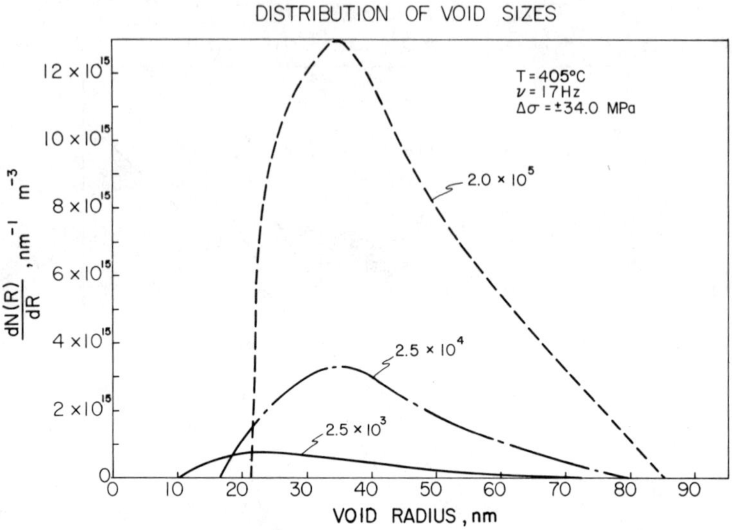

Fig. 3 - Void size distributions, dN/dR, for samples fatigued at 405°C for various numbers of cycles. Here R is void radius. (Page, Weertman and Roth (14).)

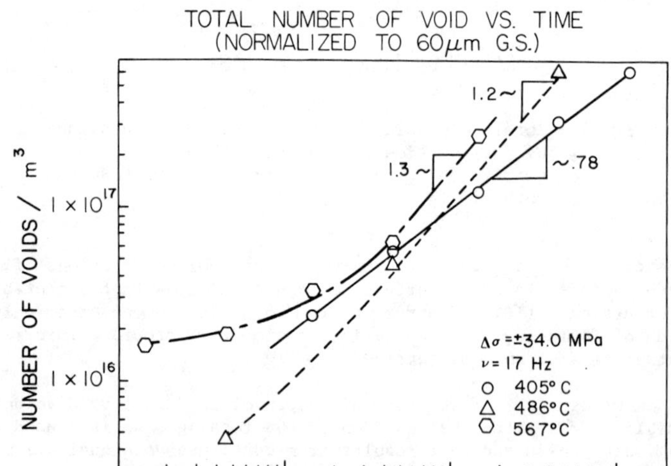

Fig. 4 - Total number of voids per unit volume as a function of fatigue time for three test temperatures, 405°C, 486°C, 567°C. (Page, Weertman and Roth (14).)

to determine the dependence of nucleation and individual void growth rates on stress, temperature, cycle frequency.

Examples of SANS in NDE

In this section we cite some examples of the use of SANS to follow processes which can be associated with material change and/or degradation. Much of the pioneering work in the use of SANS in NDE has been carried out at the Fiat Research Center (19-24). The Fiat Group has been especially interested in the development of SANS testing to the point where it can be used to (1) monitor thermal processing of complex alloys, and (2), detect damage in structural parts subjected to creep or fatigue.

Figures 5, 6, and 7 show results of an experiment (20-24) designed to demonstrate the feasibility of SANS to achieve the first of these two goals.

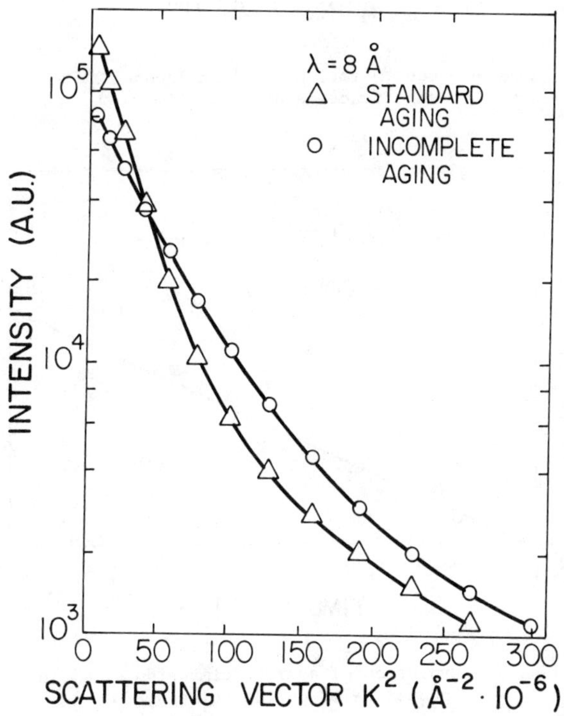

Fig. 5 - SANS from Inconel X-750 for two heat treatments (24).

Samples of the nickel base superalloy, Inconel X-750, were given one of several heat treatments designed to affect the final size of the dispersed particles of the ordered intermetallic phase, γ'. The γ' particles, which are responsible for a major portion of the small angle scattering, make a significant contribution to the strength of the alloy. The extent of this strengthening is dependent on the size of the γ'particles. Figure 5 compares

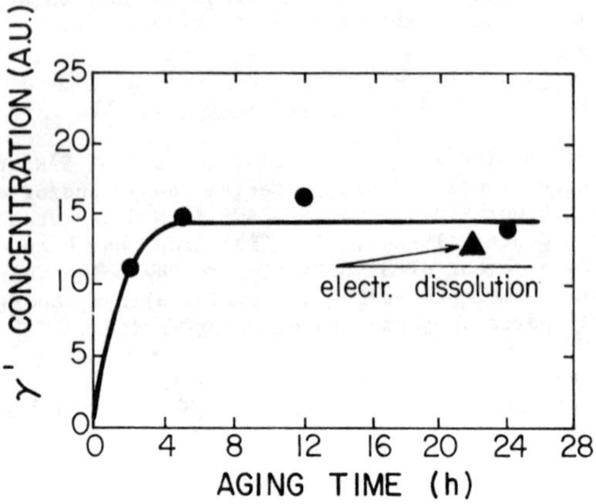

Fig. 6 - Concentration of γ' in Inconel X-750 as a function of aging time (24).

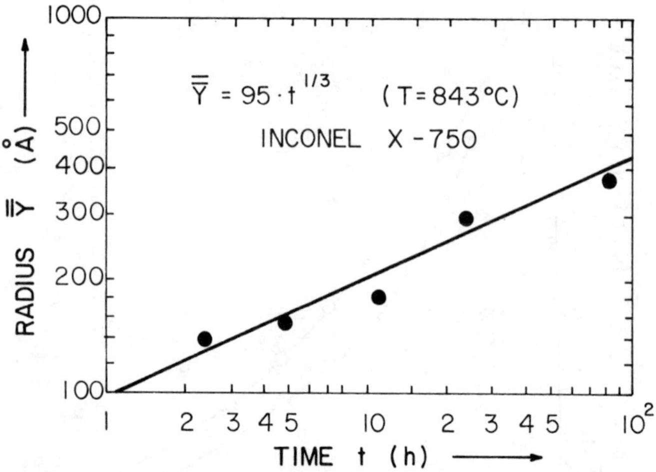

Fig. 7 - Evolution of average dimension of γ' in Inconel X-750 during aging (24).

scattering from an X-750 sample given the standard heat treatment with that from a sample in which the γ' phase is not fully developed. As would be expected, the scattering from the larger particles of the fully aged sample is pushed to smaller values of the scattering vector. Figure 6 shows the influence of aging on the volume fraction of γ'. The fraction rises rapidly at first, then stabilizes. However individual γ' particles continue to grow, as seen in Fig. 7. The evolution of particle radius with the cube root of aging time agrees with predictions of the Lifshitz-Wagner theory. It should be noted that the SANS results were verified by TEM measurements and electrochemical dissolution studies.

Figure 8 illustrates the use of SANS in following the results of different thermal treatments of a complex steel 17.4PH (22,23,24). The steel is quenched and tempered for 4 hours at the temperatures shown. The final structure contains a fraction of austenite as well as very fine Cu precipitates in a martensitic structure. While the copper precipitates lead

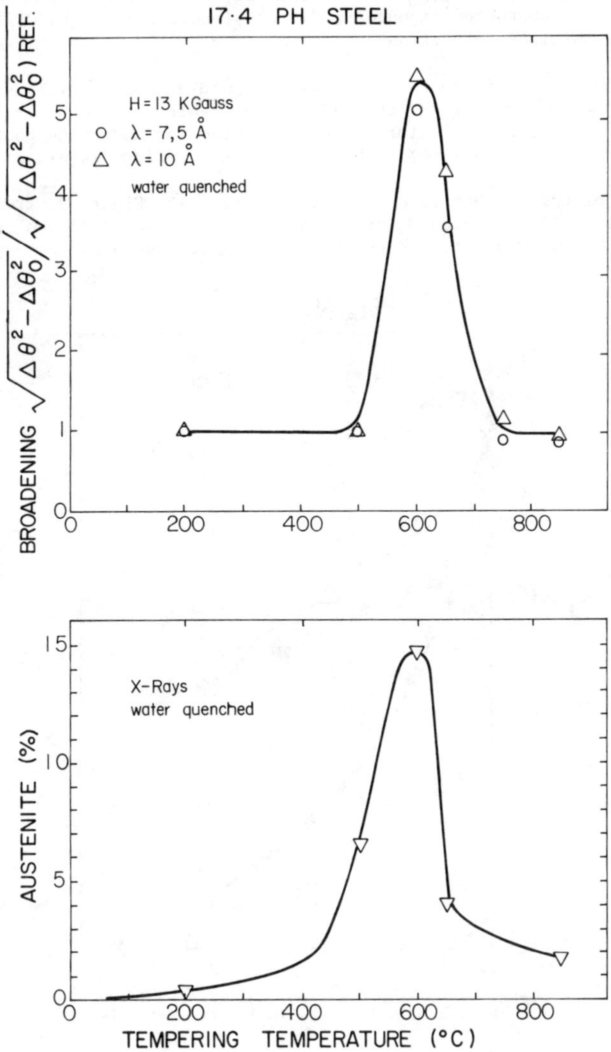

Fig. 8 - The effect of tempering temperature on line broadening in a SANS experiment and on austenite content, as determined by X-rays, in 17.4 PH steel. (After Pizzi and LaMalfa (23).)

to diffraction scattering, the austenite regions are large enough to cause multiple refraction of the neutron beam. Strong refraction effects also would arise from the presence of magnetic domains. However the variation in direction of magnetization from domain to domain has been eliminated, for the data in Fig. 8, by magnetically saturating the samples during the SANS measurements. The line broadening plotted as ordinate of the SANS data is a measure of the multiple refraction. ($\Delta\theta$ is the full width of the neutron beam at half maximum intensity.) Note that the broadening depends on the value of the neutron wave length. The form of the dependence of scattering on neutron wave length can be used to distinguish scattering by diffraction from that by multiple refraction.

It can be seen that the SANS data are in good agreement with X-ray studies. A major advantage of the SANS measurements over those by X-ray is that the bulk of the material is being sampled for austenite content (the sample thickness is 7 mm) rather than just a surface layer.

The use of SANS in NDE is not limited to metals. Figure 9, from the work of Pizzi (19), compares Guinier plots of scattering from two samples of Si_3N_4. The material from which the samples were taken had been prepared

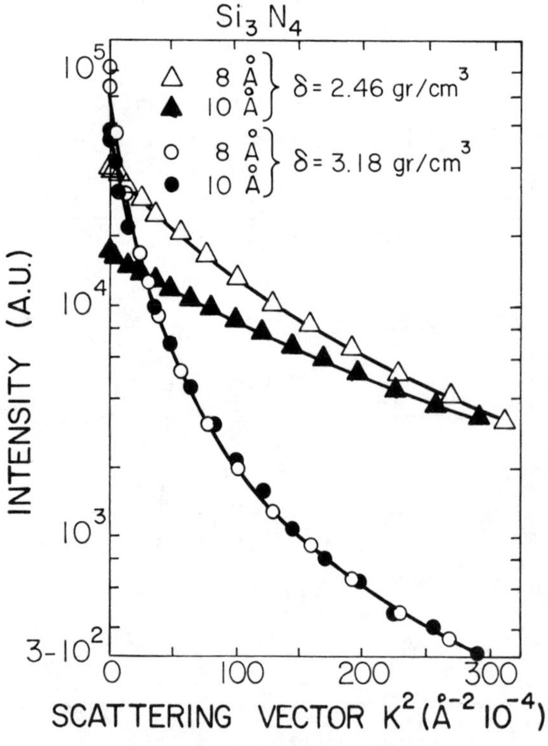

Fig. 9 - SANS from 2 samples of Si_3N_4 measured at neutron wave lengths of 0.8 and 1.0 nm. Densities of the two samples were 2.46×10^3 and 3.18×10^3 kg/m^3, respectively. (After Pizzi (19).)

by different manufacturing processes which led to products with different density. Each sample was tested at two wave lengths. It can be seen that the scattering is greatly affected by the defect structure of the silicon nitride. Multiple refraction, associated with large scattering centers, persists out to higher values of the scattering vector in the case of the less dense material.

Hot pressed silicon nitride (19) was subjected to thermal fatigue by cycling it between room temperature and 1300°C every 4 minutes. From the scattering curves taken during the course of cycling it was possible to calculate the evolution of average void radius, a measure of damage in the material (Fig. 10). The author cites good agreement with theories of void growth during thermal fatigue.

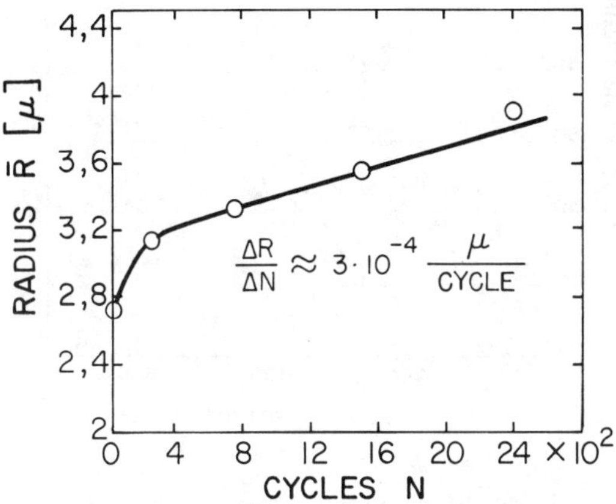

Fig. 10 - Evolution of microvoid radius in Si_3N_4 during thermal fatigue. (After Pizzi (19).)

One of the most interesting demonstrations to date of the promise of SANS in NDE is shown in Fig. 11. Pizzi, Walther and Cortese (22) have measured the effect of service life on the average γ' size in aircraft turbine blades made of the nickel base superalloy Inconel 700. As mentioned before, the ordered coherent γ' particles contribute greatly to the strength of this class of alloy, and the extent of strengthening depends on the size distribution of the particles. A ripening of the particles during service at elevated temperatures can be associated with a reduction in remaining life. In Fig. 11 we see the average γ' radius plotted as a function of distance from the root of the turbine blade up to the tip, for various service times at normal operating temperature and over temperature. The tip and root regions remain cool, and here very little ripening is seen, whereas an appreciable change in γ' size occurs in the hotter midsection. Running, even for short times, at over temperatures is especially damaging. The γ' particle sizes of Fig. 11, which were obtained from SANS measurements, were verified by TEM examination of γ' particles extracted from the matrix electrochemically. Close agreement was found between size distributions deduced from an inversion of the scattering data and the distributions directly measured by TEM (21).

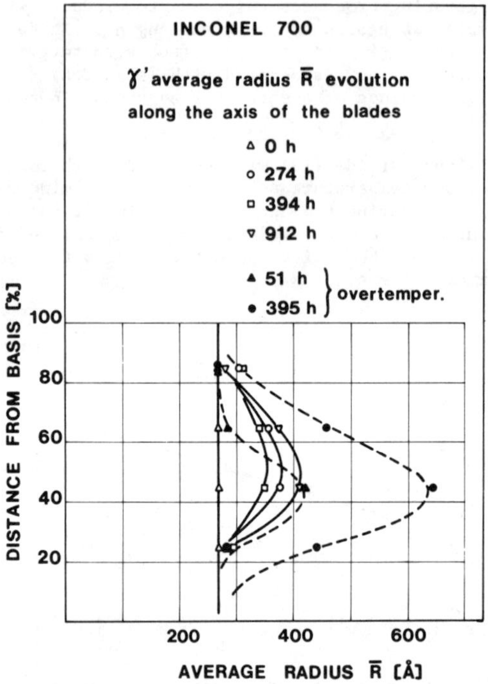

Fig. 11 - Change in γ' size in Inconel 700 turbine blades during service (Pizzi, Walther and Cortese (22).)

Another instance (24) of observation of microstructural degradation by SANS is shown in Fig. 12. Scattering curves are given for IN-100 gas turbine blades in the as-received condition and after 312 hours of service. TEM examination of the alloy shows that originally the material had a duplex structure: both large and small particles of γ' were present. After 312 hours of service the small particles had disappeared. This disappearance of the smaller scattering entities is reflected in the diminution of scattering at high values of the scattering vector.

The effect on small angle scattering caused by changes in γ' size and distribution during heat treatment and during creep has been studied in Nimonic 105 by Miller, Messoloras, Stewart and Kostorz (5). The results are complicated, but the authors conclude that SANS demonstrates an Ostwald ripening mechanism is operating during thermal treatment and during creep at low stress. However some other factor, possibly microvoid formation, modifies the scattering curves taken during high stress creep.

Galotto et al. (20) report that marked changes occur in the neutron scattering curve of 304 stainless steel as the result of welding. Both the weld area and the heat affected zone produce line broadening curves which are considerably different from that of the original 304 samples. The scattering change is attributed to refraction from ferrite particles, a supposition confirmed by electron microscopy.

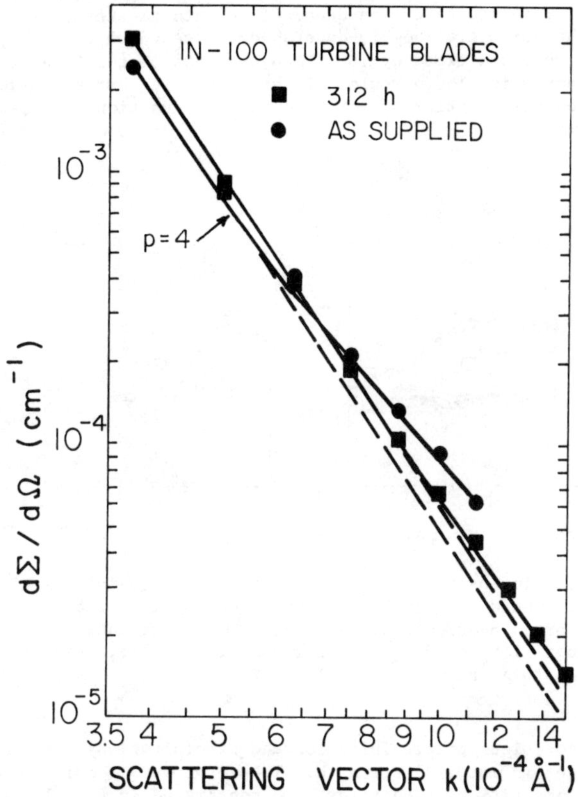

Fig. 12 - Effect of service life on SANS from IN-100 turbine blades (After Pizzi (24).)

Damage in the form of microvoid formation in irradiated material can be monitored with a high degree of sensitivity by SANS. As mentioned in the discussion of Fig. 2, the presence of minute void volume fractions can be detected by small angle scattering. Hendricks, Schelten and Schmatz (25) have used small angle neutron scattering to measure total void volume, total void surface area, and radius of gyration of voids in neutron-irradiated single crystals of aluminum. They compared the SANS values of these quantities with results obtained by precision density measurements and TEM observations, and found excellent agreement. Mook (26) also has studied void distributions in irradiated aluminum.

Faceted voids in single crystals give rise to scattering patterns which are no longer isotropic, that is, the intensity depends on $\vec{q}$ rather than simply $|\vec{q}|$. From an analysis of the scattering pattern it is possible to deduce the shape of the voids. Figure 13, taken from the work of Epperson, Gerstenberg, Berner, Kostorz, and Ortiz (27), shows iso-intensity contours of scattering from single crystals of NiAl which had been quenched from 1600°C and annealed 30 minutes at 650°C. During the anneal the supersaturation of vacancies retained after the quench form into clusters. It was concluded that the voids are rhombic dodecahedra faceted on the {110} planes. The SANS results were confirmed by TEM. Measurements of the Guinier radius, taken from the isotropic, low q region of the scattering curves, made it possible to follow the kinetics of void growth as a function of time and of annealing temperature.

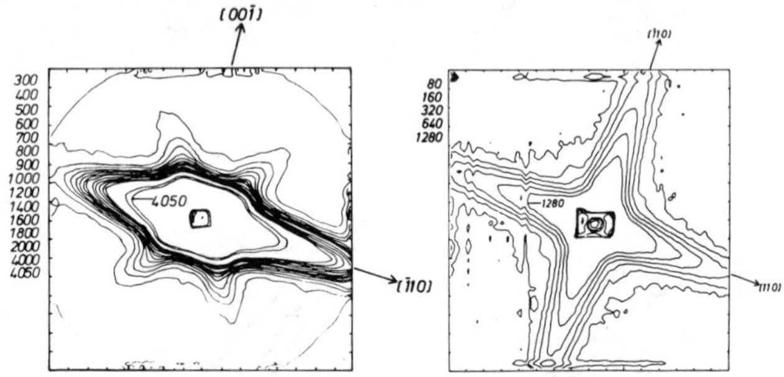

Fig. 13 - Iso-intensity contour maps for single crystals of NiAl quenched from 1600°C and annealed 30 min. at 650°C. Left: neutron beam parallel to [110]; right: beam parallel to [$00\bar{1}$]. (Epperson, Gerstenberg, Berner, Kostorz and Ortiz (27).)

Alloy decomposition (preprecipitation and precipitation) can produce significant changes in material strength. That SANS is a sensitive means for following decomposition of an alloy is demonstrated in Fig. 14, which shows scattering curves from Au-60 at.% Pt specimens quenched and aged for varying times at 550°C (Singhal, Herman and Kostorz (28)).

Dislocations are not such strong scatterers as voids, precipitates, magnetic domains, etc. However, under favorable circumstances, small angle scattering curves can be dominated by dislocation scattering. Atkinson and Hirsch (29) have calculated the scattering from a random network of dislocations in an isotropic medium. Atkinson and Hirsch found that the major contribution to scattering comes from the linear elastic strain field around edge dislocations, and that the scattering cross section is proportional to q^{-3}. Epperson, Kostorz, Ortiz, Fürnrohr and Gerstenberg (30) have investigated SANS in deformed and annealed Cu-16 at.% Al single crystals (Fig. 15). Here we see the scattering intensity plotted as a function of h^{-3} ($h = q/2$). Curve A comes from a sample of Cu-Al deformed by cold rolling through about a 100% reduction in thickness. Subsequent annealing at 300°C for increasing lengths of time lowers the slope of the curve. (According to theory (29) the scattering intensity is proportional to the dislocation density.)

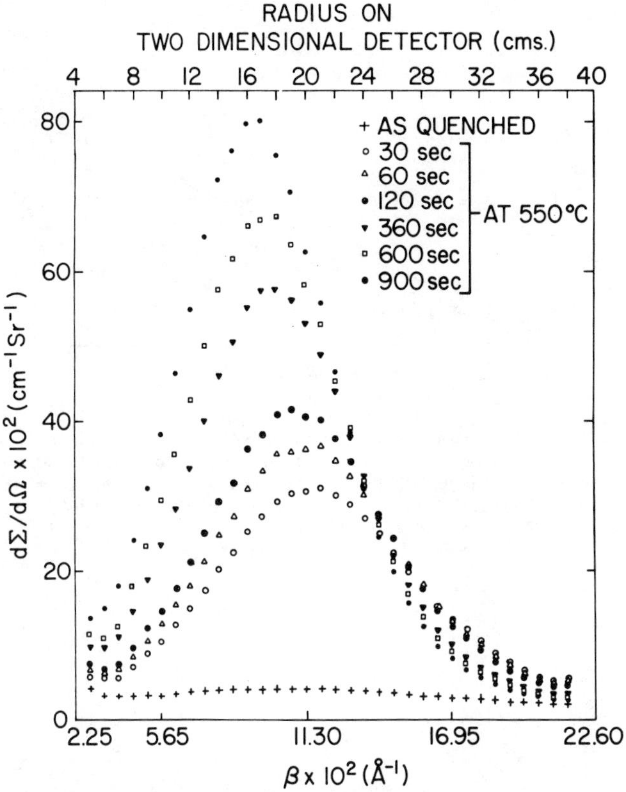

Fig. 14 - Differential scattering cross section vs. scattering vector for quenched and aged specimens of Au-60 at.% Pt. (Singhal, Herman and Kostorz (28).)

Conclusion

The foregoing examples demonstrate the sensitivity of SANS to changes in microstructure, indeed, to very small changes. The spectrum of types of microstructural alteration or defect which affect the scattering curve is broad: precipitates, G-P zones, microvoids, dislocations, growing or dissolving particles, etc. The very breadth of response can make identification of the particular phenomenon causing an observed shift in scattering difficult, especially in a complex alloy. However for NDE purposes in a given application it is only necessary to establish which patterns of change in scattering are associated with undesirable material changes. Thereafter NDE can be carried out in a routine manner.

In summary, SANS offers a powerful tool for obtaining information about changing microstructures, whether in a research situation with a model sample system or in the screening of components for potentially dangerous defects or material degradation.

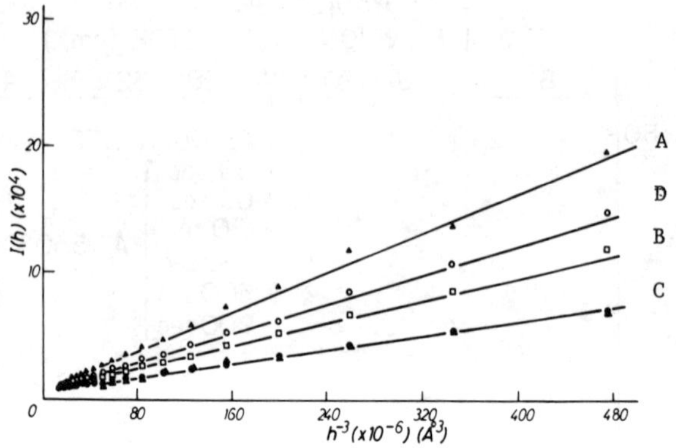

Fig. 15 - SANS from deformed and annealed samples of Cu-16 at.% Al. Curve A: sample deformed; B: deformed and annealed 20 min at 300°C; C: deformed and annealed 1 hr at 300°C; D: deformed and annealed 14 hrs at 300°C. (After Epperson et al. (30).)

References

(1) G. E. Bacon, Neutron Diffraction, Oxford University Press, London and New York, 1975.

(2) Treatise on Materials Science and Technology, Vol. 15, Neutron Scattering, ed. G. Kostorz, Academic Press, New York, 1979.

(3) H. Herman, Non-Destructive Evaluation of Materials with Cold Neutron Beams, report to Naval Air Systems Command, Contract No. N00019-77-M-0418, Washington, D.C., Dec. 1977.

(4) L. H. Schwartz and J. B. Cohen, Diffraction from Materials, Academic Press, New York, 1977.

(5) R. J. Miller, S. Messoloras, R. J. Stewart and G. Kostorz, "A Small-Angle Neutron Scattering Study of the Temperature and Stress Dependence of the Microstructure of Nimonic Alloys", J. Appl. Crystallogr., 11 (1978) pp. 583-588.

(6) A. Guinier, "La diffraction des rayons X aux tres petits angles; application l'étude de phénomènes ultramicroscopique", Ann. Phys. (Paris), 12 (1939) pp. 161-237.

(7) G. Porod, "Die Röntgenkleinwinkelstreuung von dichtgepackten kolloiden Systemen II", Kolloid Z., 125 (1952), pp. 51-57, 109-122.

(8) A. Guinier and G. Fournet, Small Angle Scattering of X-Rays (translated by C. B. Walker), Wiley, New York, 1955.

(9) I. S. Fedorova and P. W. Schmidt, "A General Analytical Method for Calculating Particle-Dimension Distributions from Scattering Data", J. Appl. Crystallogr., 11 (1978) pp. 405-411.

(10) C. G. Vonk, "On Two Methods for Determination of Particle Size Distribution Functions by Means of Small-Angle X-ray Scattering", J. Appl. Crystallogr., 9 (1976). pp. 433-440.

(11) J. H. Letcher and P. W. Schmidt, "Small-Angle X-ray Scattering Determination of Particle-Diameter Distributions in Polydisperse Suspensions of Spherical Particles", J. Appl. Phys., 37 (1966) pp. 649-655.

(12) M. Roth, "The Small-Angle Scattering of Neutrons by Surface Imperfections", J. Appl. Crystallogr., 10 (1977) pp. 172-176.

(13) B. E. Warren and R. L. Mozzi, "Multiple Scattering of X-rays by Amorphous Samples", Acta Crystallogr., 21 (1966) pp. 459-461.

(14) R. Page, J. R. Weertman and M. Roth, "Investigation of Fatigue-induced Grain Boundary Cavitation by Small Angle Neutron Scattering", Scripta Metall., 14 (1980) pp. 773-777.

(15) R. Page and J. R. Weertman, "Limitations on the Use of Intergranular Fracture Surfaces for the Determination of Cavity Spacings", Scripta Metall., 14 (1980) pp. 1029-1031.

(16) T. Saegusa, J. R. Weertman, J. B. Cohen and M. Roth, "Small-Angle Neutron Scattering from Pores Produced in High-Temperature Fatigue", J. Appl. Crystallogr., 11 (1978) pp. 602-604.

(17) R. Page and J. R. Weertman, "HVEM Observations of Grain Boundary Voids in High Purity Copper", Acta Metall., in press.

(18) A. Gittens, "The Mechanism of Cavity Growth in Copper during High-Temperature Fatigue", Metal Science J., 2 (1968) pp. 51-58.

(19) P. Pizzi, Fracture Mechanics of Ceramics, Vol. 3, eds. R. C. Bradt, D. P. H. Hasselman and F. F. Lange, Plenum Press, New York, 1978.

(20) C. P. Galotto, P. Pizzi, H. Walther, V. Angelastro, N. Cerullo and G. Cherubini, "A Cold Neutron Small Angle Scattering Device for Technological Applications", Nucl. Instrum. Methods, 134 (1976) pp. 369-378.

(21) P. Cortese, P. Pizzi, H. Walther, G. Bernardini and A. Olivi, "Non-Destructive Characterization and Examination of Turbine Blades and Nickel Alloys by Small Angle Neutron Scattering", Mater. Sci. Eng., 36 (1978) pp. 81-88.

(22) P. Pizzi, H. Walther and P. Cortese, "Non-Destructive Examination of Metals by Small-Angle Neutron Scattering", Proc. 8th World Conf. Nondestr. Test., Cannes, France, 1976, p. 3L7.

(23) P. Pizzi and U. LaMalfa, "Etude de la Microstructure des Materiaux par Diffusion des Neutrons aux Petits Angles. Comparaison avec les Etudes par Microscopie Electronique", Journées Métallurgiques d'Autommne, 18-20 Oct. 1977.

(24) P. Pizzi, "Application of Small-Angle Neutron Scattering to NDI of Materials and Manufactured Components", AGARD Meeting, Structures and Materials Panel, NDI Relationships to Aircraft Design and Materials, Norway, Sept. 1977, pp. (5-1)-(5-7).

(25) R. W. Hendricks, J. Schelten and W. Schmatz, "Studies of voids in neutron-irradiated aluminum single crystals", Philos. Mag., 30 (1974) pp. 819-837.

(26) H. A. Mook, "Neutron small-angle scattering investigation of voids in irradiated materials", J. Appl. Phys., 45 (1974) pp. 43-46.

(27) J. E. Epperson, K. W. Gerstenberg, D. Berner, G. Kostorz and C. Ortiz, "Voids formed in quenched and annealed NiAl", Philos. Mag., 38A (1978) pp. 529-541.

(28) S. P. Singhal, H. Herman and G. Kostorz, "Neutron Small-Angle Scattering Study of Phase Decomposition in Au-Pt", J. Appl. Crystallogr., 11 (1978) pp. 572-577.

(29) H. H. Atkinson and P. B. Hirsch, "The Theory of Small Angle Scattering from Dislocations", Philos. Mag., 3 (1958) pp. 213-228.

(30) J. E. Epperson, G. Kostorz, C. Ortiz, P. Fürnrohr and K. W. Gerstenberg, "Neutron Small Angle Scattering from Deformed and Annealed Cu and α-Phase Cu-Al Alloy Single Crystals", Acta Metall., 27 (1979) pp. 1363-1372.

DEFECT CHARACTERIZATION WITH POSITRON ANNIHILATION*

L. Granatelli and K. G. Lynn
Brookhaven National Laboratory
Upton, New York 11973

It has been sixteen years since the positron annihilation technique was first used to study crystal lattice defects in deformed nickel (1). Since that time it has been applied to the investigation of defect mechanisms in most pure metals and several alloy systems. This review will deal only with positron annihilation in metal crystals, although the technique has also been applied to such diverse materials as ionic solids and polymers. The positron annihilation technique has matured to the point where, for example, it is now the preferred method for measuring vacancy formation enthalpies. In addition, it has been quite successful in the study of radiation damage and has been shown to be sensitive to void formation and growth up to diameters where they first become detectable with electron microscopy. To date, positron annihilation has had its greatest success with more complex defects such as dislocations and grain boundaries when it has been used as a compliment to the more traditional methods of studying defects. Positron annihilation is just beginning to be used to study fatigue in metals. Until recently, the high kinetic energy of positrons obtained from radioactive sources has prevented positrons from being used as a surface probe. However, with the development of an energy tunable monoenergetic positron source, positron annihilation can now be applied to the study of surfaces and their associated defects.

*This work was supported by U.S. Department of Energy under contract number DE-AC02-76CH00016.

1. Introduction

In section 1 we present a brief introduction to the positron annihilation technique. In sections 2-5 the ability of the positron technique to perform microstructural characterization of four types of lattice defects is discussed. It is frequently not possible to obtain samples which contain only one type of defect in non-negligible concentrations. Such situations exist for some alloys (section 6) and for fatigued metal samples (section 7). In the final section (8) the current limitations and some future prospects of the technique will be presented.

The behavior of thermalized positrons in most metals is affected by the presence of vacancy-like defects in the crystal lattice. The positron annihilation technique is non-destructive and the positron is very sensitive to local changes in both electron density and momentum produced by these defects. In addition, relatively small samples may be used and the near surface or interior of the sample may be probed by proper choice and configuration of the positron source. For these reasons the positron annihilation technique can be the preferred method for studying crystal lattice defects in certain instances.

Positrons may be produced either in or at the surface of a specimen by the radioactive decay of a positron emitting isotope. Depending upon the isotope used, positrons can enter the sample with considerable energy: up to 1.90 MeV with ^{68}Ge. When positrons are injected into a metal sample they are rapidly thermalized (relative to their lifetime) by interactions initially with core electrons and later with both conduction electrons and phonons (2). Thermalization times are on the order of a picosecond (3,4). Once thermalized the positron will annihilate with an electron. In 99.7% of all annihilation events in metals the resulting energy is divided between two oppositely directed gamma rays each of approximately 0.511 MeV ($= m_0c^2$). The remaining 0.3% consist mainly of three quanta events which are not generally considered in lattice defect studies.

The exclusion principle allows only two conduction electrons (spin 1/2) to exist in the lowest momentum state of a crystal. The remaining electrons must, pairwise, occupy progressively higher levels. In a positron annihilation experiment only one positron exists in a sample at any moment. However, there are a large number of conduction electrons. Thus, since the positron is thermalized prior to annihilation, essentially all momentum effects on the annihilation photons can be considered to be the result of the momentum of the electron. The total energy released during annihilation is derived from the rest mass of the two particles, their mutual binding energy, the binding energy to the solid, and their kinetic energy at annihilation. Assuming that the positron is at rest and that the velocity of the electron is much less than the speed of light, c, then the energies of the two annihilation photons are approximately

$$E_{1,2} = m_0c^2 \pm \frac{cp_x}{2} \qquad (1)$$

where $p_x/2$ is the component of momentum of the center of mass of the two particle system which is parallel to the emission direction. Emission of the antiparallel gamma photons in all directions is equally probable. Thus, the energy distribution of the photons resulting from annihilation with electrons having momentum p is a continuous one symmetric* about m_0c^2 and extending from $m_0c^2 - cp/2$ to $m_0c^2 + cp/2$.

If the positron-electron pair is at rest during annihilation the two gamma photons are emitted in diametrically opposite directions. However, if only the positron is at rest then the annihilation photons will be emitted at an angle which differs from π by the value θ. For small values of θ

$$\theta \text{ (rad)} \cong \frac{p_z}{m_0c} \qquad (2)$$

where p_z is the component of electron momentum perpendicular to the direction of the emitted gamma photon.

A metallic solid is frequently depicted as a periodic three dimensional array of positively charged ion cores embedded in an electron gas. The electron gas is composed of free valence electrons with an approximately parabolic energy distribution. At absolute zero temperature this ranges from the energy at the bottom of the conduction band up to the Fermi energy. The Fermi energy for many metals has a value between 4 and 10eV. In general the core electrons have higher momentum states than do the conduction electrons. Their momentum distribution is commonly approximated as a Gaussian but can be calculated in a straightforward manner (5).

In simple metals the positron annihilation parameters are similar to those that would be expected to result from annihilations in a homogeneous

*The energy distribution is not exactly symmetric about m_0c^2 because of the binding energy of the electron and positron to the solid. However, this is a small effect.

electron gas whose density corresponds to the conduction electron density in the metal. However, in all real metals the positron wavefunction overlaps with that of the core electrons. This results in a greater annihilation rate than would be expected from the conduction electrons alone (6). The ratio of the number of core to valence annihilations is affected by the degree of lattice perfection. This ratio is however, characteristic of each metal when it is free from defects. Due to Coulombic repulsion the probability of a positron entering the core region is low but the annihilation rate there is high. Conversely, the annihilation rate with conduction electrons is low but the probability of annihilation is high. The variability of the positron lifetime in the perfect lattice among different metals is a function of these two competing factors (7,8). The lifetime of positrons in the nondefective lattice is characteristic of each particular metal. Positron lifetimes in the perfect metal lattice range from about 100 to 400 ps (Fig. 1).

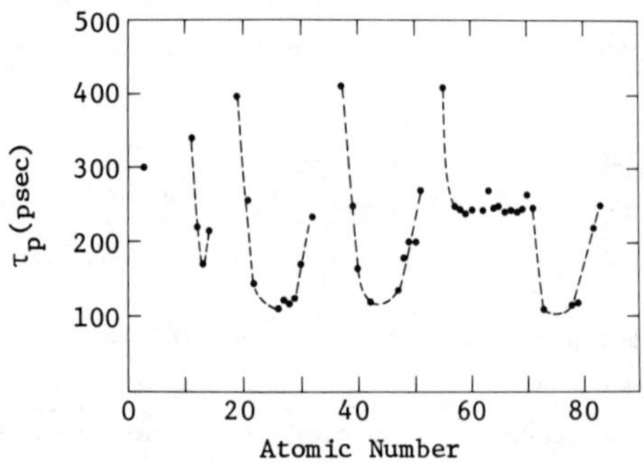

Fig. 1 - Variation of the mean positron lifetime with atomic number in the well annealed bulk lattice (8).

Positron trapping at a defect will occur when the sum of the potential and kinetic energy of the positron can be lowered by localization of the positron at that defect. The repulsive coulombic potential which a positron experiences while in a metal lattice obviously decreases with increasing distance from the positive ion cores. Open volume defects allow the positron to increase its mean distance from these positive ions (i.e. the positron wavefunction may spread out in the vicinity of these defects). This lowers the zero point energy relative to that in the perfect lattice. Even though the electron density is lower at an open volume defect than in the interstitial regions of the lattice these effects are sufficient to produce positron traps at several types of defects in many metals. Positron trapping has been reported at vacancies, voids, grain boundaries, and edge dislocations.

Due to their increased distance from the host atoms positrons which become localized at open volume defects have a reduced probability of annihilation with core electrons. The annihilation rate with conduction electrons is also reduced but the fractional reduction is less than for core electrons. Also, because the potential energy of an electron is greater in a vacancy-like region than in the interstitial region and since its total energy remains constant the momentum of the conduction electrons is reduced. These effects lead to a narrowing of the angular correlation and Doppler-broadening spectra resulting from annihilations with electrons at defects relative to that resulting from annihilations in the interstitial regions. In addition, since these positrons experience a reduced electron density their mean lifetime is increased relative to the mean lifetime of those positrons which annihilate in the non-defective lattice.

Three experimental methods are in common use for study of positron annihilations in the bulk solid. The Doppler-broadening technique, as the name implies, measures the Doppler shift of the annihilation photons. The angular correlation technique measures the angular distribution of the photons resulting from 2-γ annihilations. The lifetime technique measures the mean lifetimes of positrons in a sample.

Energy spectroscopy measurements for Doppler-broadening experiments are performed with a Ge(Li) or intrinsic Ge solid state detector. The resulting data are stored in a multichannel analyzer in terms of counts as a function of photon energy (Fig. 2). It can be determined from equation (1) that an annihilation with a 10 eV electron results in a maximum Doppler shift of the resulting γ-photon of approximately 1.5 keV. The typical resolution of a good solid state detector is about 1.1 keV at 511 keV.

Doppler-broadening data are frequently characterized with a lineshape parameter. The most commonly used lineshape parameter, usually designated "S" (9), is defined as the ratio of the number of counts in an arbitrary symmetric central region of the photopeak to the total number of counts in the peak. Thus, the S parameter is dominated by the fraction of annihilations that take place with conduction electrons. An increase in the proportion of positrons which annihilate at vacancy-type defects results in an increase in the value of the S parameter.

Angular correlation experiments are usually performed with a two photon coincidence and narrow slit arrangement. The detectors are frequently NaI(Tl) scintillators optically coupled to photomultiplier tubes. Since

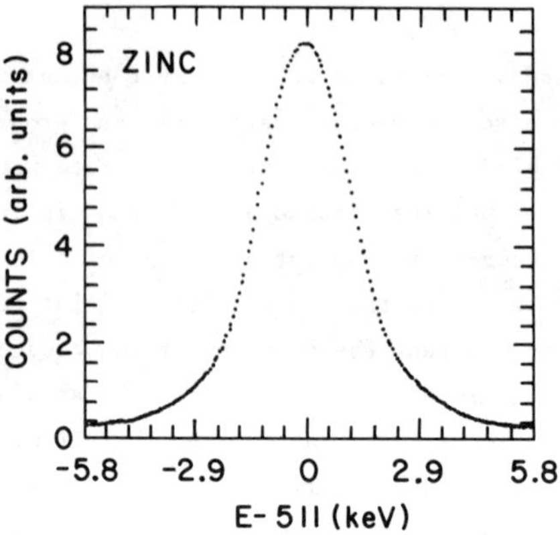

Fig. 2 - Doppler-broadening spectrum from a well annealed 99.999% Zn sample.

more than 95% of the 2-γ annihilation photons fall within $\theta \lesssim \pm 10$ mrad (Fig. 3) the sweep angle of the detectors need not be large but high angular precision is required. Angular correlation techniques restrict data collection to one momentum range at a time (except with multiple detector (10,11) apparatus). Since it is time consuming to repeatedly scan the entire angular range, angular correlation data are sometimes collected only at one angle. Data may then be analyzed in terms of a counting rate (usually at $\theta = 0$) as a function of a variable parameter (e.g. sample temperature).

Even though both Doppler-broadening and angular correlation methods provide the same information about the electron momentum distribution there are circumstances which may make one method preferable to the other. Energy spectroscopy with a solid state detector can yield poorer equivalent momentum resolution than the angular correlation method (typically by a factor of about 10), however, in addition to a lower background, energy spectroscopy has the advantages that a smaller positron source can be used ($\sim 10\mu$Ci vs. 10 mCi or more), and that data collection is much more rapid.

Positron lifetime measurements are performed using a radioactive source which emits a positron with a simultaneous gamma ray (e.g. ^{22}Na). The prompt and annihilation gamma photons are detected with scintillation counters. The time between thermalization and annihilation of the individual positron is then approximated as the time between the detection of the prompt gamma photon and the subsequent detection of one of its annihilation photons. Since positron annihilation is a statistical process the values and corresponding intensities of one or more positron lifetimes are usually

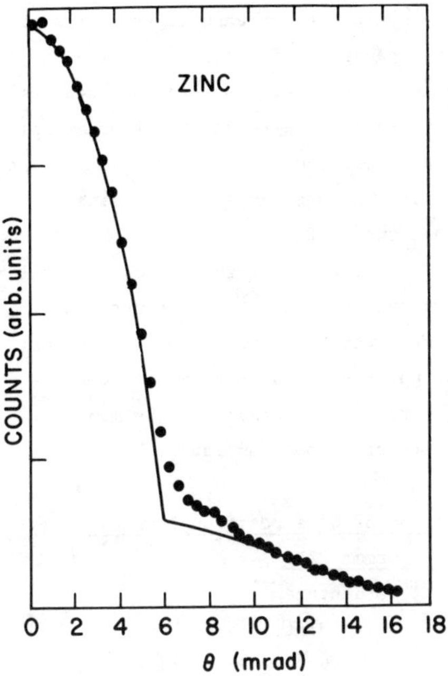

Fig. 3 - Angular correlation distribution for polycrystalline Zn [(12) and references therein].

extracted from lifetime spectra (Fig. 4) with the aid of computer programs such as Positronfit Extended (13). These programs are able to extract this information by fitting positron decay spectra to a resolution function and the sum of an appropriate number of decay terms.

Further details regarding all three techniques can be found in a number of places, including the review articles of West (12) and Doyama (14).

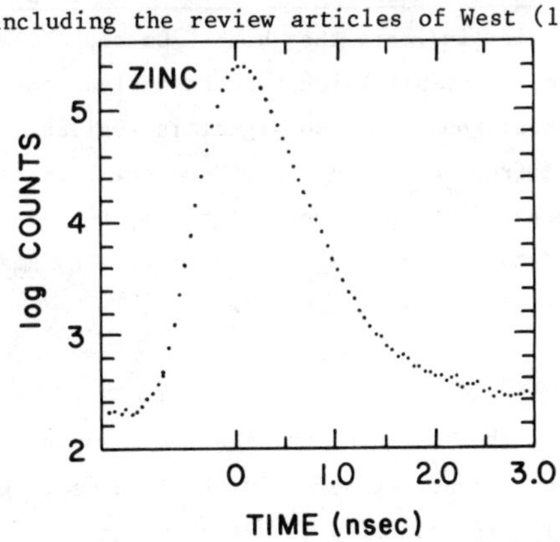

Fig. 4 - Positron lifetime spectra for well annealed 99.999% zinc single crystal

Both qualitative and quantitative information regarding the nature of defects in a sample can be obtained by measuring the angular variation of the annihilation photons, the Doppler-broadening of the annihilation photopeak and the positron lifetimes in the sample. In certain instances the same information can be deduced from all three techniques (for example vacancy formation enthalpies - see section 2). However, it should be kept in mind that to compare lifetime to either Doppler-broadening or angular correlation measurements is to compare two different physical measurements. The positron lifetime is a measure of the expectation value of the electron density of all momenta over the extent of the positron wavefunction (Bloch like in the perfect lattice). Both Doppler-broadening and angular correlation measurements, when analyzed in terms of some lineshape parameter or peak counting rate, are a measure of the momentum profile of electrons in a narrowly restricted momentum range.

Table I. Nuclear Characteristics of Some Positron Sources

Isotope	E max (keV)	Percentage of decays giving a positron	Gamma Rays: Energy (MeV)	Gamma Rays: Percentage Coincident with Positron	Production Process	Half Life
^{22}Na	545	90	1.274	90	^{24}Mg(d,α)	2.6y
^{64}Cu	656	19	1.34	0	^{63}Cu(n,γ)	12.8h
^{58}Co	474	15	0.810	15	^{55}Mg(α,n)	71 d
^{68}Ge	1900	87	1.078	1.5	^{66}Zn(α,2n)	270 d
^{44}Ti	1470	94	1.156	95	^{45}Sc(p,2n)	48 y

These characteristics indicate that both ^{64}Cu and ^{68}Ge are useless for positron lifetime measurements which require nuclear gamma rays in coincidence with emergent positrons since no signal is available to indicate the birth of a positron from these sources. However, ^{68}Ge is an excellent positron source for recent β^+-γ coincidence lifetime techniques (26) which use the positron itself to produce the start signal. Its small nuclear gamma intensity helps keep the background low.

2. Vacancies

The application of the positron annihilation technique has proven to be very successful in the quantitative study of vacancies. With present methods positron trapping in vacancies is detectable at vacancy concentrations as low as 0.1 ppm. When the vacancy concentration is such that almost all positrons are trapped and annihilate in vacancies then this technique becomes insensitive to further increases in their concentration. Saturation generally occurs at vacancy concentrations of about 100 ppm. Many experimentalists have studied particle irradiation induced or thermally generated vacancies in a large number of pure metals and several alloy systems. Vacancies in alkali metals have not as yet been shown to produce changes in the positron annihilation parameters.

As previously stated, positrons which are trapped at monovacancies decay from these traps with lifetimes which are longer than those of positrons annihilating in the bulk. Also, trapped positrons produce angular correlation and Doppler-broadening parameters which are different from those generated by positrons annihilating in the perfect lattice. Therefore, the change in the annihilation component attributable to vacancy trapped positrons is a quantitative measure of the change in vacancy concentration. By determining the changes in the vacancy concentration as a function of sample temperature the vacancy formation enthalpy can be deduced. Positron annihilation has become the most widely adopted technique for measuring vacancy formation enthalpies in metals. Indeed, the positron annihilation technique is the only method that has, to date, been successful in measuring the monovacancy formation enthalpies in the refractory metals V, Nb, and Ta (15).

Lynn et al. (16) studied the effect of temperature from 4.2 K to 1700 K on the positron lifetime in 99.999 + wt.% polycrystalline Ni. Figure 5 is a plot of the single lifetime fit (approximately the mean positron lifetime) to the data as a function of temperature. Several metals, including Ni, exhibit a "precursor" effect in the temperature range (here 900 K to 1100 K) immediately preceding the onset of detectable vacancy trapping. A number of possible causes have been suggested ranging from extrinsic defects (17) to trapping at thermally generated transient dilations occuring in the lattice (18). At present the precursor effect is not well understood.

The sharp increase in τ_m at temperatures greater than 1100 K is unquestionably attributable to vacancy trapping of positrons. Saturation occurs in the single lifetime fit analysis of Ni beyond about 1350 K. The saturation of the lifetime at high temperatures is in part a result of shortcomings in the numerical analysis methods rather than a real physical effect. Doppler-broadening lineshape parameters, which are the result of a more simplistic analysis, do not exhibit the same saturation seen in lifetime data.

Kuribayashi et al. (19) first noted that a linear relationship exists between the so-called critical temperature, T_c, of a number of fcc metals

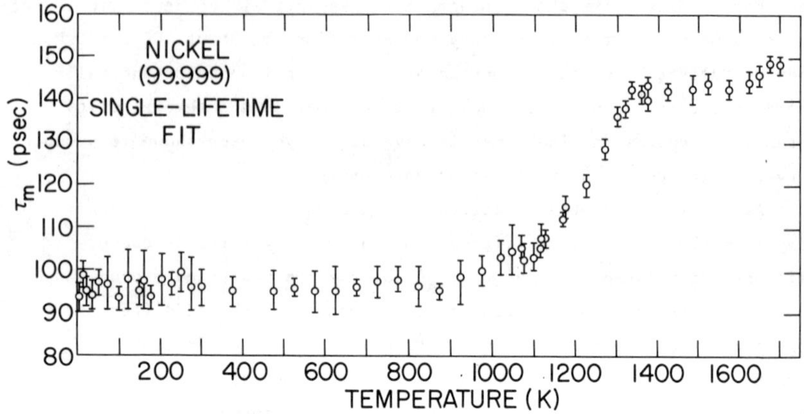

Fig. 5 - Single lifetime fit, τ_m, vs temperature for well annealed Ni

and alloys and their monovacancy formation enthalpies, H^F_{IV}. The critical temperature was defined as that temperature at which noticeable deviation occurs from the linear low temperature region. MacKenzie et al. (20) suggested an operationally simple procedure for finding the "threshold temperature", T_t. As with T_c a linear relationship exists between T_t and the monovacancy formation enthalpies of a large number of metals (practically $T_c \approx T_t$). T_t is defined as the intersection of a linear fit to the prevacancy region with a linear extrapolation back from the early vacancy dominated region. More recently Schulte and Campbell (21) produced the equation

$$H^F_{IV} = (-0.098 \pm 0.057) + (15.2 \pm 0.7)\, T_t \times 10^{-4}\ \text{eV} \qquad (3)$$

from angular correlation and Doppler-broadening measurements. Campbell et al. (22) obtained a value of $H^F_{IV} = 1.45 \pm 0.07$ eV using a similar equation and their Doppler-broadening nickel data. By applying this procedure to their lifetime data Lynn et al. (16) obtained a value of $H^F_{IV} = 1.40^{+0.06}_{-0.01}$ eV. This value is somewhat lower than the value $H^F_{IV} = 1.54^{+0.1}_{-0.2}$ eV which was obtained by fitting the two-state trapping model (23,24,25) to the lifetime data. Lynn et al. (16) state that the generally less pronounced precursor effect exhibited by lifetime data relative to angular-correlation and Doppler-broadening data resulted in the lower value of H^F_{IV} when the threshold temperature equation was applied. They therefore suggest a modified equation for lifetime data ($H^F_{IV} = 1.34\, T_t \times 10^{-3}$ eV) but warn that it is based solely on the determination of T_t for their own Ni data.

Maier et al. (15) have recently measured the monovacancy formation enthalpies for a number of refractory bcc metals (W, Ta, Mo, Nb, V) and have found that the vacancy formation enthalpies for these metals do not follow the empirical relation above which is based mostly on the characteristics of close-packed metals but are better described by the equation

$$H^F_{IV} = 17k_BT_t \qquad (4)$$

where k_B is Boltzmann's constant.

The major advantages of using the threshold temperature method are (1) it avoids the somewhat questionable approximations inherent in the trapping model (21,27) and (2) vacancy formation enthalpies can be determined without heating the sample close to its melting point. This can be a major advantage when studying refractory metals or metals exhibiting a high vapor pressure in the solid state. This method, however, inherently assumes that the sensitivity of positrons to vacancies remains constant between metals.

The ability of the positron technique to quantitatively measure defect populations is useful in non-equilibrium as well as equilibrium situations. In contrast to resistivity, which must rely on only one parameter, positron data can be interpreted in terms of several complimentary parameters (e.g. the lifetime of free (τ_f) and vacancy trapped (τ_t) positrons and their associated intensities I_f and I_t). This ability has been helpful in settling the debate over which migrating species, vacancies or interstitials, is responsible for Stage III recovery in several metals and in particular Mo (28,29,30). Eldrup et al. (29) were able to demonstrate that void formation occurred simultaneously with a decrease in the vacancy population. This was accomplished by following the intensities of the vacancy and void components in the positron annihilation spectra of electron irradiated Mo as a function of the isochronal annealing temperature. This strongly suggested that vacancies are the mobile defect.

In our own studies of 99.999 + wt.% Zn single crystals we have found that the Doppler-broadening S parameter exhibits an increase in value between the annealing temperatures 77 K and 150 K after deformation in liquid nitrogen (Fig. 6). Since this occurs in the stage III recovery temperature range for Zn (105 K to 160 K) (31) we interpret the increase as being due to the agglomeration of vacancies. Vacancy agglomeration commensurate with stage III suggests that dislocations in Zn are not efficient vacancy sinks. The knee in the curve (Fig. 6) at $\sim$200K is due to the annealing out of dislocations.

Free interstitials may combine with vacancies and so reduce the vacancy population, however, free interstitials themselves have not been observed to produce any effects on positron annihilation parameters.

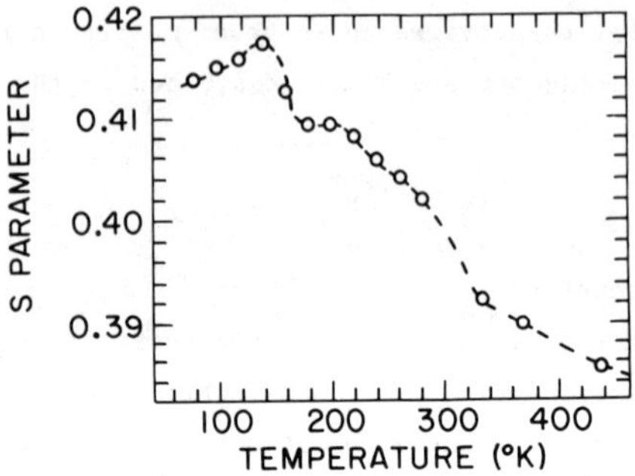

Fig. 6 - Doppler-broadening S parameter as a function of isochronal annealing temperature for Zn single crystals after deformation (10%) at 77K. All data points were taken at a sample temperature of 77K.

3. Voids

The production of voids in metals by particle irradiation is of great concern in the design of critical reactor components because of the swelling which accompanies void growth. In 1972 Mogensen et al. (32) and Cotterill et al. (33) established positron trapping at voids in Mo. Positron trapping has since been observed at voids in other metals. The positron technique is sensitive to voids of sizes well below that at which they become detectable with transmission electron microscopy. This sensitivity permits much shorter sample irradiation times. In addition, since the positron annihilation technique is non-destructive there is a possibility that it could be used to continuously monitor the void concentration in nuclear reactor components (33).

Positrons which annihilate in monovacancies do so with lifetimes which are typically 20% to 50% longer than those of positrons annihilating in the bulk. The lifetime of positrons annihilating in annealed Mo is 118 ± 2 ps (29), that of positrons annihilating in monovacancies in Mo is 193 ± 5 ps (34), and that of positrons annihilating in large voids (> 10 Å diameter) in Mo is 474 ± 10 ps (35); approximately 300% longer than that for free annihilations. Similarly, the narrowing of the angular correlation distribution and the reduction in the width of the Doppler-broadened photopeak which results from positron annihilation at voids in Mo is greater than that produced by the same fraction of positrons annihilating at vacancies.

This seems to be generally the case for positrons annihilating at voids in metals; for example in Ni (36) and in Al (37).

Positronium (electron-positron bound state) formation in voids was first considered as a possible cause for these very long lifetime and narrow Doppler-broadening distributions. However, room temperature magnetic quenching experiments performed on Mo samples containing voids (32,33) did not alter the resulting spectra. Magnetic quenching, associated with Doppler-broadening or angular correlation, is a sensitive test for positronium (Ps) only if it causes Ps transitions from a state with a broad momentum spectrum to a state with a narrow one. Since the nature of the supposed Ps state in a void is unknown, the evidence against Ps formation is not very strong.

Two models have been proposed to describe the state of positrons at a void. In the model of Hautojärvi et al. (38) the positron enters the void with its screening cloud. In the model of Hodges and Stott (39) the positron is trapped at the surface of the void. In either case both models predict that the positron lifetime should be very sensitive to changes in void dimensions for voids with diameters below about 10 Å (Fig. 7). Grynszpan et al. (40) state that the surface trap model is probably superior for larger voids whereas the model of Hautojärvi et al. (38) is probably a better representation of the physical situation in small voids.

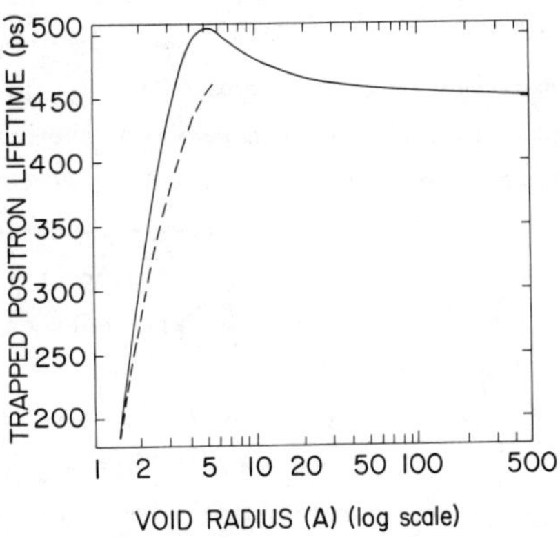

Fig. 7 - Predicted positron lifetime as a function of void radius. (---) Hautojärvi et al. (38) (——) Grynszpan et al. (40)

In any real sample there is a distribution of void sizes as well as a significant number of dislocation loops. This will result in an associated distribution of positron lifetimes. However, because of the inherent difficulties in resolving many similar lifetime components, an experimentally determined void lifetime is actually a weighted average of many lifetimes.

Petersen et al. (30) studied the effect of increasing temperature on 99.99% Mo single crystals which had been irradiated with fast neutrons (E > 0.1 MeV) to a fluence of 1.5 x 10^{18} n/cm^2 at 333 K. A similar experiment was performed on 10 MeV electron irradiated low carbon polycrystalline Mo by Eldrup et al. (29). The two studies yield results which are in qualitative agreement. Both experiments demonstrate that it is possible to correlate positron annihilation behavior with the detailed mechanisms leading to void production. They also demonstrate that positron annihilation techniques can be used to complement the results of both electron microscopy and resistivity studies. The work of Eldrup et al. (29) has shown that the positron annihilation technique can be particularly useful in complimenting resistivity studies both because of its multiparameter nature and because positron annihilation, unlike resistivity, is insensitive to isolated interstitials.

Angular correlation and Doppler-broadening data for defect studies are almost always described in terms of peak counting rates or lineshape parameters. When the full angular distribution or Doppler-broadened photopeak is analyzed in this way it is possible to overlook much of the information contained in the spectra. Computer programs such as PAACFIT (41), PARAFIT (42), or ANNIH (43) should be used whenever possible. With such programs the contributions of different types of defects can be extracted from the angular correlation or Doppler-broadening data in a manner that is analogous to that in which multiple lifetime components may be deduced from a lifetime spectrum. When the changes occurring to two types of defects simultaneously produce opposite effects on the momentum distribution a lineshape parameter or peak counting rate may exhibit only a slight change and the effect may appear small. However, when the data are analyzed with a program such as PAACFIT (41) large changes in the defect concentration and type may become evident. This was the case in the experiment on electron irradiated Mo by Eldrup et al. (29) (Fig. 8a)

In both the neutron and electron irradiation studies of Mo, τ_2 values (the lifetime component associated with voids) changed rapidly in two stages (Figure 8b). During the first stage τ_2 increased and seemed to saturate at roughly 460 ps at a temperature of about 700 K. This increase is associated with void formation. τ_2 began its second increase starting at approximately 973 K. This second increase was attributed to thermal coarsening of voids; larger voids growing at the expense of smaller ones. A later study by Thrane et al. (44) showed that, in agreement with theory, the long lifetime component in high purity Mo is independent of average void size for void radii in the range r = 9 Å to r = 45 Å. Further study

(35) associated the increase in τ_2 above ~470 ps with the migration of O, C or N impurities to void surfaces at high temperatures. This does not invalidate the basic annealing processes deduced from the irradiation damage annealing studies, however, it does demonstrate the need for more sophisticated analysis of the data. As noted by Thrane and Evans (45): "Provided such effects are (properly) identified they could add more power to the effectiveness of positrons annihilation."

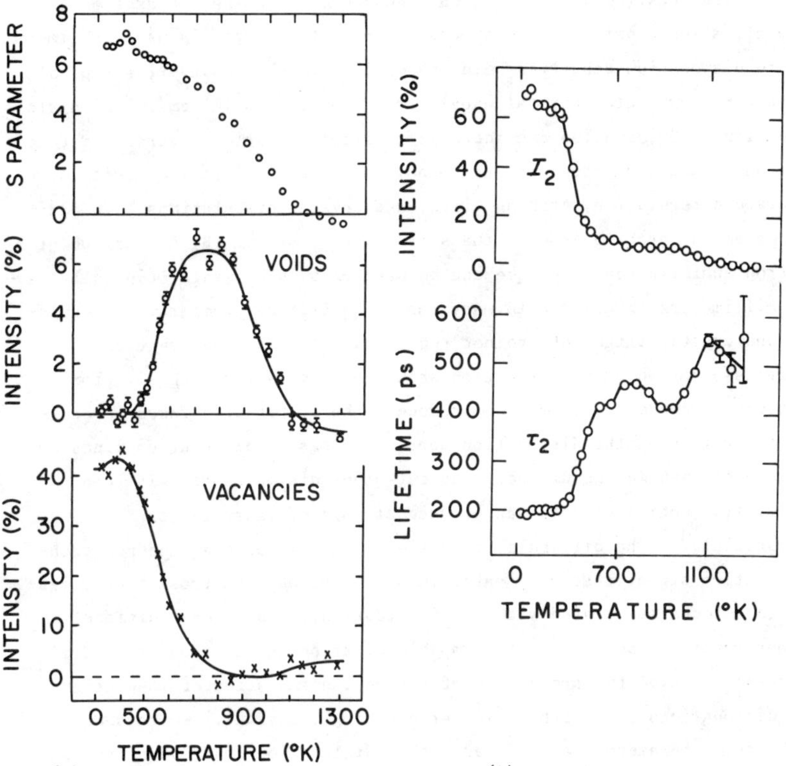

(a) (b)

Fig. 8 - Results of isochronal annealing studies of electron irradiated Mo. (Eldrup et al. (29)).

A recent paper by Schultz et al. (46) suggests thermal detrapping of positrons from void surfaces and concommitant Ps formation inside voids in Mo at high sample temperatures. The voids were characterized as having a mean diameter of ~37 Å. In addition to thermal detrapping of positrons from surfaces, slow positron studies (47) indicate that the presence of certain impurities may either reduce or increase the surface trap depth. Coupling these slow positron surface results with void studies should produce a quantitative description of positron trapping at voids.

4. Dislocations

The most common techniques used to study the effects of plastic deformation on the production of dislocations are transmission electron microscopy (TEM), X-ray topography, and etch pitting at emergent dislocations. TEM is destructive since the samples must be thinned to several hundred angstroms. In addition, because of thinning, relaxation of the dislocation substructure may occur. X-ray topography is usually limited to dislocation densities below about 10^6 cm/cm^3. However, a typical laboratory grown metal crystal usually contains 10^4 to 10^6 cm/cm^3 of dislocation line. Deformation can increase this concentration several orders of magnitude. Chemical etch pitting techniques frequently rely on critical temperature regulation, etching times, and crystal orientations. In addition, chemical etching affects the surface under study. By comparison, the positron annihilation technique can be used on bulk crystals (typically several millimeters thick) and in many instances critical temperature regulation and crystal alignment are not required. It is thought that the dilated side of an edge dislocation should provide a trapping site for positrons. However, deformation produces other types of defects in addition to increasing the dislocation density. Jogs in dislocation lines can be produced when moving dislocations cut other dislocations. The non-conservative motion of jogs can produce strings of vacancies or interstitials. The difficulty then lies in distinguishing the positron annihilation signal that originates in dislocations from those that originate in defects associated with the dislocation. Since screw dislocations are not generally believed to be capable of trapping positrons the ideal situation would be the production of a high concentration of long straight edge dislocations in an otherwise perfect single crystal. In this way the annihilation parameters associated with a dislocation could be established (48). Failing this, what is usually done is to deform a single or polycrystalline metal sample and then take advantage of the fact that point defects can generally be annealed out of a sample at lower temperatures than can dislocations.

Hautojärvi et al. (49) performed an experiment of this type on 99.998 wt.% and commercial grade polycrystalline iron. In this experiment the samples were rolled to 40% of their initial thickness. Deformation to this extent produces large numbers of dislocations. After several days (presumably at room temperature) a series of isochronal anneals were performed at increasing temperatures. The lifetime and Doppler-broadening data which were taken resulted in the curves of Figure 9. Vacancies in high purity iron are known to become mobile at 220 K (50) and are therefore not expected to seriously affect the results. A sharp recovery is observed between 673 K and 873K. Recrystallization is known to occur in this temperature range and therefore the recovery is ascribed to dislocation movement to grain boundaries. Polygonization is the suggested cause of the slight recovery beginning at 573 K in the high purity sample.

In an attempt to produce dislocations without the associated debris resulting from deformation Cotterill et al. (51) quenched 99.9999% polycrystalline Al samples from high temperatures to freeze in vacancies and then annealed at 353 K to produce dislocation loops.

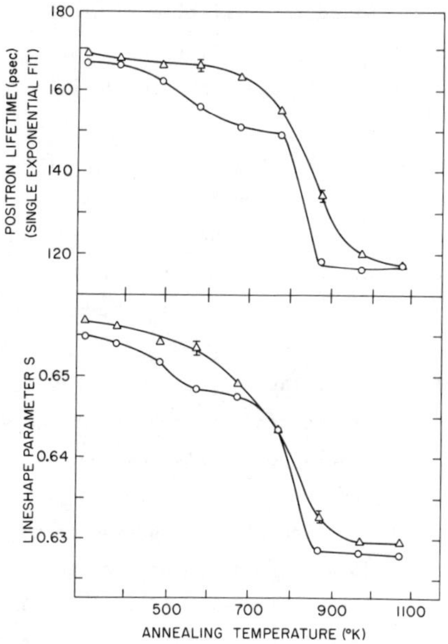

Fig. 9 - Positron lifetime and Doppler-broadening recovery curves produced by isochronal annealing after deformation of polycrystalline iron (49).
Δ Fe 99.86% ○ Fe 99.998%

The lifetime of positrons annihilating at vacancies and dislocation loops was found to be 246 ± 4 ps and 250 ± 30 ps, respectively. A free positron lifetime of 166 ± 2 ps was measured. The similarity of positron lifetimes at vacancies and dislocations exists in a number of metals. This has lead several investigators to suggest that positrons may be trapped by dislocations but ultimately annihilate at vacancies which are themselves trapped by the dislocation (52) or at jogs with vacancy-like character (53,58). Hjelmroth et al. (54) have noted that quenching and subsequent annealing of vacancies in Al is known to produce a mixture of voids and dislocation loops (55). The presence of voids could have contributed to the results of this experiment since voids in Al are known to be deep traps for positrons (37,56,57).

Another experiment which attempted to isolate the response from positrons trapped at dislocations was performed by Dannefaer et al. (48). In this experiment Ni, Co and a series of Ni-Co alloys were deformed and then subjected to a series of isochronal anneals at increasing temperatures. In most of the samples the lowest trapped positron lifetime observed after annealing was 140 ps. The fact that this same value is obtained from materials with widely differing stacking fault energies and hence different dislocation core configurations implies that the positrons are not annihilating in the dislocations. Since this lifetime is close to the 142 ps Ni monovacancy lifetime (16) it is suggested that the annihilations are taking place at vacancies or at jogs with vacancy-like character.

Hjelmroth et al. (54) proposed a model consisting of two types of traps in close physical proximity (nominally (A) dislocations and (B) jogs). In this model either defect may trap positrons but once trapped a positron may not escape into the bulk. However, positrons trapped at the A-type traps may escape or diffuse to the deeper B-type traps. There is no detrapping from B-type traps.

Smedskjaer et al. (53) have produced a similar model. In this theory the positron binding energy to a perfect edge dislocation is small ($\sim$0.1 eV). The positron signal from the dislocation is not easily resolved from the bulk signal. The presence of dislocations results in changes in the positron lifetime primarily because of the trapping of positrons by point defects associated with dislocations (e.g. jogs). Positron binding to these point-like defects is presumed to be high (a few eV). Figure 10 (53) shows a schematic representation of the trapping model. This version also takes into account detrapping from dislocations with a rate δ. The model predicts the possibility of a temperature dependent trapping rate to shallow traps (i.e. dislocations). It also predicts, for positrons already localized at a dislocation, a high probability of trapping into deep traps along the dislocation line when the deep trap "density" is $\gtrsim 10^{-4}$ Å^{-1}. The authors suggest that the low temperature effects that have been observed by several experimentalists (17,59,60) might be explained by their model.

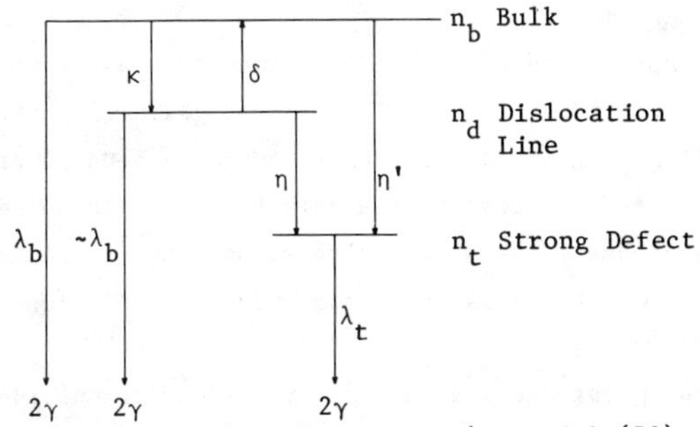

Fig. 10 - Schematic representation of the trapping model (53). κ, η, and η' are trapping rates to the states shown. δ is the detrapping rate from dislocations. λ_b and λ_t are annihilation rates.

Although the exact nature of positron trapping at dislocations is clearly not well understood at the present time this need not preclude the "applied" use of positrons for dislocation related defect studies. Lynn et al. (61) have demonstrated that positron annihilation techniques can be used to complement X-ray particle size studies during fatigue in Ni and Ni-Co alloys. During this experiment the mean positron lifetime was observed to increase monotonically (Fig. 11) and saturate at ∽7% of the fatigue life. The changes in mean positron lifetime saturated earlier than did changes in X-ray particle size (Fig. 12) and thus the mean positron lifetime is a more sensitive indicator of early fatigue damage.

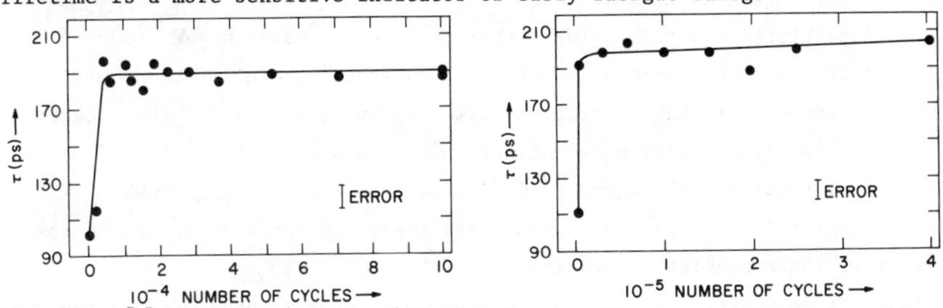

Fig. 11 - (a) The mean positron lifetime in pure Ni as a function of cyclic deformation. (b) in Ni - 66.5% Co (61).

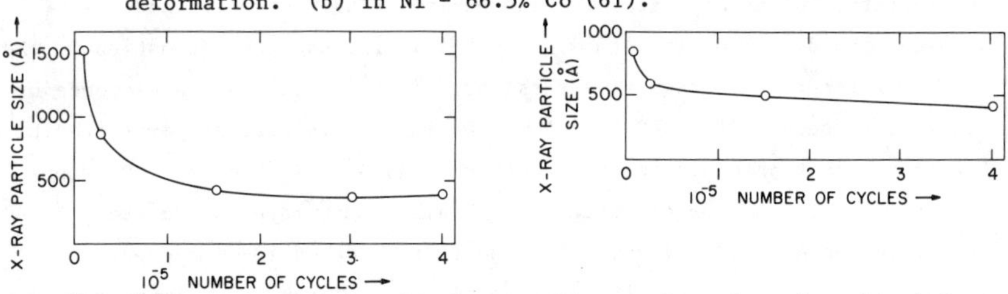

Fig. 12 - (a) X-ray particle size in pure Ni as a function of cyclic deformation. (b) in Ni-66.5% Co (61).

5. Grain Boundaries

Several experiments have been performed to measure positron annihilation characteristics as a function of grain size (62,63,64). Positron trapping at grain boundaries is expected because of the lower atomic density there relative to the bulk lattice. As noted by McKee et al. (62) it is probable, as with dislocations, that a spectrum of lifetimes favoring the deepest traps is associated with annihilations at grain boundaries.

The trapping rate to a defect may be either transition limited or diffusion limited. In the transition limited case the trapping rate is solely determined by the defect density and the positron transition time between the free and trapped states. In the diffusion limited case the trapping rate is determined by the positron's ability to get to the defect. (See McKee et al. (62) and references therein.) McKee et al. (62) find a monotonically decreasing mean lifetime and Doppler-broadening S parameter with increasing grain size (0.38μm to 1.9 μm) in fine grain Zn-22 wt.% Al alloy. Using the diffusion limited model they measured a positron diffusion coefficient in Zn-22 wt.% Al at 293°K of $D_{e+} = 0.6\ cm^2\ sec^{-1}$; a value of both theoretical and practical importance. Slow positron measurements on high purity Al at room temperature yield a value of $D_{e+} = 0.4\ cm^2\ sec^{-1}$ (see section 8).

6. Alloys

The positron annihilation technique is often used in the study of lattice defect mechanisms in pure metals, thus avoiding many of the complications inherent in the study of alloys. However, because of their technological importance lattice defect mechanisms in alloys are worthy of similar consideration. Al-Cu and Fe-C form the basis for a large number of technologically important alloys and are therefore interesting candidates for positron annihilation studies.

The positron annihilation technique has recently been used to study the formation of precipitates in high purity Al samples which had been alloyed with 0.5, 2.0, and 4.0 wt.% Cu (65). The eventual formation of the Cu rich incoherent tetragonal θ phase precipitate occurs via a sequence of metastable phases GP1 → GP2 → θ' → θ. During the initial stage of precipitation coherent precipitate particles known as Guinier-Preston (GP) zones are formed. The GP1 precipitate is a single atomic layer of Cu atoms formed in the Al (100) planes. GP2 zones are ordered Al-Cu platelets up to about 25 layers thick. The θ' phase forms as a partially incoherent precipitate which, upon further growth, finally results in the fully incoherent tetragonal θ phase precipitate.*

Gauster and Wampler (65) homogenized their alloyed samples at 823 K and then quenched to 200 K. A series of isochronal anneals were then performed at successively increasing temperatures. A 99.995% Al sample underwent similar treatment for comparison. It was quenched to 200 K from 773 K. All of the positron measurements were taken at 85 K. They interpret the results (Fig. 13) as follows: The large value of the S parameter for the as-quenched samples (relative to the annealed state) is attributable to positron trapping at quenched-in thermally generated vacancies. The increase in the value of the S parameter immediately after the 200 K anneal is ascribed to vacancy agglomeration. Close association of vacancies to Cu allows for less agglomeration with increasing Cu content. The drop in the value of the S parameter between 230 K and 300 K is related to a decrease in the total number of vacancies. Electron microscopy provides evidence of GP zone formation by 300 K. S parameter analysis shows that positrons are not sensitive to the further growth (GP1→GP2) of the precipitate between 300 K and 450 K. In this temperature range the S parameter for the 2.0 and 4.0 wt.% Cu alloy is lower than that for pure annealed Al. This indicates that the copper rich GP zones are being preferentially sampled by positrons.** Vacancies are thought to be associated with these zones for temperatures up to 450 K. Positron trapping at vacancies associated with the precipitate would provide a mechanism for preferential sampling of the Cu rich regions.

For annealing temperatures below ∿450 K the GP1 and GP2 zones remain coherent with the Al lattice, however, growth beyond this stage results in lattice strain too great to be accommodated coherently. The θ' phase is formed as an ordered tetragonal precipitate of composition near $CuAl_2$. It grows in the form of thin platelets whose faces are coherent with the Al lattice but whose edges have misfits which are accommodated with dislocations. In addition, vacancy clusters may be associated with these interphase boundaries. Positron trapping at defects associated with the θ' precipitates would result in preferential sampling of the Cu rich areas.

---

*Gauster and Wampler (65) cite a number of references on θ phase precipitation.

**Gauster and Wampler (65) find that the S parameter in their analysis is 19% larger for pure annealed Al than for pure annealed Cu.

Gauster and Wampler (65) suggest that the creation of the lattice defects associated with the θ' precipitate formation would produce positron trapping which results in the sharp reduction in the S parameter immediately after the 450 K anneal for the 2% and 4% Cu samples. Annealing at temperatures above ∿650 K produces the fully incoherent θ phase. Since these precipitates are large and widely separated the probability of positron trapping is reduced to the precipitate volume fraction.

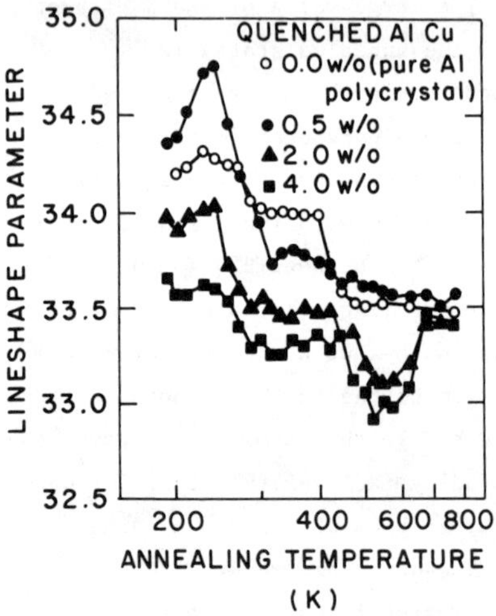

Fig. 13 - Doppler-broadening lineshape parameter S (x100) as a function of isochronal annealing temperature (log scale) (65).

The usual situation in lattice defect studies is that positron trapping results in an increase in both the S parameter and the mean lifetime. In many cases the longer second lifetime and its associated intensity can be deduced from the data. The results of this study, however, suggest that lattice defect trapping of positrons results in a decrease in the S parameter due to preferential sampling of one component of the alloy. Gauster and Wampler (65) did not report any lifetime data. It might be interesting to see if two positron lifetimes could be resolved. One could, however, imagine running into prohibitive numerical problems since, assuming that vacancy and dislocation trapped positron lifetimes are approximately equal, the positron lifetime in well annealed Cu and Al differ by only about 45 ps (120 ps Cu vs. 165 ps Al) and the lifetimes of positrons annihilating in bulk Al and Cu vacancies (∿180 ps) are nearly the same.

Carbon in Fe is one of the oldest, most frequently studied, and most technologically important alloys. Yet, a number of questions remain unanswered with regard to defect mechanisms in α-Fe (66). Specifically, the mechanism responsible for stage III recovery in α-Fe has been the subject of much controversy (50,66). Recently Hautojärvi et al. (50) have used the positron annihilation technique to study the interaction of monovacancies and interstitial carbon impurities in electron-irradiated α-Fe. In this experiment pure Fe samples and samples doped with 50 and 750 ppm carbon were annealed at 1023 K, quenched to 273 K, and electron irradiated at 20 K with 3 MeV electrons to produce isolated vacancies. Positron lifetime measurements were then performed after a series of isochronal anneals at increasing temperatures up to 600 K. The results of the study are interpreted by the authors to indicate that vacancy-carbon pair formation occurs at 220 K resulting from vacancy capture by immobile carbon atoms. The pair still exhibits positron trapping with a lifetime of 160 ps; only slightly shorter than the positron lifetime in vacancies (175 ps) for pure Fe. This suggests that, in agreement with theoretical calculations (67), carbon is located rather far off center of the vacancy. Above 350 K interstitial carbon becomes mobile further decorating the vacancies and inhibiting positron trapping.

Positron trapping is not generally believed to occur at isolated substitutional impurities (68,69), however, the results of recent experiments (70,71) suggest the contrary for Al-Mg alloys. In these experiments 99.9999% Al was alloyed from the melt with 100 ppm and 1000 ppm Mg. The samples were annealed in vacuum and surface impurities were chemically polished away. Results of the analysis of lifetime data (70) were shown to be inconsistent with positron trapping at nonequilibrium defects (such as Mg-vacancy pairs). The results seem to suggest positron trapping at isolated substitutional Mg. Angular correlation studies on the same samples (71) support the previous lifetime studies. As noted by the authors these results are important because, in addition to showing unexpected trapping at substitutional impurities, the possibility of positron trapping by substitutional elements may alter the interpretation of positron studies of the Fermi surfaces of alloys.

Bernardin and Dupasquier (71) state only that their Al-Mg samples were alloyed in the melt and annealed in vacuum. Experiments performed by Alam (72) have demonstrated that positrons are sensitive to MgO which is produced in the bulk of an Al-500 ppm Mg sample if it is annealed in the presence of even a small amount of O_2 (N_2 containing 500 vol. ppm O_2). Also, muon depolarization measurements (73) have suggested the possible existence of small clusters of Mg or even MgO in annealed Al-1000 ppm Mg alloy. Thus, an alternative interpretation, and one more consistent with the general concepts of positron trapping, may be possible depending upon the specific sample preparation conditions.

7. Fatigue

From a technological point of view detailed knowledge of the defect mechanisms associated with the fatigue of metals is obviously highly desirable. However, fatigue is a complex process involving point, line, and surface defects. The ability of the positron technique to distinguish both defect type and concentration should be of use when applied to the study of this process.

Since the positron technique is nondestructive and since the gamma ray detectors need not be in physical contact with the sample, dynamic fatigue experiments can be performed involving temperature, ultimate stress and strain or even composition of the ambient atmosphere as variable parameters. Dynamic fatigue measurements are probably not amenable to the usual sample-source-sample sandwich arrangement. In certain instances "massless" sources (74) may be applicable or corrections for annihilations outside the sample may be necessary when the positron source cannot be sandwiched between two samples.

By using isotopes which emit positrons at different energies and through the use of an energy tuneable positron beam (see section 8) a variable depth of penetration may be attained. This would allow the sample to be studied in layers without the need to polish away surface material. To date, this method has not been applied to fatigued samples. However, it has been used in a study of Al clad aluminum alloy (75) (Fig. 14).

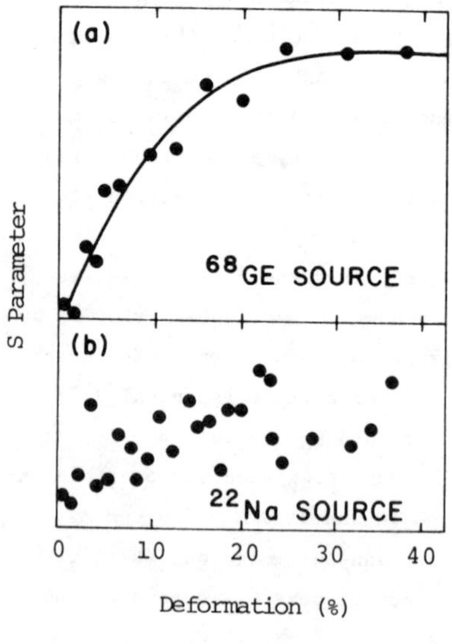

Fig. 14 - Variations of the Doppler-broadening annihilation line width with deformation in rolled Al clad 2024 samples irradiated with (a) ^{68}Ge positrons (b) ^{22}Na positrons (J. Eady et al. unpublished).

Despite these possibilities very few positron studies of fatigued metals have been performed, possibly owing to supposed difficulties in analysis and interpretation of data. In one of the few positron studies of fatigued metals (76) the mean positron lifetime in fatigued 304 stainless steel attained, early on in the fatigue process, a value which was $\sim$7% greater than that in a heavily deformed sample of the same material. This result indicates the presence of vacancy agglomerates which are apparently produced during cyclic deformation but not during monotonic deformation.

In another study (77) the difference in the S parameter, ΔS, between a fatigued and annealed sample of a low carbon steel was found to have a square dependence on the maximal applied stress for a fixed number of cycles. If the ΔS value is assumed to be proportional to the defect density then this dependence will reflect the relation between defect density and applied stress.

Hjelmroth et al. (54) studied the effects on positron annihilation of fatigue in 99.9999% Al single crystals. Only the free positron lifetime was found for fatigue below 60 cycles. During a later stage (100-350 cycles) in addition to the free lifetime a 215 ± 15 ps lifetime was also measured. Still later (>350 cycles) a lifetime of 230 ± 15 ps increasing with the number of fatigue cycles to 260 ± 15 ps was measured. Under the experimental conditions a relatively untangled dislocation population should be produced during the initial stage of fatigue (<60 cycles) and subsequently rearrange ($\sim$100 to 350 cycles) to form subgrain boundaries whose walls are composed of dense (10^9 -10^{11} cm/cm^3) heavily jogged dislocations. This implies that untangled dislocations are not efficient positron traps. Fatigue for $\sim$100 to $\sim$350 cycles produces heavily jogged dislocation tangles. During this period a 215 ps lifetime is measured and ascribed to positron trapping and subsequent annihilation at jogs. Fatigue beyond $\sim$350 cycles produces a rearrangement of dislocations which precipitates the production of a large number of point defects. During this "third stage" a third lifetime component (230 to 260 ps) is detected and assigned to vacancies or voids.

8. Future Prospects and Present Limitations on the Positron Annihilation Technique

The positron annihilation technique provides the researcher with one or more annihilation components each of which may be associated with annihilations in the bulk or at vacancy-like defects. The combination of insensitivity to interstitials and a unique "signal" resulting from annihilations at each type of vacancy-like defect promises to make positron annihilation useful in the study of complex recovery or recrystallization mechanisms. This is particularly so under conditions where several types of defects may exist simultaneously. By following the intensity and character of the annihilation components, defect types, mechanisms, and kinetics may be determined. However, with presently available experimental techniques it is usually impossible to resolve more than two or at best three annihilation components. Further complications can arise when the data contain annihilations from defects such as grain boundaries and dislocations. Annihilations at these defects probably result in a range of very similar annihilation components which are very difficult to resolve numerically.

The large mean energy of positrons (50 keV-MeV range) obtained from radioactive isotopes has not allowed positrons to be used in the study of surface or near surface defects. Typically less than 1% of the particles injected into a solid diffuse back to the surface before annihilation. This is clearly an advantage for bulk studies but makes surface studies practically impossible. Recently, however, it has become possible to greatly increase this percentage by using an energy tuneable (1 eV to 10 keV) monoenergetic positron beam (78). The variable energy beam allows the initial implantation depth to be varied and therefore permits surface as well as near-surface positron annihilation studies. In addition to conventional Doppler-broadening measurements the condition of the surface of a sample may be studied by measuring the fraction, energy, and direction of reemitted positrons and the fraction of positrons which are reemitted as positronium. Very recently it has become possible, by pulsing the slow positron beam (79), to perform lifetime measurements as well.

The positron annihilation technique is a useful laboratory tool for, among other things,* basic research in the field of crystal lattice defect mechanisms. In many instances its greatest value, to date, has been when used in conjunction with other techniques such as resistivity or electron microscopy. While the positron annihilation technique is not yet mature enough to be used for nondestructive evaluation of samples in the field, there is certainly the promise of many practical applications in applied as well as basic research settings; particularly in regard to slow positron studies of defects at metal-semiconductor and semiconductor-semiconductor oxide interfaces.

---

*The positron annihilation technique has also been used to study the electronic structure of metals.

Acknowledgements

We would like to thank A. N. Goland for his critical reading of the manuscript and continued interest. One of us (L.G.) would also like to thank J. C. Bilello for useful discussions and encouragement.

References

(1) I. Ya. Dekhtyar, D.A. Levina, V.S. Mikhalenkov, Soviet Phys.-Dolk. 9 (1964) p. 492.

(2) S. M. Kim, A. T. Stewart, Phys. Rev. B11 (1975) p. 2490.

(3) P. Kubica, A. T. Stewart, Phys. Rev. Lett. 97 (1975) p. 852.

(4) G. E. Lee-Whiting, Phys. Rev. 97 (1955) p. 1557.

(5) P. Hautojärvi ed., Positrons in Solids Springer-Verlag, New York, N.Y. 1979.

(6) R. M. Nieminen, M. J. Manninen, "Positrons in Imperfect Solids: Theory," p. 145 in Positrons in Solids, P. Hautojärvi, ed., Springer-Verlag, New York, N.Y. 1979.

(7) O. Johnson, Phys. Stat. Sol. b99 (1980) p. 745.

(8) D. O. Welch, K. G. Lynn, Phys. Stat. Sol. b77 (1976) p. 277.

(9) I. K. MacKenzie, J. A. Eady, R. R. Gingerich, Phys. Lett. 33A (1970) p. 279.

(10) W. Triftshauser, Proceedings of the Second Int. Conf. Positron Annihilation, Kingston, Canada, p. 477.

(11) R. N. West, (to be published in J. of Phys. C).

(12) R. N. West, Advances in Physics, 22 (1973) p. 312.

(13) P. Kirkegaard, M. Eldrup, Comp. Phys. Comm. 7 (1974) p. 401.

(14) M. Doyama, "Studies of Lattice Defects and Phase Transformations by Positron Annihilation," p. 3 in Progress in the Study of Point Defects, M. Doyama, S. Yoshida, eds., University of Tokyo Press, Tokyo, Japan, 1978.

(15) K. Maier, M. Peo, B. Saile, H. E. Shaefer, A. Seeger, Phil. Mag. A40 (1979) p. 701.

(16) K. J. Lynn, C. L. Snead, Jr., J. J. Hurst, J. Phys. F: Met. Phys. 10 (1980) p. 1753.

(17) L. C. Smedskjaer, D. G. Legnini, R. W. Siegel, J. Phys. F: Met. Phys. 10 (1980) p. L1.

(18) P. C. Lichtenburger, C. W. Schulte, I. K. MacKenzie, Appl. Phys. 6 (1975) p. 305.

(19) K. Kuribayashi, S. Tanigawa, S. Nanao, M. Doyama, Sol. State Comm. 12 (1973) p. 1179.

(20) I. K. MacKenzie, P. C. Lichtenberger, Appl. Phys. 9 (1976) p. 311.

(21) C. W. Schulte, J. L. Campbell, Appl. Phys. 19 (1979) p. 269.

(22) J. L. Campbell, C. W. Schulte, J. A. Jackman, J. Phys. F: Met. Phys. 7 (1977) p. 1985.

(23) W. Brandt, p. 155 in Positron Annihilation, A. T. Stewart, L. O. Roellig; Academic Press, New York, N.Y. 1967.

(24) B. Bergersen, M. J. Stott, Solid State Comm. 7 (1969) p. 1203.

(25) D. C. Connors, R. N. West, Phys. Lett. A30, (1969) p. 24.

(26) K. Maier, R. Myllylä, "Positron Lifetime Spectrometer with β^+-γ Coincidence," p. 829 in Proc. 5th Int. Conf. Positron Annihilation, R. R. Hasiguti, K. Fujiwara eds; Japan Institute of Metals, Japan, 1979.

(27) V. H. C. Crisp, D. G. Lock, R. N. West, J. Phys. F: Met. Phys. 4 (1974) p. 830.

(28) L. Stals, J. Nihoul, J. Cornelis, P. De Meester, Phys. Stat. Solidi 18 (1973) p. 283.

(29) E. Eldrup, O. E. Mogensen, J. H. Evans, J. Phys. F: Met. Phys. 6 (1975) p. 499.

(30) K. Petersen, N. Thrane, R. M. J. Cotterill, Phil. Mag. (1973) p. 9.

(31) H. Vandenborre, J. Nihoul, L. Stals, Crystal. Lattice Defects 5 (1974) p. 89.

(32) O. Mogensen, K. Petersen, R. M. J. Cotterill, B. Hudson, Nature 239 (1972) p. 97.

(33) R. M. J. Cotterill, I. K. MacKenzie, L. Smedskjaer, G. Trumpy, J. H. O. L. Träff, Nature 239 (1972) p. 99.

(34) M. Eldrup, B. T. A. McKee, A. T. Stewart, J. Phys. F: Met. Phys. 9 (1979) p. 637.

(35) N. Thrane, J. H. Evans, Appl. Phys. 12 (1976) p. 182.

(36) M. Hasegawa, T. Suzuki, Rad. Effects 21 (1974) p. 201.

(37) W. Trifthäuser, J. D. McGervey, R. W. Hendricks, Phys. Rev. B9 (1974) p. 3321.

(38) P. Hautojärvi, J. Heinio, M. Manninen, R. Nieminen, Phil. Mag. 35 (1977) p. 973.

(39) C. H. Hodges, M. J. Stott, Sol. State Comm. 12 (1973) p. 1153.

(40) R. Grynszpan, B. T. A. McKee, A. T. Stewart, "Relation between Void Size and Positron Lifetime in a Surface Trapping Model," p. 717 in Proc. 5th Int. Conf. Positron Annihilation, R. R. Hasiguti, K. Fujiwara eds.; Japan Institute of Metals, Japan, 1979.

(41) P. Kirkegaard, O. E. Mogensen, Riso report (1973).

(42) P. Kirkegaard, O. E. Mogensen H12 in Proc. 4th Int. Conf. Positron Annihilation, G. Trumpy ed.; 1976.

(43) V. Bantchev, T. Troev, p. 833 in Proc. 5th Int. Conf. Positron Annihilation, R. R. Hasiguti, K. Fujiwara, eds.; Japan Institute of Metals, Aoba Aramaki, Sendai 980, Japan, 1979.

(44) N. Thrane, K. Petersen, J. H. Evans, E19 in Proc. 4th Int. Conf. Positron Annihilation, G. Trumpy ed., 1976.

(45) N. Thrane, J. H. Evans, E19b in Proc. 4th Int. Conf. Positron Annihilation, G. Trumpy ed.; 1976.

(46) P. J. Schultz, K. G. Lynn, I. K. MacKenzie, Y. C. Jean, C. L. Snead, Jr., Phys. Rev. Lett. 44 (1980) p. 1629.

(47) K. G. Lynn, Phys. Rev. Lett. 43 (1979) p. 391.

(48) S. Dannefaer, D. P. Kerr, S. Kupca, B. G. Hogg, J. O. Madsen, R. M. J. Cotterill, Can. J. Phys. 58 (1980) p. 207.

(49) P. Hautojärvi, A. Hehanen, V. S. Mikhalenkoo, Appl. Phys. 11 (1976) p. 191.

(50) P. Hautojärvi, S. Johnson, A. Vehanen, J. Yli-Kauppilla, Phys. Rev. Lett. 44 (1980) p. 1326.

(51) R. M. J. Cotterill, K. Petersen, G. Trumpy, J. Traff, J. Phys. F: Met. Phys. 2 (1972) p. 459.

(52) M. Doyama, R. M. J. Cotterill, p. 89 in Proc. 5th Int. Conf. Positron Annihilation, R. R. Hasiguti, K. Fujiwara, eds.; Japan Institute of Metals, Japan, 1979.

(53) L. C. Smedskjaer, M. Manninen, M. J. Fluss, J. Phys. F: Metal Physics 10 (1980) p. 2237.

(54) H. E. Hjelmroth, J. G. Rasmussen, J. Träff, Ch. Nuovo Cimento 41B (1977) p. 284.

(55) S. Yoshida, M. Kiritani, Y. Shimomura, Lattice Defects in Quenched Metals, p. 173; New York, NY, 1965.

(56) D. Segers, M. Dorikens, L. Dorikens-Vanpnaet, O. Dauwe, E16 in Proc. 4th Int. Conf. Positron Annihilation, G. Trumpy ed. 1976.

(57) G. Kogel, J. Weinter, W. Trifthauser, p. 707 in Proc. 5th Int. Conf. Positron Annihilation, R. R. Hasiguto, K. Fujiwara eds.; Japan Institute of Metals, Aoba Aranaki, Sendai 980, Japan, 1979.

(58) B.T.A. McKee, S. Saimoto, A. T. Stewart, M. J. Stott, Can. J. Phys. 52 (1974) p. 759.

(59) D. Herlach, H. Stoll, W. Trost, H. Metz, T. E. Jachnan, K. Maier, H. E. Schaefer, A. Seeger, Appl. Phys. 12 (1977) p. 59.

(60) P. Rice-Evans, I. Chaglar, F. El-Khangi, Phys. Rev. Lett. 40 (1978) p. 716.

(61) K. G. Lynn, C. M. Wan. R. W. Ure, J. G. Byrne, Phys. Stat. Solidi 22 (1974) p. 731.

(62) B. T. A. McKee, G. J. C. Carpenter, J. F. Watters, R. J. Schultz, Phil. Mag. 41 (1980) p. 65.

(63) K. G. Lynn, R. Ure, J. G. Byrne, Acta Metal. 22 (1974) p. 1075.

(64) B. T. A. McKee, A. T. Stewart, L. Morris, H. Sang, "Positron Diffusion and Trapping at Grain Boundaries in Al," p. 169, Proc. 5th Int. Conf. Positron Annihilation, R. R. Hasiguti, K. Fujiwara, eds.; Japan Institute of Metals, Japan 1979.

(65) W. B. Gauster, W. R. Wampler, Phil. Mag. A 41 (1980) p. 145.

(66) H. E. Schaefer, K. Maier, M. Weller, D. Herlach, A. Seeger, J. Diehl, Scripta Metall. 11 (1977) p. 803.

(67) R. A. Johnson, A. C. Damask, Acta Metall. 12 (1964) p. 443.

(68) D. G. Lock, R. N. West, J. Phys. F: Metal Phys. 4 (1974) p. 2179.

(69) M. J. Stott, P. Kubica, Phys. Rev. B11 (1975) p. 1.

(70) M. Bernardin, A. Dupasquier, Lett. Nuovo Cimento 26 (1979) p. 161.

(71) A. Dupasquier, A. Gallone, C. Regazzoni, Solid State Comm. 35 (1980) p. 415.

(72) M. A. Alam (1980) Ph.D. Thesis, University of East Anglia, England.

(73) W. J. Kossler, A. T. Fiory, W. F. Lankford, K. G. Lynn, R. P. Minnich, C. E. Stronach, Hyperfine Inter. 6 (1979) p. 295.

(74) M. J. Fluss, L. C. Smedskjaer, p, 407, in Proc. 5th. Int. Conf. Positron Annihilation, R. R. Hasiguti, K. Fujiwara, eds.; Japan Institute of Metals, Japan, 1979.

(75) J. A. Eady, I. K. Mackenzie, R. R. Gingerich, W. C. Winegard (unpublished).

(76) K. Nishiwaki, N. Owada, K. Hinode, S. Tanigawa, K. Shibata, T. Fugita, M. Doyama, "The Study of Fatigued Stainless Steel by Positron Annihilation," p. 177, Proc. 5th Int. Conf. Positron Annihilation, R. R. Hasiguti, K. Fujiwara, eds.; Japan Institute of Metals, Aoba Aramaki, Sendai 980, Japan, 1979.

(77) W. Gostaw, T. Goworek, J. Wawryszczuk, Phys. Stat. Sol. 47 (1978) K141.

(78) K. G. Lynn, H. Lutz, Rev. Sci. Instrum. 51 (1980) p. 977.

(79) A. Mills, L. Pfiefer, Phys. Rev. Lett. 43 (1979) p. 1961.

ULTRASONIC CHARACTERIZATION AND MICROSTRUCTURE OF STAINLESS STEEL WELD METAL*

D.S. Kupperman, K.J. Reimann and D.I. Kim**

Materials Science Division, Argonne National Laboratory
Argonne, Illinois

One of the most difficult materials to inspect or characterize ultrasonically is austenitic stainless steel weld metal. The problems arise because of the preferred orientation associated with the columnar grains that are present. This structure is highly anisotropic and, as a result, the sound velocity and attenuation in the weld can vary with propagation direction, by as much as ∿ 20% for longitudinal and ∿ 100% for shear waves. Experimental results compare favorably with model calculations. The phenomenon of beam deviation in anisotropic materials is also discussed. Theoretical predictions and experimental results for velocity, attenuation, and the deviation of longitudinal and shear waves in several stainless steel weld metal samples are compared. Predicted deviations of up to ∿ 20° for longitudinal waves and ∿ 45° for shear waves have been confirmed experimentally. The consequences for attenuation and flaw detection are discussed.

*Work supported by the U.S. Department of Energy, Division of Materials Science.
**On leave of absence from Inha University, Incheon, Korea.

Introduction

The microstructure of austenitic stainless steel weld metal is significantly different from that of ferritic weld metal. During cooling, ferritic welds undergo a phase transformation not seen in stainless steel welds; this transformation results in a relatively homogeneous grain structure. In contrast, slow cooling of stainless steel welds leads to the formation of a coarse-grain dendritic structure. These grains tend to grow with a <100> orientation and essentially along the lines of heat dissipation, producing an acoustically anisotropic microstructure. Thus, the ultrasonic inspection of austenitic stainless steel weld metal is difficult, and the acoustic properties are not easily understood.

Many investigators, some of them mentioned below, have studied the propagation of ultrasonic waves in stainless steel weld metal and the implications of the greater anisotropy of austenitic welds for flaw detection. In a 1954 report, Musgrave (1) described the propagation of ultrasonic waves in anisotropic media and presented calculations for the deviation of the beam from the wave normal. In 1963, Holmes (2,3) pointed out that in austenitic stainless steel weld metal, the erroneous detection of flaws and inability to detect known defects were due to the unpredictable propagation of ultrasonic waves. The high ultrasonic attenuation in the weld metal and the increase in this attenuation with grain size were also noted at that time. Pilseneer and Louis (4) noted variations in sound velocities and problems in locating and sizing defects in stainless steel weld metal because of apparent beam deviations.

Serabian (5) pointed out that the polarization of shear waves propagating through the weld metal parallel to the weld-pass direction greatly influenced the attenuation. Below 3 MHz, with polarization parallel to the columnar grains, the attenuation was lowest (above 3 MHz, the attenuation appeared to be independent of polarization). For waves propagating across the weld-pass direction (but still perpendicular to the long axis of the grains), no variation in attenuation with polarization was reported. More recently, Mech and Michaels (6) have studied the problems of inspecting stainless steel weld metal and have shown that a frequency moment algorithm, utilizing refracted longitudinal waves, helps discriminate defect-notch signals from grain noise.

Yoneyama et al. (7) have studied the causes of false indications sometimes observed in austenitic stainless steel welds and have demonstrated that shear-wave signals can be reflections caused when the acoustic beam deviates from the anticipated path and follows the lines of grain growth. Baikie et al. (8) showed the orientation dependence of longitudinal wave velocity and demonstrated that the apparent attenuation of longitudinal waves at 2 MHz in stainless steel weld metal varied with propagation direction relative to the columnar grain axis. The minimum attenuation occurred for waves propagating at about 45° to the columnar grain axis (the direction of maximum wave velocity). More recently, Tomlinson et al. (9) interpreted Baikie's data on the basis of beam deviation, concluding that the variation in apparent attenuation may be primarily due to the variation in beam width (and deviation) with propagation direction, with the smallest beam width observed for directions of maximum velocity. In a recent report, Adler et al. (10) discuss anisotropic structural materials (stainless steel welds and centrifugally cast pipe) and illustrate how anisotropy must be accounted for when employing frequency analysis for flaw characterization. Data on velocity of sound waves (longitudinal and shear) as a function of propagation direction are presented for a single-pass weld of Type 308 stainless steel and an Inconel

82 multipass weld. In addition, some data are presented on the relative apparent attenuation of longitudinal waves as a function of propagation direction, in qualitative agreement with Baikie (8).

References 11-19 describe additional efforts on inspection of austenitic stainless steel welds. These include an examination of the variation in frequency spectra with microstructure (including polarization effects) by Kupperman et al. (19). Some of the work described in the present paper is also discussed in Ref. 20.

The main factors that contribute to ultrasonic attenuation in stainless steel welds are (a) grain-boundary scattering, (b) mode conversion, and (c) the phenomenon of beam steering, in which the ultrasonic beam is guided towards the direction of maximum ultrasonic velocity in the weld (9). Spurious signals, which cannot be correlated with component geometry or actual flaws, contribute to the difficulties in interpreting the results. The difficulties encountered, as well as the high attenuation in the weld metal, have been discussed in the literature (2-12, 14, 17, 18). Fewer grain boundaries are encountered along the path of the long axes of the grains, and thus the scattering attenuation of the ultrasonic waves may vary with wave propagation direction.

Since beam scattering depends upon anisotropy and the ratio of grain size to wavelength, the severity of the ultrasonic attenuation due to scattering in stainless steel weld metal may be reduced by welding techniques that tend to produce a finer and more isotropic crystallographic structure. However, for coarse-grain materials, proper selection of the acoustic mode and an understanding of the acoustic propagation characteristics of the weld metal may enhance the reliability and sensitivity and lead to an improved ultrasonic testing method.

In the present paper, a model for the weld metal is described and predictions of sound velocity and beam deviation, based on this model, are compared with experimental results. In addition, data on microstructure, sound, velocity, and attenuation are compared for samples obtained from four gas-tungsten-arc (TIG) V-type welds. Weld-metal samples of the same type, with coarse and fine grain structure, are compared with respect to sound velocity and attenuation. Frequency spectra and flaw detection results are also discussed.

Theory

From the columnar grain character and anisotropy of stainless steel weld metal (see next section), one may assume that the vertical grains are long cylindrical single crystals randomly oriented about the vertical axis in the <100> direction; i.e., the samples has orthotropic symmetry (23). X-ray diffraction studies carried out at Argonne on weld-metal coupons show strong <100> preferred texture along the grain axis and much less <100> texture for orthogonal directions. Sound-velocity measurements confirm these x-ray results for the bulk. The effectiveness of the model can be checked by comparing the predicted and measured ultrasonic velocities in various propagation directions. The following discussion (a) indicates how the velocity in various directions can be predicted for the weld metal and (b) describes the phenomenon of beam steering, which can have a great effect on apparent attenuation of both longitudinal and shear waves.

Calculations of sound velocity and deviations (i.e., skewing of the maximum energy at the wave front from the wave-front normal) in the (110) plane for single-crystal Type 304 stainless steel were performed by Musgrave's method (1, 13, 15), using elastic constants from single-crystal specimens (20). The deviation is considered positive if it is in the same direction as the increment of the angular-velocity direction cosine and negative if it is in the opposite direction. The results show that the quasilongitudinal velocity can vary from 6.4 x 10^3 to 5.1 x 10^3 m/s and the shear-wave velocities from 4.0 x 10^3 to 2.1 x 10^3 m/s. The angle between velocity vector and energy ray can amount to 22° for longitudinal waves, 50° for vertically polarized (SV) shear waves, and 35° for horizontally polarized (SH) shear waves in single crystals of stainless steel. The results for single crystals are presented for comparison with the weld-metal results. Columnar grain structures exhibit symmetry around one axis (the z-axis), and there will be five independent elastic constants (16):

$$\bar{C}_{11D} = \bar{C}_{22D} = \bar{C}_{11} + \frac{3\gamma C}{20} \qquad \bar{C}_{33D} = \bar{C}_{11} + \frac{2\gamma C}{5}$$

$$\bar{C}_{44D} = \bar{C}_{55D} = \bar{C}_{44} - \frac{\gamma C}{5} \qquad \bar{C}_{66D} = \bar{C}_{44} + \frac{\gamma C}{20} = \frac{1}{2}\left(\bar{C}_{11D} - \bar{C}_{12D}\right)$$

$$\bar{C}_{13D} = \bar{C}_{23D} = \bar{C}_{12} - \frac{\gamma C}{5} \qquad \bar{C}_{12D} = \bar{C}_{12} + \frac{\gamma C}{20}$$

Here, $C = C_{11} - C_{12} - 2C_{44}$ (16) where C_{11}, C_{12}, C_{44} are single-crystal elastic constants; $A = 2C_{44}/(C_{11} - C_{12})$, and A = 1 has been assumed for polycrystalline material; and $\bar{C}_{ij}$ is the elastic constant for polycrystalline materials (Voigt average) (10,16). The texture anisotropy factor γ can be derived from the following considerations:

(a) In the (001) plane (cross section of a grain), the ultrasonic velocity should be independent of direction of incidence.

(b) The velocities parallel to the z-axis (the axis of symmetry) should be the same as given by C_{11} and C_{44} in a single crystal.

These restrictions establish the value $\gamma = 1$. Using the relations $\bar{C}_{ijD}$ and the single-crystal elastic constants of Ref. 20, velocity surfaces, wave surfaces, and deviation were calculated for columnar grain structures of Type 304 stainless steel. The results are plotted for the (110) plane in Figs. 1 and 2. In the (001) plane, the plots of velocity and wave surfaces would be concentric circles and the deviation would be zero. Examination of the velocity curves indicates that in the (110) plane the quasilongitudinal velocity can vary between 6.2 x 10^3 and 5.1 x 10^3 m/s, while the shear-wave velocities can assume values between 4.0 x 10^3 and 2.4 x 10^3 m/s (SV) and 4.0 x 10^3 and 3.2 x 10^3 m/s (SH). The deviation of longitudinal waves can be as large as 21°, larger than the calculated maximum value of $\sim$18° and experimental maximum of $\sim$ 15° given in Ref. 9, while that of shear waves can reach values of 44° or 13°, depending on polarization. (The difference between the current results and those of Ref. 9 is due to different values of the elastic constants used in the calculations.) Experimental and calculated values are compared in the following section.

The above theoretical considerations lead to a better understanding of consequences that will be encountered in the practical application of ultrasonic testing of structures with preferred orientations, such as stainless steel welds. The deviation of the beam from the incident direction can lead to errors in flaw location. Bending along the axis of symmetry can result in a focusing or defocusing of the beam, making defect sizing very difficult.

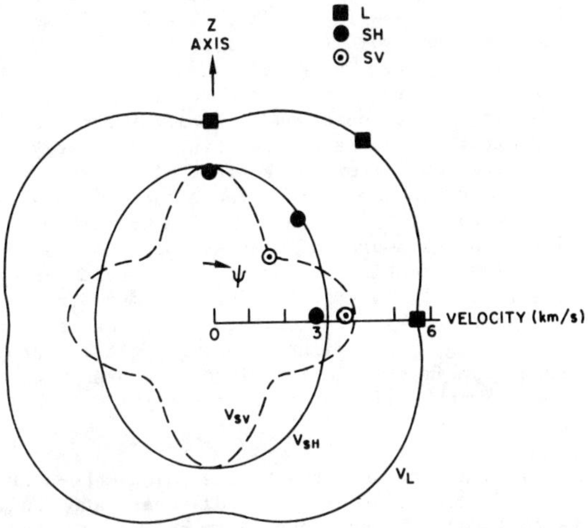

Fig. 1 - Velocity-surface intersection with the (110) plane for columnar grain structure of Type 304 stainless steel. ψ is propagation angle relative to the columnar grain axis. Curves represent calculated results for three types of sound waves; symbols indicate experimetal data for Type 308 SS weld-metal samples.

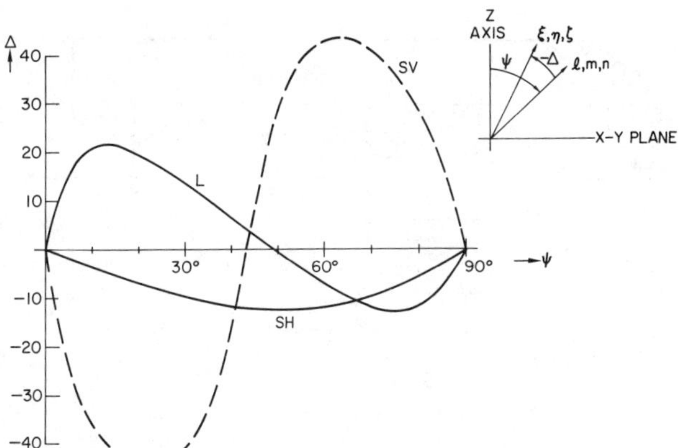

Fig. 2 - Energy-ray deviation (Δ) vs propagation angle relative to columnar grain axis (ψ) in the (110) plane for columnar grain structure of stainless steel.

Specimens and Experimental Methods

Specimens were cut from Type 308 stainless steel welds in Type 304 stainless steel plates. The welds were "V"-type, with a 77° included angle. Ultrasonic examinations were performed on a 24 x 25 x 38-mm coupon from a multipass shielded-metal arc weld (Fig. 3). The microstructure of this weld sample is presented in Fig. 4. The coupon was etched after removal from the welded plate. Grains up to 10 mm in length and 1 mm in width, essentially parallel to the z-axis, can be observed. The x-ray results indicate a strong preferred <100> orientation in the z direction. (The x-ray diffraction data are, of course, obtained only from the surface of the interrogated specimen.) A similar sample was used to demonstrate the effect of shear-wave polarization on defect detection (20). Other Type 308 SS weld-metal coupons (TIG V-type welds) were the source of sound-velocity and attenuation data for various propagation directions relative to the columnar-grain axis. They are shown schematically in Fig. 5. Four coupons were prepared with a 13-mm-square cross section, with the orientation of one end rotated 45° with respect to the other (Fig. 5b). Figure 6 shows optical micrographs of one end of each of these specimens; similar microstructures with columnar grains are evident. Samples with different microstructures may well have different ultrasonic characteristics.

Sound-velocity data were acquired by the pulse-echo-overlap method (20,21). Normal-incidence 2.25-MHz and 5-MHz shear- and longitudinal-wave transducers were employed. These transducers and an Aerotech UTA3 pulser-receiver system (except as noted below) were used for flaw detection as well as velocity and attenuation measurements. A shear-wave couplant was used to couple shear waves directly into the weld metal. An x-ray diffractometer 2θ scan method was used to evaluate the crystallographic texture in the weld samples.

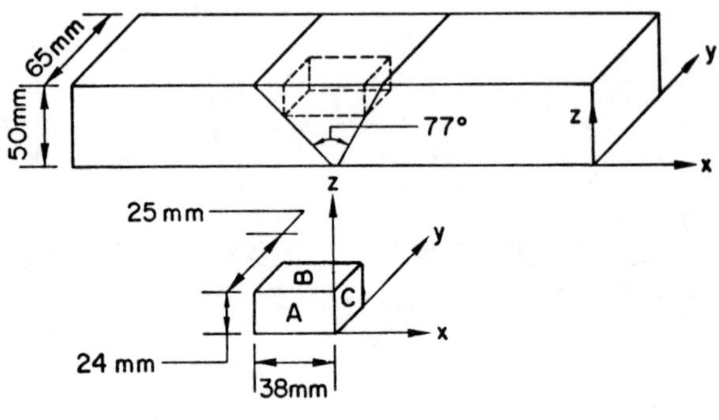

Fig. 3 - Type 308 SS weld-metal coupon for studies of acoustic properties. Base metal is Type 304 SS.

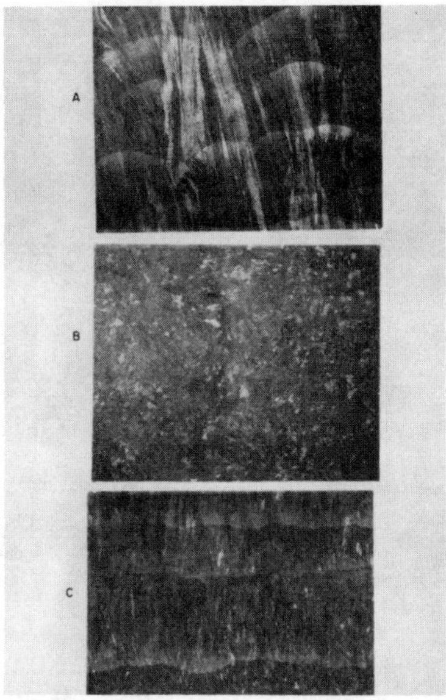

Fig. 4 - Micrographs of ∿ 1-cm^2 areas of surfaces A, B, and C of weld-metal coupon shown in Fig. 3.

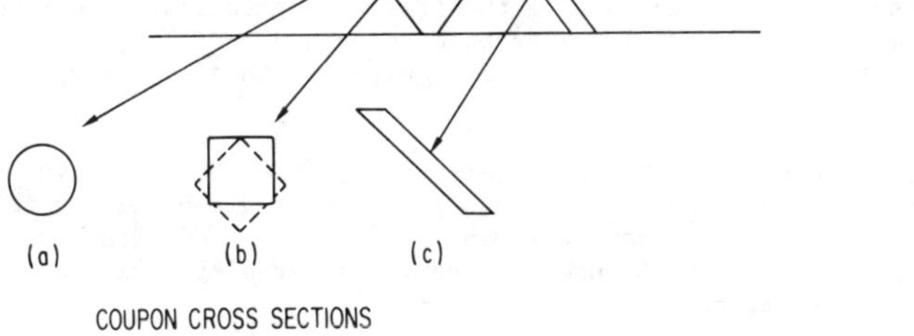

Fig. 5 - Cross sections of several types of coupons removed from Type 308 SS V-type weld for use in studies of (a) beam deviation; (b,c) sound velocity and attenuation.

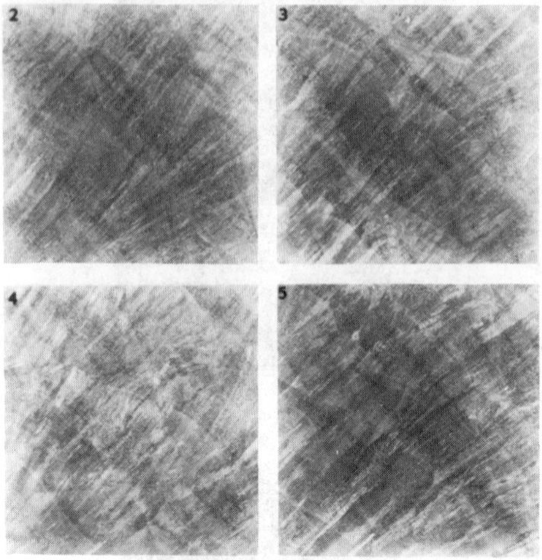

Fig. 6 - Micrographs of 308 SS weld-metal coupons (Fig. 5b).

Results

Sound Velocity and Attenuation

The sound velocities measured in three different propagation directions in the weld-metal samples (Figs. 3 and 5b) are shown in Fig. 1 along with the theoretical calculations. Our results for longitudinal-wave velocity are consistent with those of Refs. 9 and 10 but only qualitatively consistent with the shear-wave results of Ref. 10. (The present data show greater anisotropy.) A variation in velocity of 14 percent for longitudinal and 70 percent for shear waves is observed here. The maximum velocity for longitudinal waves is in a direction ∿ 45° to the columnar grain axis, whereas this is the direction of minimum shear velocity. The maximum velocity for shear waves is in a direction parallel to the columnar grain axes or perpendicular to the columnar grains with polarization parallel to the axes (these are the directions of the lowest longitudinal velocities). Velocity measurements parallel, perpendicular and at 45° to the grains were sufficient to verify the theoretical results, as these are the directions of minimum and maximum velocity.

Sound velocity (in a specific direction, i.e., 0, 45 and 90° to the columnar grain axis) varies by less than 2.5% for longitudinal and 4% for shear waves among the four samples shown in Fig. 6. This suggests that welds made under similar conditions should have very similar ultrasonic wave propagation characteristics.

As discussed earlier, the variation in velocity of sound with direction of propagation through a material is a measure of anisotropy. Grain structures that are finer and more homogeneous, and less anisotropic, should be associated with less velocity variation than coarse structures. Samples with low anisotropy and fine grains should also produce less apparent attenuation (i.e., less beam steering and less Rayleigh-scattering attenuation). To test these predictions, the velocity and relative apparent attenuation of shear waves were measured in three weld coupons (Fig. 5b) with different grain sizes, designated A, B, and C. The velocity of sound was determined for three propagation directions: z (parallel to columnar grain growth); y (parallel to the weld-pass direction); and x (orthogonal to y and z). In each case, the shear waves were polarized in two orthogonal directions perpendicular to the direction of propagation. The small size ($\sim 1\ cm^3$) and coarse grain structure of the samples made the velocity of sound difficult to measure accurately. In most cases, through-transmission techniques had to be used, as there was often only one weak ultrasonic echo. A Panametrics normal-incidence 2.25-MHz shear-wave transducer was employed in conjunction with a Panametrics Model 5052 PRX pulser-receiver and Tektronix oscilloscope with a delaying time base. The results are shown in Table I. The relatively fine-grained sample B (with grains about 20% as large as those of sample A) showed the smallest variation in velocity, as might be expected; sample A, with the coarsest structure (similar to that of Fig. 5), showed the greatest variation. Sample B, as expected, also produced much less attenuation than the other two samples. In Fig. 7, the echoes from waves propagating in the weld-pass direction (y) with polarization parallel to the columnar grain axis (z) are shown for the three samples. Sample B (Fig. 2b) gives three clear echoes, whereas samples A and C each indicate at most one weak echo.

Table I. Comparison of Shear-wave Velocities for Three Type 308 SS Weld-metal Coupons

Sample	Relative Grain Size	V_{max},[a] 10^5 cm/s	V_{min},[a] 10^5 cm/s	V_{avg}[b]	$(V_{max}-V_{min})/V_{avg}$
A	Coarsest	3.7 y(z)	2.8 x(y)	3.2	0.28
B	Fine	3.3 y(z)	3.2 z(y)	3.2	0.03
C	Coarse	3.7 y(z)	3.0 z(x)	3.3	0.21

[a]The directions of propagation and polarization (in parentheses) for which the maximum and minimum velocities were obtained are indicated below the velocity values.

[b]Average of 6 velocity measurements (3 propagation directions with 2 polarizations each).

A

B

C

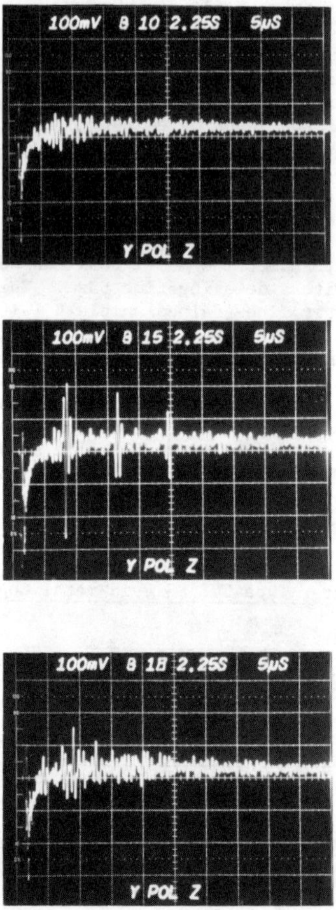

Fig. 7 - Comparison of 2.25-MHz ultrasonic shear-wave echoes from three 1-cm^3 stainless steel weld-metal samples with different microstructures. Wave propagation is in the weld-pass direction, with polarization parallel to the columnar grains. The relative grain sizes are (A) coarsest, (B) fine, and (C) coarse.

These results show that microstructural variations have a significant effect on the ultrasonic propagation characteristics of weld metal. Welding procedures which produce a uniform, fine-grained texture, and thus reduce the anisotropy, should greatly alleviate ultrasonic inspection problems in stainless steel weld metal.

The variation of attenuation with wave propagation direction can be seen in Fig. 8. Here the apparent attenuation of longitudinal ultrasonic waves in each of four propagation directions (the directions of expected minimum and maximum attenuation) is plotted for the four different samples of Fig. 6. Attenuation is slightly greater for propagation across the columnar grain axis than parallel to it (possibly because there are fewer grain boundaries to scatter waves propagating parallel to the grains). Waves propagating at approximately 45° to the columnar grain axis exhibit the least attenuation (of the order of 1.5-3 dB/cm less than waves propagating parallel or perpendicular to the grains), in general agreement with Ref. 9. The low attenuation at 45° to the grains can be explained by consideration of the beam-deviation phenomenon. The calculated energy-ray deviation for longitudinal waves is indicated in Fig. 2. In this plot, maximum convergence with no beam deviation occurs where $\Delta = 0$ with negative slope, i.e., at $\sim$ 50°. This implies that a beam propagating along this direction would not deviate from the straight path and would experience a focusing effect. The experimental results indicating maximum beam intensity along a direction $\sim$ 45° to the columnar grains are consistent with the prediction that the beam will tend to focus when traveling along a direction of maximum velocity.

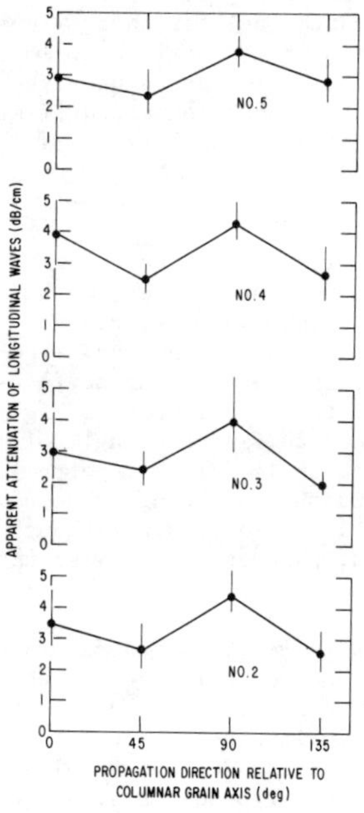

Fig. 8 - Apparent attenuation vs propagation direction relative to long axis of grains for longitudinal waves at 2.25 MHz. Results are for the four weld samples shown in Fig. 5.

The results for shear waves are not as clear. Maximum shear velocities occur for waves propagating (a) parallel to the grain axes and (b) perpendicular to the grain axes, with polarization parallel to them. The minimum velocity is along a direction 45° to the columnar grains with polarization in the (110) plane. The current experimental observation is that shear waves propagating across the columnar grains with polarization parallel to the grains have at least 3 dB/cm less attenuation than all other signals, including those from waves propagating at 45° to the grains. (This is qualitatively consistent with the theoretical predictions.) For shear waves traveling parallel to the grain axis, one observes a difference in attenuation for various polarizations which is not predicted. This may be due to variation in mode-conversion characteristics (the grains are not exactly parallel to the wave-propagation direction) or to a grain structure which is not as symmetric about the z axis as assumed. For shear waves traveling perpendicular to the grain axis, grain-boundary scattering may be an important factor, since the acoustic-impedance mismatch across the grains varies considerably for different polarizations and thus different wavelengths. The mismatch is least for waves polarized along the columnar grains and greatest for waves polarized across them. In the latter case, scattering would help explain the larger-than-expected variation in attenuation with polarization for shear waves propagating across the grains. The importance of wavelength (Rayleigh scattering) for shear waves can be seen from the ultrasonic frequency spectra presented in Ref. 18. The frequency spectra vary considerably with polarization and thus with wavelength and velocity (for a given frequency), as would be expected if Rayleigh scattering were a major factor in the apparent attenuation.

Beam Deviation

In order to help establish the adequacy of the model presented earlier to predict beam deviations for shear waves, we have attempted to measure these deviations by means of two normal-incidence 2.25-MHz shear-wave transducers in a pitch-catch mode. A couplant and 1-cm^2 curved plexiglass "shoes" were used to attach the transducers to the surface of a cylindrical specimen (Fig. 9, top) cut from the weld coupon of Fig. 3. The transducers can be moved around the cylinder to find the angle of maximum signal transmission and thus the beam-deviation angle (Fig. 9, bottom). In practice, the transmitting probe was fixed and the receiving transducer moved manually until the maximum received signal was located. Although many difficulties are encountered with this method, the data do show significant beam deviations, as shown below.

The experimental results for longitudinal-wave deviations are presented in Fig. 10 along with the theoretical predictions. As the columnar grains are $\sim$ 1 mm in diameter, we can assume the beam crossed a minimum of 20 grains in traversing the specimen. At least three data points were obtained for each transmitter location indicated. The results (except for those obtained at 67° to the grain axis) were reproducible; deviations as large as 22° were observed for wave propagation at $\sim$ 22° and $\sim$ 67° to the grain axis, while only a small deviation was observed for wave propagation at $\sim$ 45° to the grain axis.

The data for shear waves were more difficult to acquire. A solid couplant (epoxy) was required; also, the deviations were generally smaller than for longitudinal waves and SH and SV shear waves were often evident at the same time in the oscilloscope trace. These modes had to be separated by time-of-flight measurements since, as shown by the wave-velocity curves of Fig. 1, either the SH or SV waves can have a higher velocity depending on the wave propagation direction relative to the columnar grain axis. The beam-deviation curves for SV waves are shown in Fig. 11 and the corresponding results for SH waves are shown in Fig. 12. A comparison of these figures shows that the angle of deviation is better defined, and the results are in closer agreement with the theoretical results, for SV than for SH waves. Deviations from predicted results are due to variations from the symmetry assumed in the calculations.

Visualization of Beam Distortion

The distortion of an ultrasonic beam as it propagates through weld metal can be visualized by use of a Beam Profile Analyzer (courtesy of Sonoscan, Inc., Bensenville, IL). Figure 13 shows the schematic arrangement for this commercially available device. A transducer and sample are placed in a water bath. The ultrasonic beam travels through the specimen and strikes a gold-plated plastic plate. A laser beam scans the gold film and, by use of an optical interferometer, monitors the intensity of the beam in the plane of the film. The sample used here was a 4-mm-thick plate, half 308 SS weld metal and half 304 SS base metal (Fig. 5c). As discussed previously, longitudinal waves are expected to converge when propagating along a direction 45° to the grain axis in SS weld metal. Figure 14 (left) shows a comparison of the same 2.25-MHz longitudinal wave traveling through the weld metal (at 45° to the grain axis) and through the virtually isotropic* base metal. The incident beam was normal to the specimen surface in both cases. The gain of the system was adjusted (only 1 dB) to make the maximum intensity of the beam the same for each half of the specimen. Despite the short path, a focusing effect can be observed for the beam passing through the weld metal.

The right side of Fig. 14 shows the splitting of the beam through refraction in the weld metal when the plate is tilted slightly (10° and 20°).

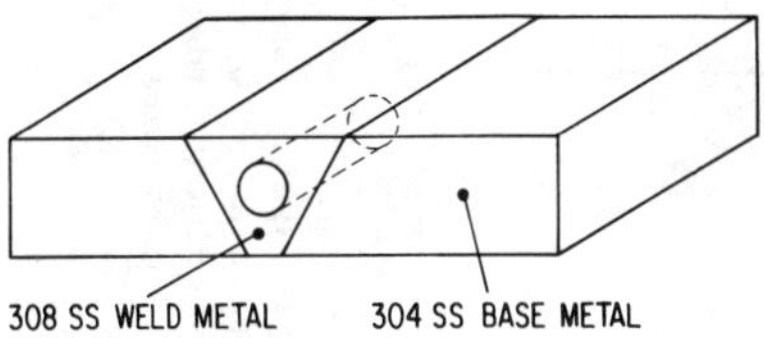

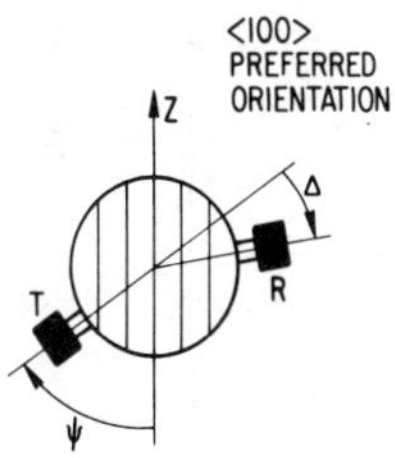

Fig. 9 - Schematic representation of system for measuring beam deviation. The cylinder used (2.5 cm long x 2 cm diameter) was cut from the weld coupon of Fig. 3. The transmitted pulse (T) is propagated at an angle ψ with respect to the columnar grain axis. The maximum received signal indicates the deviation angle Δ.

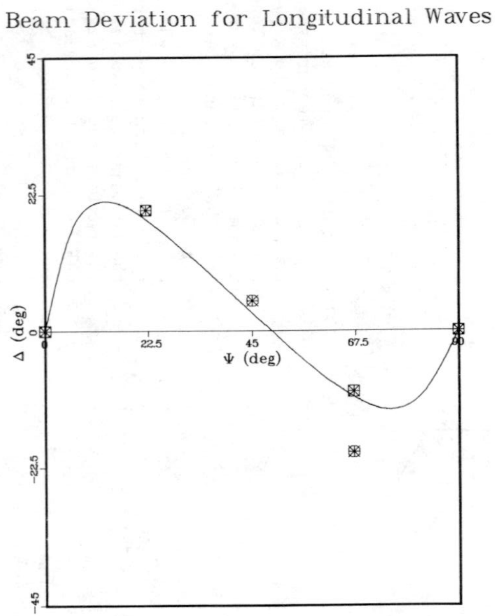

Fig. 10 - Beam deviation (Δ) vs propagation angle relative to columnar grain axis (ψ) for longitudinal waves. Experimental data (symbols) and calculated curve are shown.

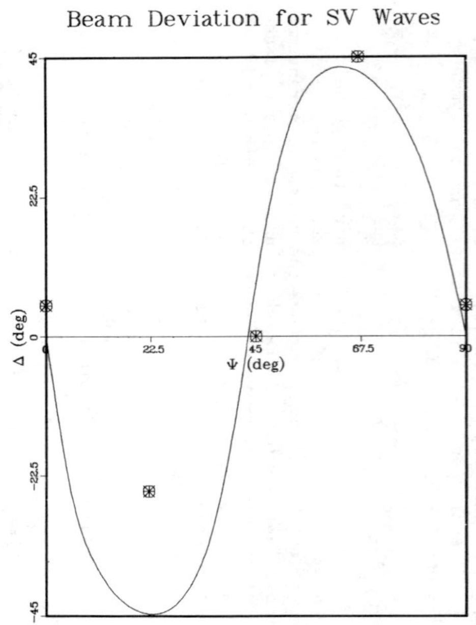

Fig. 11 - Beam deviation (Δ) vs propagation angle relative to columnar grain axis (ψ) for SV waves. Experimental data (symbols) and calculated curve are shown.

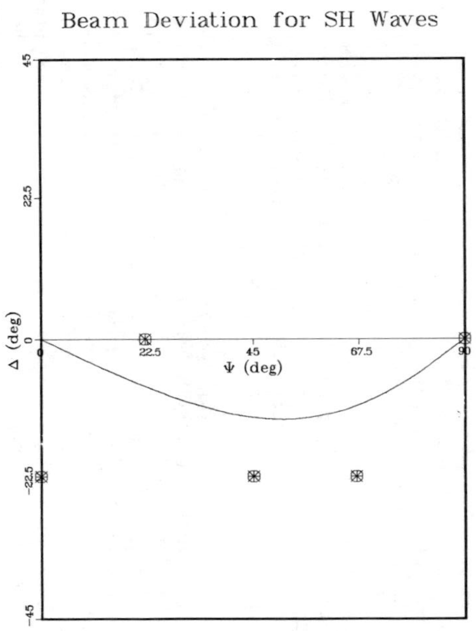

Fig. 12 - Beam deviation (Δ) vs propagation angle relative to columnar grain axis (ψ) for SH waves. Experimental data (symbols) and calculated curve are shown.

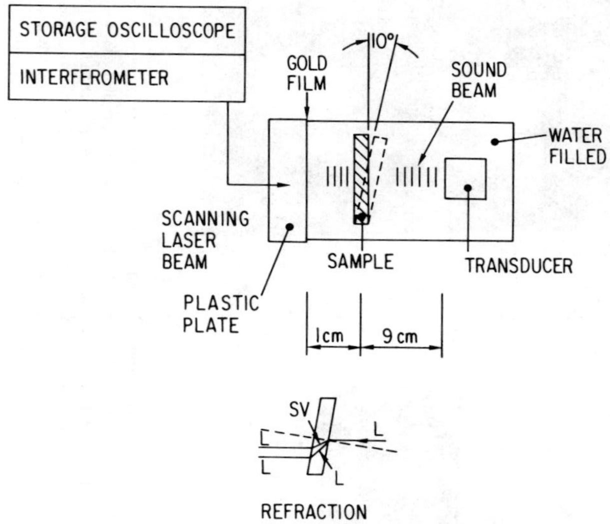

Fig. 13 - Schematic arrangement of Beam Profile Analyzer with ultrasonic beam passing through weld-metal sample.

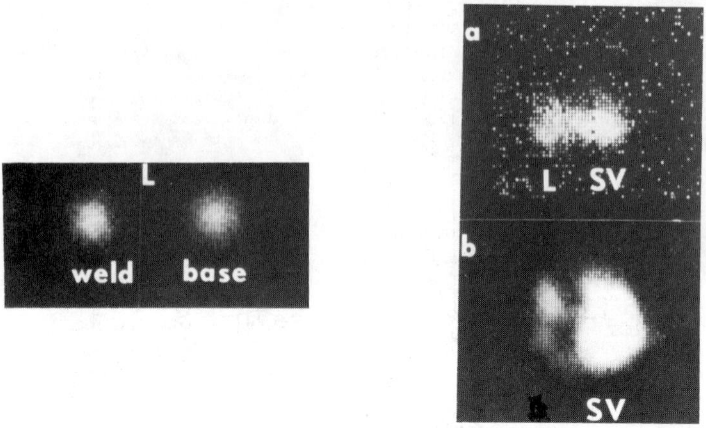

Fig. 14 - Cross sections of 2.25-MHz longitudinal-wave ultrasonic beam after passing through a 4-mm-thick sample containing 308 SS weld metal and 304 SS base metal, at 45° to the weld-metal grain axis. (Left) specimen surface normal to incident beam; (right) specimen tilted (a) 10° and (b) ∿ 20° from normal. Photo on right shows refraction of beam into longitudinal (L) and shear (SV) beams as it passes through the weld metal.

Flaw Detection

The detectability of flaws in austenitic SS weld metal is affected by wave mode (longitudinal or shear) and by variations in polarization of the shear waves used. For example, Kupperman and Reimann (18) were unable to detect a side-drilled hole (∿ 6 mm beyond the weld fusion line) in a weld-metal sample similar to that shown in Fig. 3 by conventional SV waves, but detected the same hole with SH waves propagating in a direction normal to the weld fusion line. A longitudinal-wave transducer also enabled detection of the artificial reflector. This result is consistent with those presented earlier. The ultrasonic wave propagates parallel to grains for a distance of ∿ 3 mm (the grains are initially perpendicular to the fusion line) and then at ∿ 45° to the grains for another 3 mm. In both regions, the SH waves undergo the least attenuation; attenuation of the longitudinal waves is greatest for the parallel and least for the skewed grains.

A comparison of SV shear waves with longitudinal waves leads to the conclusion that longitudinal-wave testing may be the better method for defect detection in stainless steel weld metal. However, the difference between SH shear waves and longitudinal waves is not as apparent. To determine whether these relationships are dependent upon the direction of acoustic-wave propagation and microstructure of the weld metal, as suspected, another hole was drilled in the same weld. In this case, the hole was placed in the center of the weld metal and normal-incidence waves propagating parallel to the columnar grain axis were used for detection. The radio-frequency (RF) signal obtained from the hole using SH shear waves was clearly distinguishable from the noise. No reflection from the hole was apparent when SV shear waves were used, indicating at least for this example that orthotropic symmetry cannot always be assumed for a particular propagation direction. If the sample is orthotropic, there should be no difference between SH and SV waves for the "z" direction. For longitudinal waves the signal was also distinguishable from the noise, but not as clearly as in the case of the SH shear waves. Longitudinal waves propagating parallel to the dendrites may be defocused because of beam deviation, but this propagation direction is favorable for shear waves polarized in the weld-pass direction (y).

Summary

Data on sound velocity and attenuation have been acquired for stainless steel weld-metal samples and compared with theoretical predictions and efforts of other investigators. Sound velocity has been observed to vary with propagation direction by as much as 70 percent for shear waves and 14 percent for longitudinal waves.

A simple model that describes the propagation of ultrasonic waves in Type 308 stainless steel weld metal is generally but not always consistent with sound-velocity data and apparent attenuation for both longitudinal and shear waves. The model assumes that the weld consists of a collection of long single-crystal grains with their long axes in the <100> crystallographic direction. These grains are arranged parallel to each other with a random orientation about the long axis [that is, the axis has orthotropic symmetry (20)]. Velocity as well as energy-ray deviation curves have been calculated for stainless steel single-crystal and coarse-grain structures. These results compare favorably with experimental results. Velocity varied by only a few percent between samples welded under similar conditions. Focusing of longitudinal waves propagating at $\sim$ 45° to the columnar grain axis was visualized by use of a commercially available ultrasonic beam profiling instrument. Inspections of welds with ultrasonic waves at an angle of 45° to the grain axes are more successful with SH shear or longitudinal waves than with conventional SV shear waves. For inspections with ultrasonic waves oriented parallel or perpendicular to the grains, shear waves of the proper polarization are more effective than longitudinal waves.

Acknowledgments

We wish to thank H.W. Knott for the x-ray diffraction work, N.F. Fiore for the stainless steel single crystals, G.M. Dragel for the metallography, and D. Yuhas for help in obtaining the beam profiles.

References

(1) M. P. J. Musgrave, "On the propagation of elastic waves in aeolotropic media. I: General principles; II: Media of hexagonal symmetry," Proc. R. Soc. London, Ser. A, 226 (1964) pp. 339 and 356.

(2) E. Holmes, "Ultrasonic behavior in austenitic stainless steel," Appl. Mater. Res. (July 1963) p. 181.

(3) E. Holmes and D. Beasely, "The influence of microstructure in the ultrasonic examination of stainless steel welds," J. Iron Steel Inst. London, 200 (1962) p. 283.

(4) J. P. Pelseneer and G. Louis, "Ultrasonic testing of austenitic steel castings and welds," Br. J. Non-Destr. Test., 16 (1974) p. 107.

(5) S. Serabian, "Influence of attenuation upon the weld interrogation distance amplitude curve," Mater. Eval., 32(12) (1976) p. 265.

(6) J. J. Mech and T. E. Michaels, "Development of ultrasonic examination methods for austenitic stainless steel weld inspection," Mater. Eval., 35(7) (1977) p. 81.
(7) H. Yoneyama, S. Shibata, and M. Kishigami, "Ultrasonic testing of austenitic stainless steel welds," NDT Int., 11 (1978) p. 3.
(8) B. L. Baikie, A. R. Wagg, M. J. Whittle, and D. Yapp, "Ultrasonic inspection of austenitic welds," J. Br. Nucl. Energy Soc., 15(3) (1976) p. 257.
(9) J. R. Tomlinson, A. R. Wagg, and M. J. Whittle, "Ultrasonic inspection of austenitic welds," p. 64 in Nondestructive Evaluation in the Nuclear Industry, Conf. Proc., R. Natesh, ed.; ASM, Metals Park, OH, 1979.
(10) L. Adler, K. V. Cook, W. A. Simpson, D. K. Lewis, and D. W. Fitting, "Ultrasonic flaw detection and characterization in structural materials by spectral analysis," Oak Ridge Nat. Lab. Rep. ORNL/TM-6456, Oct. 1978.
(11) E. Neumann, M. Romber, T. Just, E. Nable, K. Mattheis, and E. Mundry, "Development and improvements of ultrasonic testing techniques for austenitic nuclear components," paper presented at Second Int. Conf. on Nondestructive Evaluation in the Nuclear Industry, Salt Lake City, UT, Feb. 13-15, 1978.
(12) S. Matsumoto and K. Kumura, "The relationship between grain size and ultrasonic attenuation coefficient in austenitic stainless steel and iron," Trans. Natl. Res. Inst. for Metals, 14(4) (1972) pp. 155-164.
(13) M. P. J. Musgrave and G. F. Miller, "The propagation of elastic waves in aeolotropic media. III: Media of cubic symmetry," Proc. R. Soc. London, Ser. A., 235 (1956) p. 352.
(14) D. S. Kupperman and K. J. Reimann, "Microstructural effects and signal-enhancement techniques in ultrasonic examination of stainless steel," Argonne Nat. Lab. Rep. ANL-76-115, Sept. 1976.
(15) M. P. J. Musgrave, "The propagation of elastic waves in crystals and other anisotropic media," Rep. Prog. Phys., XIII (1959) p. 74.
(16) H. Pursey and H. L. Cox, "The correction of elasticity measurements on slightly anisotropic materials," Philos. Mag. (Ser. 7), 45(362) (1954) p. 295.
(17) I.N. Ermolov and B.P. Pilin, "Ultrasonic inspection with coarse grain anisotropic structures," NDT Intl., 9(6) (1976) p. 275.
(18) D. S. Kupperman and K. J. Reimann, "Effect of shear wave polarization on defect detection in stainless steel weld metal," Ultrasonics, 16(1) (1978) p. 21.
(19) D. S. Kupperman, K. J. Reimann, and N. F. Fiore, "Role of microstructure in ultrasonic inspectability of austenitic stainless steel welds," Mater. Eval., 36(5) (1978) p. 70.
(20) D. S. Kupperman and K.J. Reimann, "Ultrasonic wave propagation and anisotropy in austenitic stainless steel weld metal," IEEE Trans. Sonics Ultrason. SU-27(1) (1980) p. 7.
(21) E. P. Papadakis, "New, compact instrument for pulse-echo-overlap measurements of ultrasonic wave transit times," Rev. Sci. Instr., 47(7) (1976) pp. 806-813.
(22) M. C. Mangabek and N. F. Fiore, "Orientation dependence of dislocation damping and elastic constants in Fe-18Cr-Ni single crystals," Trans. AIME, 252 (1968) p. 2363.
(23) B. R. Dewey, L. Adler, R. T. King, and K. V. Cook, "Application of anisotropic elasticity to centrifugally cast piping," Oak Ridge Nat. Lab. Rep. ONRL/TM-5994, Oct. 1977.
(24) L. C. Lynnworth, "Ultrasonic probes using shear wave crystals -- principles," Mater. Eval., 25 (1967) p. 12.

BOUNDARIES BETWEEN ISOTROPIC AND ANISOTROPIC SOLIDS AND THEIR EFFECT ON QUANTITATIVE NONDESTRUCTIVE EVALUATION*

W. A. Simpson
Metals and Ceramics Division
Oak Ridge National Laboratory
Oak Ridge, Tennessee 37830

L. Adler
Department of Welding Engineering
The Ohio State University
Columbus, Ohio 43220

T. K. Bolland
Department of Physics
The University of Tennessee
Knoxville, Tennessee 37916

Theories seeking to describe the scattering of ultrasound from discontinuities generally assume the host material to be isotropic. However, in several cases of importance (e.g., the scattering from discontinuities in austenitic welds) the host material exhibits pronounced anisotropy. In addition, for such a system the incident energy must often propagate across an interface between an isotropic parent material and the anisotropic host (weld) material. A treatment of the propagation of elastic waves in such a binary system is presented as well as some results concerning the quantitative effect that propagation across such an interface has on the characterization of discontinuities.

---

*Research sponsored by the Office of Basic Energy Sciences, U.S. Department of Energy, under contract W-7405-eng-26 with the Union Carbide Corporation.

In quantitative NDE (nondestructive evaluation), as in virtually all areas of physics, certain simplifying assumptions must be made before the problems become mathematically tractable. For example, in studying the propagation of elastic waves in solids or the scattering of such waves from discontinuities embedded in solids, it is generally assumed that the host material is linear, homogeneous, and isotropic. While these assumptions were often adequate in the past, the use of austenitic material in welds and the generally more stringent requirements placed on NDE techniques have prevented this simplistic approach. For example, Fig. 1 depicts a common NDE situation in which a discontinuity in a weld must be interrogated from the base metal (i.e., through the solid-solid interface). If one assumes that acoustic properties for both materials are appropriate to the respective base metals, then subsequent metallographic examination of the specimen can, for austenitic welds and certain other materials, lead to finding no discontinuity at or even near the predicted location. The reason, of course, is that the assumption of isotropy for the weld metal does not hold in this instance. Not only is the weld velocity different from that obtained in a corresponding sample of base material (thus leading to range errors), but skewing of the beam on entering the weld produces angulation errors as well.

When it was first apparent that a possible anisotropy in the weld itself could be the source of many of the observed difficulties, a study was begun to more fully characterize the weld metal (1). This study involved measuring the ultrasonic parameters of attenuation and velocity as a function of direction within the weld. Only the velocity results will be considered here. Several large welds were available from which samples composed entirely of weld metal could be cut. Figure 2 shows how these samples were chosen. The cylindrical specimens are not relevant to the current discussion and will not be considered. Numbering of axes was arbitrary; for reference, the 3-axis is the <100> crystallographic direction.

The cubical specimens were next prepared by grinding the surfaces and adding small flats to the edges, which were marked to identify the axes (Fig. 3). These flats also identified additional directions within the cubes along which velocity was measured further.

In addition to the ultrasonic studies being performed, several of the samples were chosen for concurrent metallographic analysis. Figures 4–6 depict the results in the 1-3, 2-3, and 1-2 planes, respectively. Note that traditional ultrasonic immersion testing of welds is performed in the 1-3 plane.

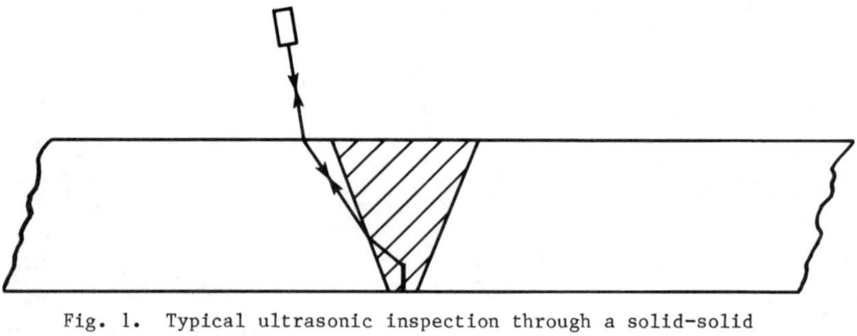

Fig. 1. Typical ultrasonic inspection through a solid-solid interface.

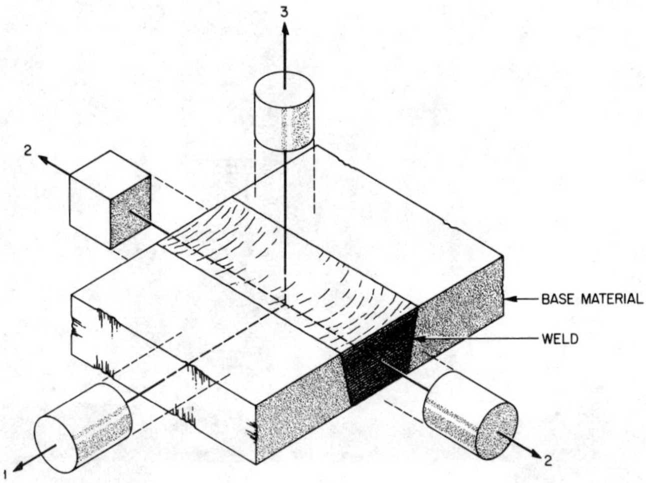

Fig. 2. Coordinate system for welding showing generation of experimental samples.

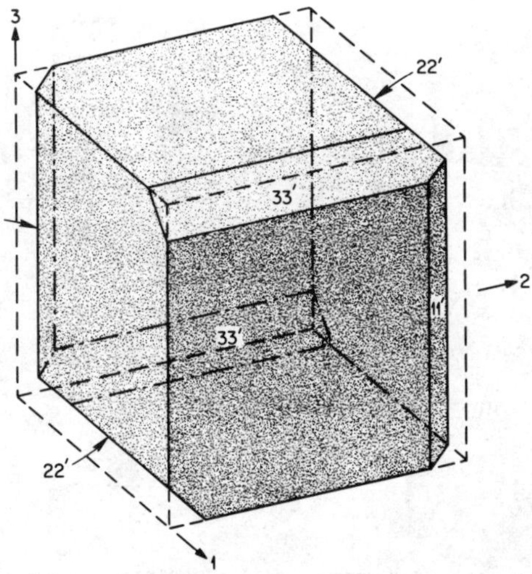

Fig. 3. Preparation of weld specimen for velocity measurements.

Fig. 4. Metallographic section of weld in the 1-3 plane.

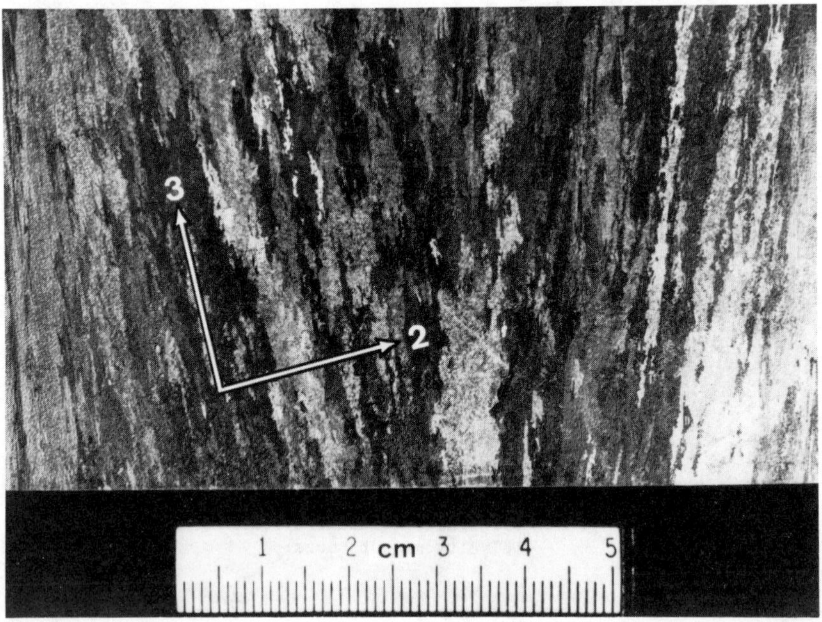

Fig. 5. Metallographic section of weld in the 2-3 plane.

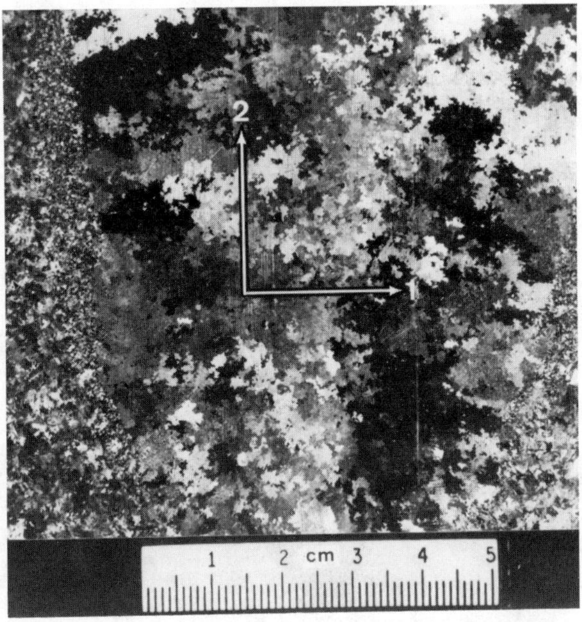

Fig. 6. Metallographic section of weld in the 1-2 plane.

As indicated in Figs. 4 and 5, the most obvious feature of the weld is a columnar grain structure strongly oriented along the 3-axis. No great imagination is required to anticipate that the acoustic properties of the weld will be strongly direction dependent in both the 1-3 and 2-3 planes. Figure 6 suggests, however, that in the 1-2 plane the grain structure, and presumably the acoustic properties, will be much more nearly isotropic. From these results and from preliminary velocity measurements, we proposed an orthorhombic or a transversely isotropic (hexagonal) model for the weld (1).

Figure 7 shows through-transmission velocity being measured on one of the cubical specimens. Longitudinal wave velocities and transverse wave velocities having various polarizations were determined for propagation along the three orthogonal axes. Since one can determine five elastic constants at most from such measurements and nine are required for orthorhombic symmetry, some of the samples were cut further to allow waves to be propagated at selected angles with respect to the axes. Figure 8 shows the preparation of one of the samples.

The velocity measurements in the 1-3 plane are given in Fig. 9. Recall that conventional ultrasonic testing is usually performed in this plane. The middle panel depicts the results for the familiar SV-wave, (i.e., the mode-converted shear wave that would be generated in immersion testing). The right panel gives the values for an SH-wave, which would not be generated during immersion testing. Note that the maximum change in the SV-wave velocity occurs at a 45° angle with respect to the 3-axis, which is the traditional value for angle-beam shear-wave inspection of welds. Since this variation was as large as 40% for some of the samples,

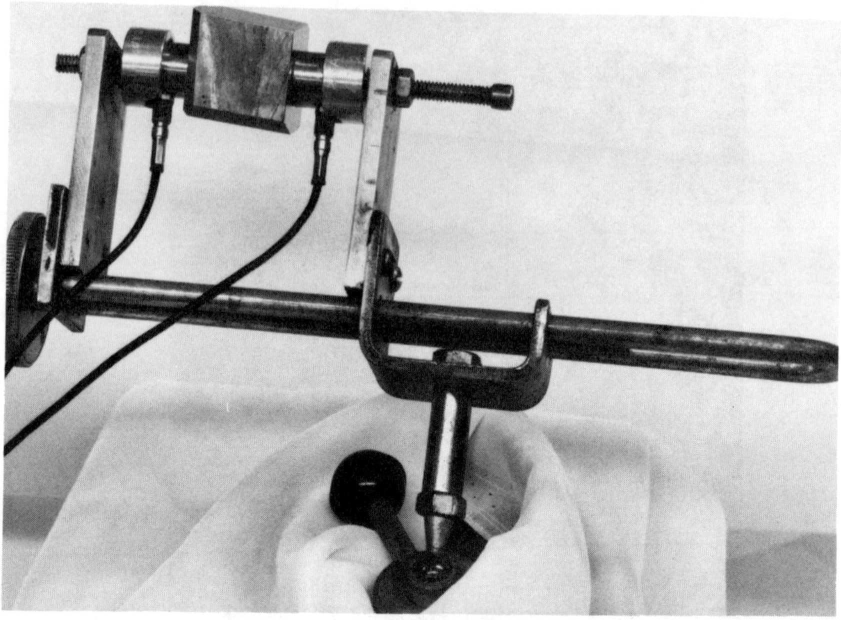

Fig. 7. Fixturing for through-transmission velocity measurements.

failure to account for the effects of anisotropy can obviously lead to gross errors in both locating and sizing discontinuities in austenitic welds. Note also that the SH-wave is considerably less affected by the weld anisotropy. The theoretical curves in Fig. 9 are based on an orthorhombic model.

The results in the 2-3 and 1-2 planes are given in Figs. 10 and 11, respectively. In the latter figure the data are shown for an Inconel 82 rather than a stainless steel weldment, but the results are typical for many austenitic materials. As anticipated, the results in the 2-3 plane are quite similar to those in the 1-3 plane; however, Fig. 11 would appear to show that the 1-2 plane was one of isotropy. In fact, a small but measurable variation in this plane is not easily discernible in the figure.

The velocity measurements were in good agreement with the assumed orthorhombic model, but the large variation in the SV-wave velocity (as large as 40% for some of the samples) indicated that neglecting the anisotropy of the weld material would lead to severe errors in calculating the position and size of defects from the ultrasonic scattering data. It thus became obvious that wave propagation in such media would have to be considered, particularly the effect of propagation across an interface between isotropic and anisotropic media.

Musgrave (2) studied elastic wave propagation in anisotropic media. Our approach was similar to his, except we considered a single Fourier component (by transforming the spatial dependence of the incident wave along the interface) since this approach lends itself well to extension

Fig. 8. Progressive cuts in weld metal sample for velocity measurements.

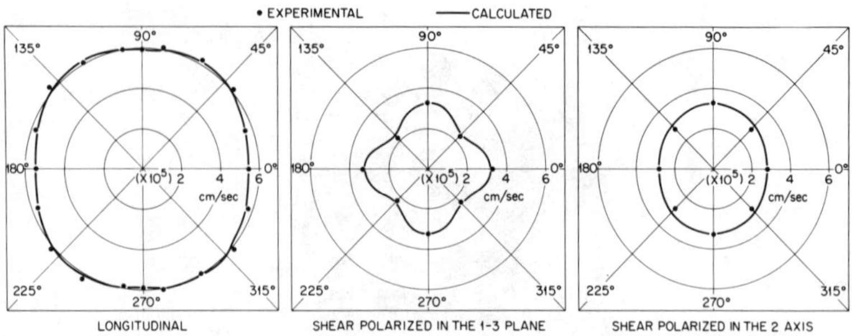

Fig. 9. Ultrasonic velocities in the 1-3 plane.

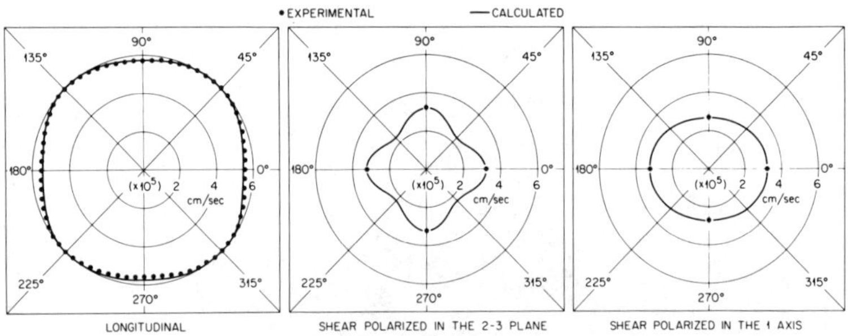

Fig. 10. Ultrasonic velocities in the 2-3 plane.

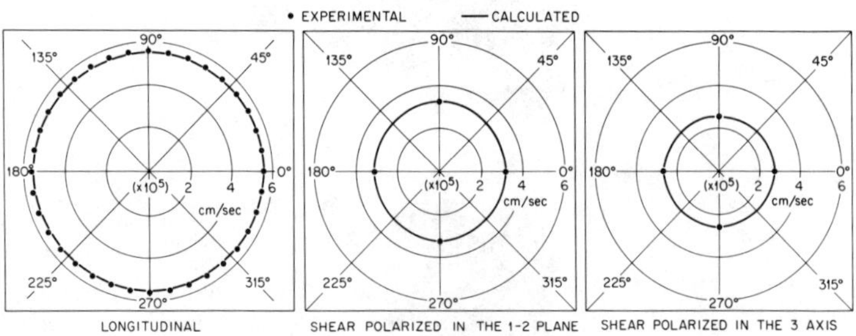

Fig. 11. Ultrasonic velocities in the 1-2 plane.

to finite-beam effects, a future consideration in our study. We began with the usual linearity assumptions in which the stress and the strain are related through the elastic stiffness matrix, C:

$$\{\sigma\} = [S]^{-1} \{\varepsilon\} = [C] \{\varepsilon\} , \tag{1}$$

where $\{\sigma\}$ and $\{\varepsilon\}$ are the stresses and strains, respectively, and [C] is the elastic stiffness matrix. Since the elastic compliance matrix, [S], is often used in the literature, it is included for completeness and is simply the inverse of the stiffness matrix. A material exhibiting orthorhombic symmetry has nine independent constants so that the elastic stiffness matrix becomes

$$[C] = \begin{pmatrix} C_{11} & C_{12} & C_{13} & 0 & 0 & 0 \\ C_{12} & C_{22} & C_{23} & 0 & 0 & 0 \\ C_{13} & C_{23} & C_{33} & 0 & 0 & 0 \\ 0 & 0 & 0 & C_{44} & 0 & 0 \\ 0 & 0 & 0 & 0 & C_{55} & 0 \\ 0 & 0 & 0 & 0 & 0 & C_{66} \end{pmatrix} . \tag{2}$$

Love (3) established this result, but the derivation proceeds through the artifice of the strain-energy function. In addition, the above result is true only for a particular choice of axes — the crystallographic axes in orthorhombic material. For arbitrary axes the stiffness matrix can be considerably more complex. Since the matrix for noncrystallographic axes may be required in the present case, the form and transformation properties for arbitrary axes were studied in somewhat more detail.

Since the elastic stiffness matrix is a fourth-rank tensor, the transformation equations for various axes become

$$C'^{kl}_{ij} = \frac{\partial Xm}{\partial X'_i} \frac{\partial Xn}{\partial X'_j} \frac{\partial X'k}{\partial X_o} \frac{\partial X'l}{\partial X_p} C^{op}_{mn} . \tag{3}$$

For all choices of axes in the present case, there is no distinction between covariant and contravariant components; thus the superscripts may be lowered. From these equations, which are well suited for computer evaluation, the elastic constants may be obtained for any coordinate system. In addition, by requiring that

$$C'_{ijkl} = C_{ijkl} \tag{4}$$

for a given rotational symmetry operation, constraints on the elastic constants imposed by the particular operation may be derived. By evaluating all possible rotations and passing to the Voigt notation, the following symmetry constraints may easily be derived:

I

If $\theta = \pi$:

$$C_{14} = C_{15} = C_{24} = C_{25} = C_{34} = C_{35} = C_{46} = C_{56} = 0 . \tag{5}$$

II

If $\theta = \pi/2$:

$$C_{14} = C_{15} = C_{24} = C_{25} = C_{34} = C_{35} = C_{36} = C_{45} = C_{46} = C_{56} = 0;$$

$$C_{11} = C_{22};\ C_{13} = C_{23};\ C_{16} = -C_{26};\ C_{44} = C_{55}\ . \tag{6}$$

III

If $\theta = 2\pi/3$:

$$C_{16} = C_{26} = C_{34} = C_{35} = C_{36} = C_{45} = 0;$$

$$C_{11} = C_{22};\ C_{13} = C_{23};\ C_{44} = C_{55};\ C_{66} = 1/2(C_{11} - C_{12});$$

$$C_{14} = C_{56} = -C_{24};\ C_{15} = -C_{25} = -C_{46}\ . \tag{7}$$

IV

For any other angle:

$$C_{14} = C_{15} = C_{24} = C_{25} = C_{16} = C_{26} = C_{34}$$

$$= C_{35} = C_{36} = C_{45} = C_{46} = C_{56} = 0;$$

$$C_{11} = C_{22};\ C_{13} = C_{23};\ C_{44} = C_{55};\ C_{66} = 1/2(C_{11} - C_{12})\ . \tag{8}$$

Additional constraints may be established for reflection symmetry.

When these constraints are applied to a material exhibiting orthorhombic symmetry, the elastic stiffness matrix has the form shown in Eq. (2), and the transformation equations give these constants in any desired coordinate system. Substituting these results in the equation of motion then yields a determinantal equation for the velocities, the Christoffel relation, shown in Eq. (9) below:

$$\begin{vmatrix} \lambda_{11} - \rho V^2 & \lambda_{12} & \lambda_{13} \\ \lambda_{12} & \lambda_{22} - \rho V^2 & \lambda_{23} \\ \lambda_{13} & \lambda_{23} & \lambda_{33} - \rho V^2 \end{vmatrix} = 0\ , \tag{9}$$

where

$$\lambda_{11} = l^2C_{11} + m^2C_{66} + n^2C_{55} + 2mnC_{56} + 2nlC_{15} + 2lmC_{16}\ ,$$

$$\lambda_{12} = l^2C_{16} + m^2C_{26} + n^2C_{45} + mn(C_{46} + C_{25}) + nl(C_{14} + C_{56}) + lm(C_{12} + C_{66})\ ,$$

$$\lambda_{13} = l^2C_{15} + m^2C_{46} + n^2C_{35} + mn(C_{45} + C_{36}) + nl(C_{13} + C_{55}) + lm(C_{14} + C_{56})\ ,$$

$$\lambda_{22} = l^2C_{66} + m^2C_{22} + n^2C_{44} + 2mnC_{24} + 2nlC_{46} + 2lmC_{26}\ ,$$

$$\lambda_{23} = l^2C_{56} + m^2C_{24} + n^2C_{34} + mn(C_{44} + C_{23}) + nl(C_{36} + C_{45}) + lm(C_{25} + C_{46})\ ,$$

$$\lambda_{33} = l^2C_{55} + m^2C_{44} + n^2C_{33} + 2mnC_{34} + 2nlC_{35} + 2lmC_{45}\ ,$$

Here l, m, and n are the direction cosines of the wave normal in the anisotropic material. The salient features of this result are (1) since the equation is cubic in velocity, generally three waves are propagating in the anisotropic material rather than two, as in an isotropic solid; (2) the velocity of propagation depends on the direction; and (3) the particle displacement will, in general, be neither parallel nor perpendicular to the wave normal. This last characteristic is profoundly important in that it affects both the apparent amplitude and apparent direction of propagation in the anisotropic medium. It also gives rise to the names "quasi-longitudinal" and "quasi-transverse" for the waves propagating in the material.

Analytical solutions were obtained for infinite elastic waves propagating across an interface between an isotropic and an arbitrary anisotropic material. The crystallographic axes of the anisotropic component may have any orientation with respect to the interface between the two media. A diffusion-bonded sample was obtained for concurrent experimental studies of how such propagation affects the scattering of ultrasound from discontinuities. Figure 12 shows the sample, fabricated by Rockwell Science Center, in a goniometer that was also designed and constructed there. The sample consists of two half-cylinders, one titanium and the other type 1020 steel, diffusion-bonded along their common plane parallel to the axis. This arrangement provides an excellent, uniform metallurgical bond so that propagation and scattering can be studied without interference from irregular bonds.

Fig. 12. Steel-titanium diffusion-bonded sample.

Future studies will include the effects of such interfaces. While we are indeed fortunate to have obtained such a sample, cost considerations and constraints on the selection of materials that can be diffusion-bonded by established procedures dictated the current choices, neither of which exhibits macroscopic anisotropy. Thus, while this sample is quite useful for studying scattering from discontinuities across a solid-solid interface, additional samples will be required to provide anisotropy.

The reflected and transmitted ultrasonic wave amplitudes are being measured experimentally on the titanium-steel sample to characterize the materials for scattering studies. Figure 13 shows the results obtained for the longitudinal components. Measurements are not complete at this time on the shear components.

Calculations have been made on a steel-graphite complex, assuming hexagonal symmetry for the graphite. The crystallographic axis was assumed perpendicular to the interface between the components, and the results for the quasi-longitudinal and one of the quasi-transverse waves are given in Fig. 14. The angle indicated is that between the wave normal and the normal to the interface. The figure does not show that, for most of the angular range included, the energy flow is not along the wave normal, and this must be taken into account when comparing theory and experiment.

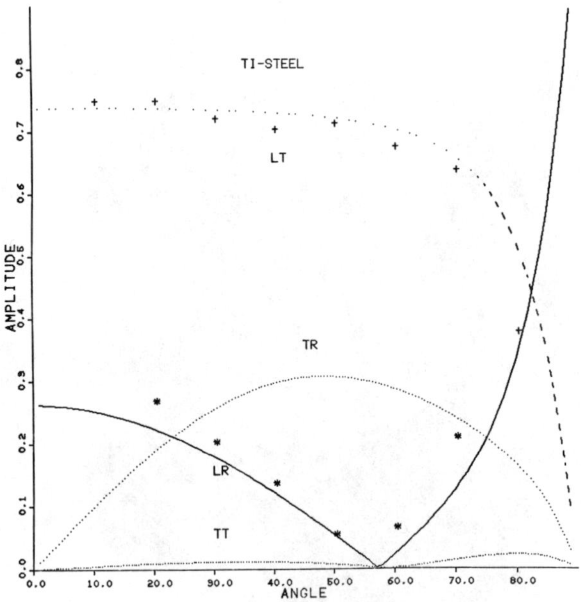

Fig. 13. Theoretical (curves) and experimental (*,+) results on ultrasonic amplitudes in a titanium-steel diffusion-bonded sample.

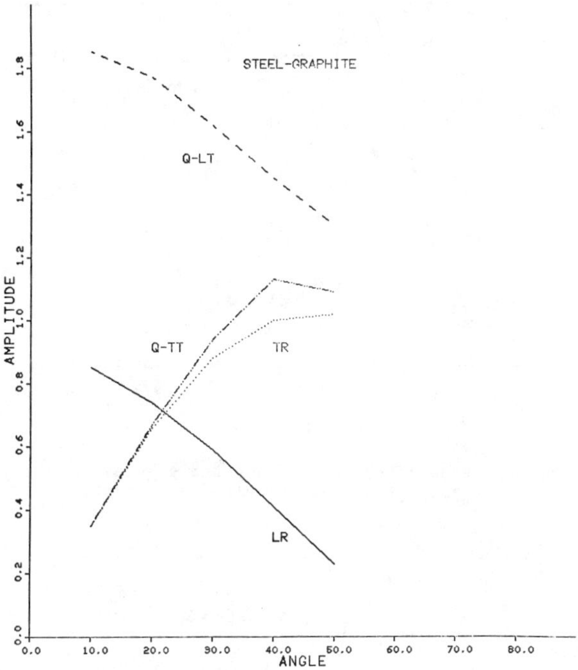

Fig. 14. Theoretical ultrasonic wave amplitudes for steel-graphite interface.

Although a solid-solid interface sample containing an anisotropic component is not yet available for study, we have several specimens that are entirely anisotropic and that contain machined discontinuities. When tested by immersion methods, the system contains a liquid-solid interface, which exhibits most of the problems attendant to the solid-solid interface. One such sample is a 72-mm-diam by 38-mm-long cylinder cut entirely from a large type 308 stainless steel weld. The cylinder axis is along the columnar grain axis (3-axis or [100] direction). A flat-bottom hole 6 mm in diameter was placed in the center of the specimen with the end of the hole lying in the 1-2 plane of near-anisotropy. A longitudinal wave was incident normally on the hole (i.e., the wave propagated along the 3-axis), and the diffracted longitudinal energy was measured at several off-axis angles. For each angle, predicted frequency-domain characteristics were calculated by using the elastic-analog Keller theory (4). These calculations were corrected for the anisotropy of the sample by using the computer programs described earlier. Figures 15 and 16 show the comparison between theory and experiment for scattering angles of 15 and 36°, respectively. Note that the agreement in amplitude as well as frequency is quite good. These corrected results provide sizing estimates whose errors are comparable to those obtainable in isotropic media.

Examples of elastic wave scattering from discontinuities embedded in an isotropic host medium have been given, and data was presented to support the contention that failure to account for this anisotropy can

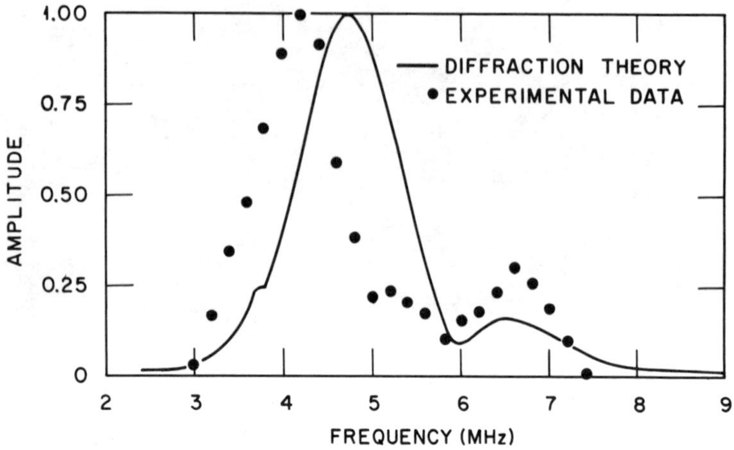

Fig. 15. Spectrum of scattered longitudinal wave in stainless steel. Scattering angle, 15°.

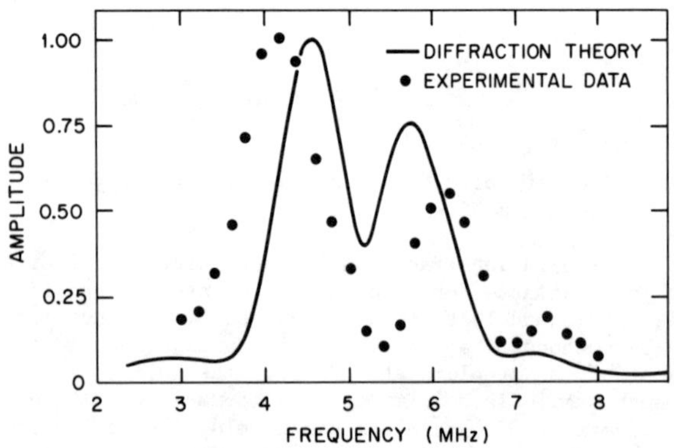

Fig. 16. Spectrum of scattered longitudinal wave in stainless steel. Scattering angle, 36°.

lead to gross errors in locating and sizing these flaws. Preliminary results on a limited number of samples have indicated that proper correction for the anisotropy of the host medium can yield sizing errors comparable to those obtainable in isotropic media.

Acknowledgments

The authors would like to extend their sincere thanks to J. C. Lewis and D. W. Fitting for their aid in performing the measurements, to W. E. Deeds and J. H. Smith for reviewing the manuscript, to B. G. Ashdown for editing, and to J. L. Bishop and S. G. Frykman for preparing the article for publication.

References

1. L. Adler and K. V. Cook, Liquid Metal Fast Breeder Reactor Materials Development Program Quart. Prog. Rep. June 30, 1976, ORNL-5218, pp. 200–202.

2. M.J.P. Musgrave, "On the Propagation of Elastic Waves in Aeolotropic Media, I. General Principles," Proceedings of the Royal Society of London, A226 (1954) pp. 339–55.

3. A.E.H. Love, A Treatise on the Metallurgical Theory of Elasticity, Dover, New York, 1944.

4. L. Adler and D. K. Lewis, "Diffraction Model for Ultrasonic Frequency Analysis and Flaw Characterization," Materials Evaluation, 35 (1) (1977) pp. 51–56.

QUANTITATIVE MODELS FOR EDDY CURRENT ANALYSIS OF METAL SOLIDIFICATION

R. M. Siegfried, J. P. Wallace, G. A. Boser, D. C. Kunerth
Department of Chemical Engineering and Materials Science
University of Minnesota
Minneapolis, Minnesota 55455

There are physical and analytic limitations to eddy current measurements for both ferrous and non-ferrous casting. Typical physical constraints can involve mold shielding of the melt, dimensional changes of the mold, and thermal transients which are common. These contributions can be estimated by modeling the effects individually in the solution of the electromagnetic boundary value problems. The basis for analysis is that the eddy current signals measured are sensitive to electrical conductivity changes within the casting. The major analytic limitation is the rapid and accurate data analysis which provides information in a form suitable for process control or evaluation. To improve the analysis of multifrequency eddy current data a transform technique is introduced to qualitatively estimate internal structural changes and their accurate spatial positions. This reduction of raw eddy current data is accomplished without resorting to data comparison of a modeled solution of the electromagnetic boundary value problem.

This work was supported by a grant from the Magnetic Analysis Corporation, Mt. Vernon, New York.

Introduction

Eddy current data obtained during a test is dependent upon the geometry of the specimen, the electrical conductivity through the sample, and the placement of the detector and transmitter of the eddy current signal.[1] The detected signal contains information on the spatial conductivity variation between the mold and sample. The difficulties in using eddy current data to monitor ingot or continuous casting can be separated into two categories. First, one must take high quality data in a difficult environment, and secondly, this data must be both reducible and analytically descriptive of the physical events taking place within the mold of the caster. For single frequency eddy current testing, these experimental problems can be mastered by the use of coils that produce an electromagnetic field that can be mathematically modeled. However, this approach constrains the number of dimensions in the analysis to a minimum, preferably one. The experimental problem of single frequency measurement has been mastered, and reduced to laboratory practice (2). The more difficult problem is handling the large quantity of information that results and its interpretation. Our initial approach was to model the solidifying metal or alloy by calculating the eddy current signal from the electromagnetic boundary value problem for a set of probable physical situations (3). The results thus obtained can be compared to experimental results and this allows equipment to be easily calibrated, by the insertion of samples of known geometries and conductivities. This type of calibration, which allows a scaling of experimental response to calibrated response, is particularly useful in estimating the equipments ability to resolve structures or defects whose responses can be modeled. This method of analysis is adequate where one has a limited number of events that must be identified.

The solidification phenomena that we have been able to detect in the laboratory using eddy current measurements are listed in Table I. The detection of the direction, extent and rate of solidification has been identified in both lead alloys and aluminum. This is not a difficult measurement if the signal is not attenuated beyond detection by thick, highly-conductive molds. Steels have been the most difficult systems in which to detect solidification because of the small conductivity difference between the liquid and solid states. Subsequently, solidification in continuously cast steels can be more easily followed below the mold than by monitoring eddy current signals through the casting mold. The passage of nonconducting inclusions such as gas bubbles have been monitored both when gas has been fed through a melt, and when gas is rejected from the melt during solidification. Their signals are characteristic of transients moving rapidly past the detector. When trying to monitor solidification through a highly conductive mold, the mold becomes a strong signal source due largely to thermal transients which may occur within the mold. In well-chilled molds, these transients are of short duration, and may be due either to liquid metal bleeding through a break in the ingot's frozen skin or to physical deformation of the mold. Proper data analysis will allow these three events to be distinguished. Ferromagnetic transitions provide very strong signals, as does the cooling of a carbon steel. Such a ferromagnetic transformation will dominate all other signals. In the 18-8 stainless steels, ferromagnetic transformations can be detected in the small quantities of retained δ-ferrite present. The segregation of precipitates in the liquid has been detected in lead as well as solid state precipitation of carbide in 18-8 stainless steels. The only requirement for detecting these events is that the change in the sampled volume's mean conductivity be sufficient to produce a signal greater than the mean detected noise level. Quantitative estimates of the detection thresholds can be made for specific physical events, and in practice, they agree well with experimental work.

Table I

Experimental Measurement With Eddy Currents.

The following events have been experimentally monitored and analytically identified.

(1) Solidification Morphology

(a) Position of the plane front and the direction of its motion.
(b) Extent of Dendritic Solidification and dimensions of mushy zone.
(c) Total volume fraction solidified.

(2) Solidification Rates

(3) Ingot-Mold Separation

(4) Formation of Inclusions

(a) Gas evolving upon solidification
(b) Oxides
(c) Precipitates in the liquid melt.

(5) Post-Solidification Phenomena

(a) Carbide Precipitation (in 18-8 stainless steels)
(b) Magnetic Transformations

(6) Principle Thermal Gradients

Table II

Resistivities of Some Mold Materials at 40°C and Some Solid and Liquid Metals at their M. P.

(ρ in $\mu\Omega$cm; $d\rho/dT$ in $\mu\Omega$cm $^{o}K^{-1}$)

Mold	ρ	$d\rho/dT$	Melt	M.P. (^{o}K)	ρSOL	ρLIQ
Cu (5)	1.8027	3.93×10^{-3}	Steel (Fe)(8)	1808.2	127.5	139
Al (6)	2.902	3.9×10^{-3}	Al (6)	933.2	10.95	24.2
			Pb (9)	600.5	54.76 (320°C)	96.7 (330°C)
C(graphite) (7)	950	4×10^{-7}	Cu (5)	1356.2	11.5	
			Si (10,11)	1693	1700	81
			Ge (10,11)	1209	800	63

Table III

Ratio of Response to 1°C Temperature Rise in Mold to Total Solidification Response.

Mold	Melt	Ratio
Cu	Pb	1.10×10^{-2}
Al	Al	8.79×10^{-3}
Cu	Steel	2.82×10^{-1}
C	Cu	9.57×10^{-2}

To demonstrate the types of responses one can model, we will consider four examples; steel in a copper mold, aluminum in an aluminum mold, copper in a graphite mold, and lead in an aluminum mold. The thermal and electrical properties of these components are listed in Table II. The ingot diameter will be held constant at 0.20 meters, so that the calculated eddy current responses may be compared. The solidification responses for the semiconductors silicon and germanium are also calculated, assuming the absence of a mold, as is the practice with these materials.

The detailed solution of the one-dimensional, boundary value problem in cylindrical coordinates is described elsewhere in detail (2), and it involves solving the problem of multilayer wave propagation into the material, as illustrated in Figure 1. The waves propagating inward and outward are equated, and are required to be continuous at each boundary. The phase and amplitude of the reflected signal a_2 which is detected, carries information on the structure of the forming ingot. The data that is presented is in graphical form, illustrating the phase difference between the returning signal and the source. This is a two-dimensional polar plot, whose radial component is scaled as a log function. The use of this type of boundary value solution can be used to isolate solidification morphology; however, here only the relative magnitude of the responses for complete solidification in these different systems are of interest. The calculations are performed for three frequencies - 100Hz, 500Hz and 1000Hz - and the first datum is the complete response for solidification within the mold (see Figure 2). The two-dimensional vector plot represents the magnetude and phase shift change in the reflected signal due to solidification. In each case, the frequency affects both the phase and relative amplitude of the response signals. Included in this figure is a detection threshold that is typical for current equipment. In figure 3 are the maximum responses to solidification at 100Hz for a range of mold thickness. As the mold thickness is reduced the relative sensitivity of the system is much greater, and thus more detailed information, can be obtained. Consequently, a detector coil placed below the mold in a continuous caster could be used to more accurately determine skin thickness and local freezing rates. More obviously, defect detection, inclusions, shrinkage cavity formation, and hot tears are more easily detected because of the increased sensitivity below the mold. The effect of the mold material on solidification measurements are also detailed in Figure 3 with copper showing the greatest attenuation. The molds not only introduce signal attenuation, but thermal fluctuations which alter their electrical conductivity and act on separate signal sources. A $1^{\circ}C$ temperature rise in the 0.5 cm thick mold is recorded in Table III. The materials with higher values of $d\rho/dT$ show particularly strong responses. Where thermal fluctuations within the mold are an experimental problem, a testing frequency can be selected that puts the information being monitored 90° out of phase with respect to these fluctuations. Also, mold deformation and deflection must be considered since they can contribute significantly to the eddy current response; however, these signals are quite characteristic and can be separated from thermal and solidification responses. This type of modeled analysis works quite well and provides a set of standards to which data can be compared.

Multifrequency Analysis

In an attempt to provide a more general analysis of eddy current data taken during solidification, we considered the use of multifrequency data, which can be taken during the course of an experiment. Multifrequency eddy current data would provide more information on geometry and conductivity than one could gain from a single frequency measurement. The multifrequency data can be sorted by using the solution to the modeled boundary

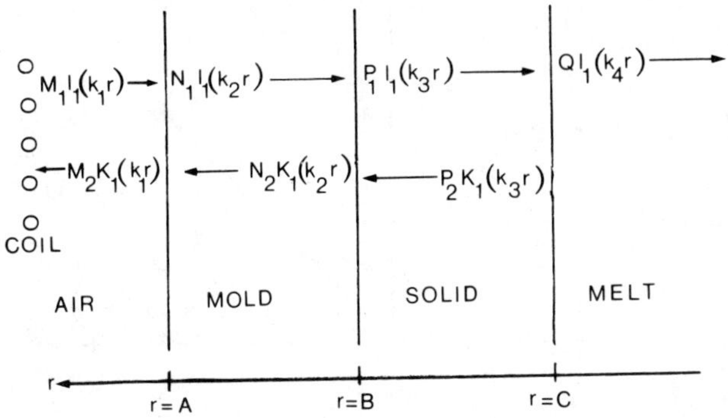

$$M_1I_1(K_1A) + M_2K_1(K_1A) = N_1I_1(K_2A) + N_2K_1(K_2A)$$
$$M_1I_1'(K_1A) + M_2K_1'(K_1A) = N_1K_1'(K_2A) + N_2K_1'(K_2A)$$
$$N_1I_1(K_2B) + N_2K_1(K_2B) = P_1I_1(K_3B) + P_2K_1(K_3B)$$
$$N_1I_1'(K_2B) + N_2K_1'(K_2B) = P_1I_1'(K_3B) + P_2K_1'(K_3B)$$
$$P_1I_1(K_3C) + P_2K_1(K_3C) = QI_1(K_4C)$$
$$P_1I_1'(K_3C) + P_2K_1'(K_3C) = QI_1(K_4C)$$

Illustrated above is the modeled electromagnetic boundary value problem used to interpret eddy current data taken in cylindrical geometry. The solution of the six equations for M_2, the measurable quantity, provides the relations between the signal and the spatial conductivity variations in the ingot.

Fig. 1. Boundary conditions.

value problems, as was done for the single frequency analysis. The multifrequency data is particularly useful when ambiguous interpretation is possible for a single measurement. This approach to interpretation of multifrequency data increases the effort required for analysis in proportion to number of frequencies. A more general technique of handling eddy current data is desirable which directly provides information on the conductivity profile. This would make final interpretation less difficult and allow machine interpretation. One can approach this problem by considering the detection of differential changes in conductivity of a volume under test. In particular any conductivity variation can be represented as a spatial sum of these differential changes in conductivity. If under some specific conditions, an integral transform exists for a sum of signal contributions, one can express the spatial conductivity variation as an integral over a continum of multifrequency data. Such an analytical approach would eliminate analysis in the complex signal plane and result in spatial conductivity information alone. Locating a small fluctuation in conductivity within a body can be expressed rather simply. A schematic of the electromagnetic boundary value problem is shown in Figure 4. The perturbation in conductivity between x and x' is expressed as a change in the propagation vector $k' = k+\delta k$.

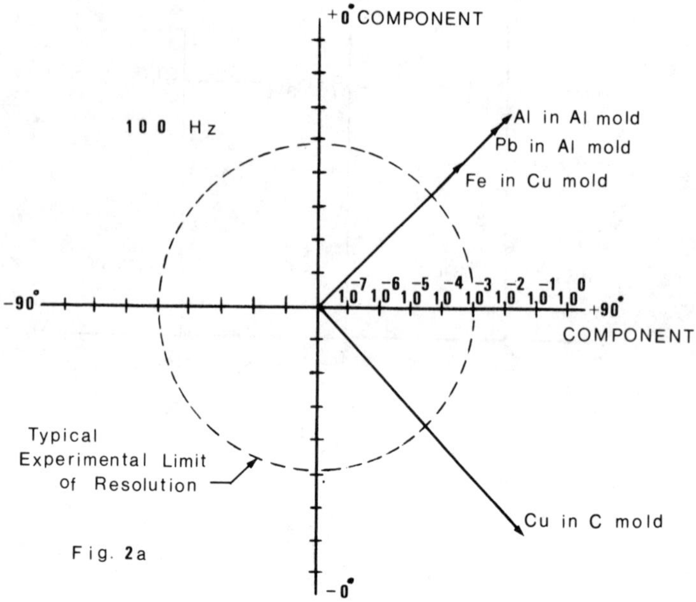
+0° COMPONENT
100 Hz
Al in Al mold
Pb in Al mold
Fe in Cu mold
-7 -6 -5 -4 -3 -2 -1 0
-90°
+90° COMPONENT
Typical Experimental Limit of Resolution
Cu in C mold
-0°

Fig. 2a

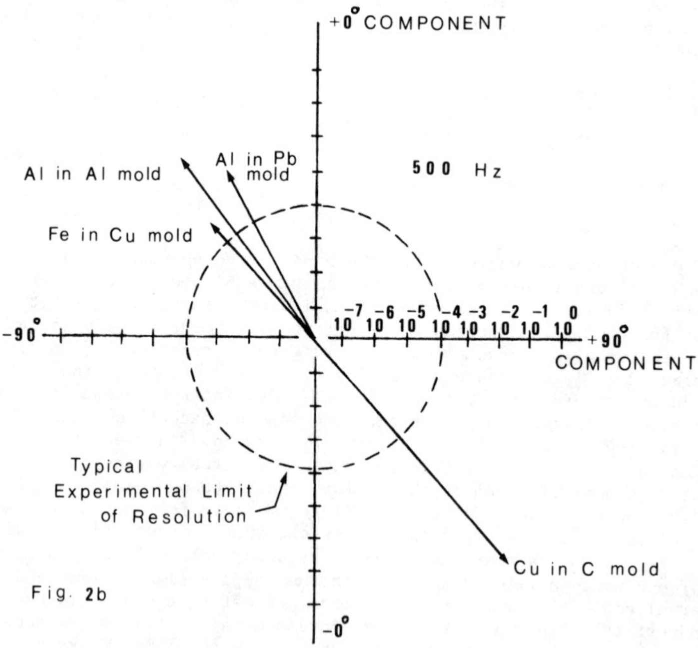
+0° COMPONENT
Al in Al mold
Al in Pb mold
500 Hz
Fe in Cu mold
-7 -6 -5 -4 -3 -2 -1 0
-90°
+90° COMPONENT
Typical Experimental Limit of Resolution
Cu in C mold
-0°

Fig. 2b

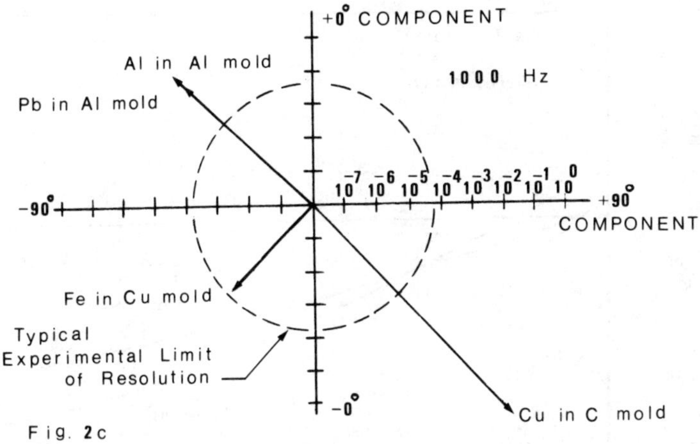

Fig. 2c

The calculated eddy current signal response for solidification of four materials in different molds is presented for three frequencies of test. The typical limit of experimental resolution is indicated by the broken line. The data is presented in polar form similar to how the experimental data would be taken in a mold .5 cm thick. The origin in all three plots represents complete liquid in the mold at the melting point.

Fig. 2. Calculated eddy current signal response.

The four equations comprising the boundary value problem are:

$$b_1e^{ikx} + db_2e^{-ikx} = C_1e^{ik'x} + C_2e^{-ik'x} \tag{1}$$

$$b_1e^{ikx} - db_2e^{-ikx} = \frac{k'}{k}\left(C_1e^{ik'x} - C_2e^{-ik'x}\right) \tag{2}$$

$$C_1e^{ik'x'} + C_2e^{-ik'x'} = f\ e^{ikx'} \tag{3}$$

$$C_1e^{ik'x'} - C_2e^{-ik'x'} = \frac{k}{k'}\ f\ e^{ikx'} \tag{4}$$

Solving the equations, and eliminating:

C_1, C_2, and f produces:

$$b_1e^{ikx} + db_2e^{-ikx} = \frac{k}{k'}\ \frac{\frac{k'+k}{k'-k}\ e^{-2ik'dx} + 1}{\frac{k'+k}{k'-k}\ e^{-2ik'dx} - 1}\left\{b_1e^{ikx} - db_2e^{-ikx}\right\} \tag{5}$$

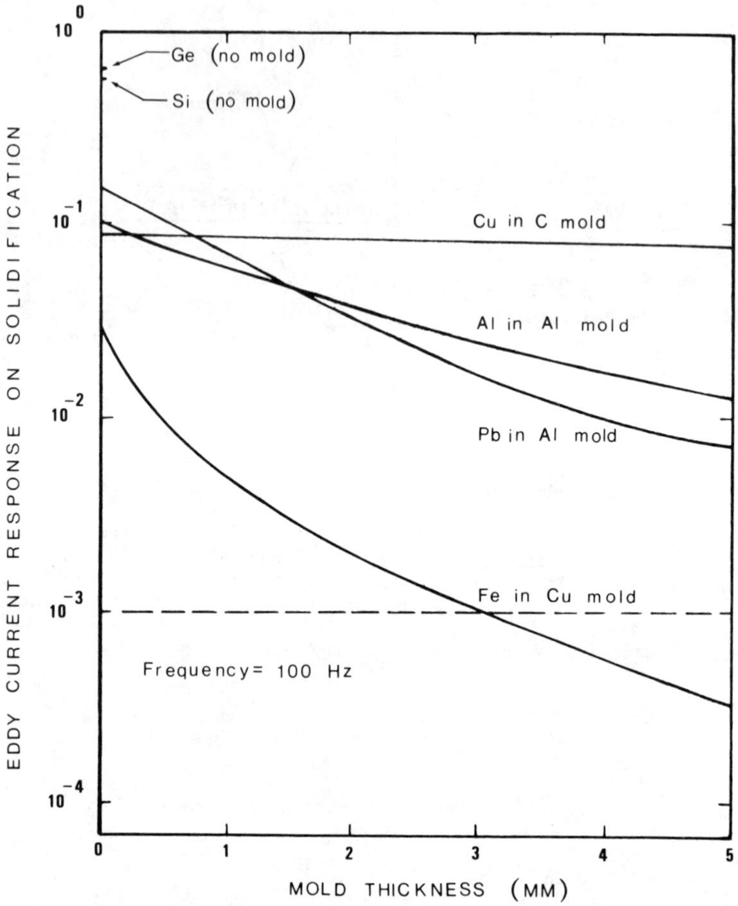

Fig. 3. The magnitude of the eddy current response is plotted as a function of mold thickness for six materials. The limit of experiment resolution is indicated by the broken line.

where $dx = x' - x$. Assuming $2k'dx$ is small, we can expand those terms containing $2k' dx$; the result is:

$$b_1 e^{2ikx} \left[2idxk'(k-k')\right] = db_2 \left[(k-k') + (k'+k)(1-2ik'dx)\right] \quad (6)$$

when $dk = k'-k$ is very small

$$db_2 \simeq b_1 \; i \; dkdxe^{2ikx} \quad (7)$$

These small responses can be summed to calculate the total contribution to the reflected signal:

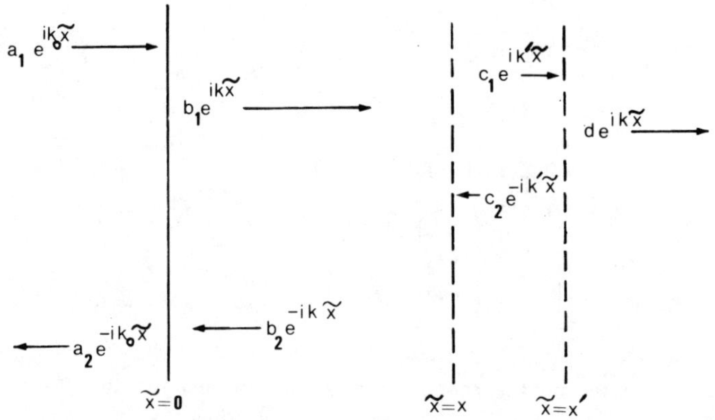

Fig. 4. Illustrated above is the one-dimensional planar perturbation model used to derive the Laplace transform from which the conductivity profile is calculated.

$$b_2 = \sum_j (db_2)_j = i \sum b_1 dx_j dk_j e^{i\,2kx_j} \qquad (8)$$

This sum is an approximation based on the assumption that the initial propagation signal b_1 is not significantly affected by these small conductivity perturbations. The dependence on the mean propagation vector k and the input signal b_1 can be removed by rewriting:

$$\frac{b_2}{ib_1k} = \sum \frac{\delta k_j}{k} e^{2ikx_j} dx \qquad (9)$$

This summation can be replaced by an integral over the region of measurement

$$\frac{b_2}{ib_1k} = \int_0^\infty \frac{\delta k(x)}{k} e^{2ikx} dx \qquad (10)$$

where $k = (1+i)\sqrt{\frac{\omega\mu\sigma}{2}}$ and $\frac{\delta k(x)}{k} = \frac{\delta\sigma(x)}{2\sigma}$ which result in:

$$\frac{b_2}{ib_1k} = \int_0^\infty \frac{\delta\sigma(x)}{2\sigma} e^{2ikx} dx \,. \qquad (11)$$

Letting p = -2ik the above integral can be written in a form that allows the use of the Laplace transform theorem when in the form:

$$\frac{-2b_2}{b_1 p} = \int_0^{\infty} \frac{\delta\sigma(x)}{2\sigma} e^{-px} dx \tag{12}$$

providing a spatial conductivity dependence:

$$\frac{\delta\sigma(x)}{\sigma} = \frac{1}{2\, i} \int_{\gamma - i\infty}^{\gamma + i\infty} \left\{ \frac{-2b_2}{b_1 p} \right\} e^{px} dp \tag{13}$$

After replacing p and noting that $b_2 = o$ for Re $k<o$ and setting $\gamma = o$, the spatial conductivity variation becomes useful in the form:

$$\delta\sigma(x) = \frac{2\sigma}{\pi i} \int_0^{\infty} \frac{b_2(k)}{b_1 k} e^{-2ikx} dx \tag{14}$$

This integral can be evaluated with the use of the corresponding contour integral, shown in Figure 5, where the integral over the entire contour, I, can be expressed as a sum:

$$I = I_{rR} + I_{RB} + I_{BA} + I_{Ar} \tag{15}$$

For a planar conductor whose propagation vector changes from k to k' at a depth, c, the above contour integral shows a singularity at $k = o$. Evaluating this argument for Re $k \geq 0$ and Re $k' \geq 0$ where

$$k' = \sqrt{\frac{\sigma'}{\sigma}} \; k$$

results in the expression:

$$\frac{b_2}{b_1 k} = \frac{1}{k} \left(\frac{k-k'}{k+k'} \right) e^{2ikC} \tag{16}$$

Evaluating the integral over separate segments of the contour:

$$I_{RB} = \lim_{a \to o} \frac{2\sigma}{\pi i} \int_a^{\pi/4} \frac{R-R'}{R+R'} \, i \, e^{2iRe^{i\theta}(C-x)} \, d\theta \tag{17}$$

where $k = Re^{i\theta}$, $k' = R'e^{i\theta}$ and $dk = Ri\, e^{i\theta} d\theta$

As $R \to \infty$, $I_{RB} \to 0 \; (x < c)$

For x > c the integral in equation 17 become unstable and oscillatory producing the major source of uncertainty in the inversion attempt. However, the rate at which equation 17 for x > c becomes unstable is bounded for values of x less than the skin depth of the lowest test frequency used in practice.

For:

$$I_{Ar} = \lim_{r\to 0} \frac{2\sigma}{\pi i} \int_{\pi/4}^{0} \frac{r-r'}{r+r'}\, ie^{2ire^{i\theta}} (C-x)\,d\theta \qquad \text{taking} \quad \Delta\sigma = \sigma' - \sigma \tag{18}$$

$$I_{Ar} = \frac{\Delta\sigma}{8} \tag{19}$$

where, I_{Ar} will be a small real number. As a result I_{Ar} will be the source of error in the inversion of multifrequency data. Setting I_{Ar} = o we have:

$$I \simeq I_{rR} + I_{BA} = 0 \tag{20}$$

$$I_{rR} = -I_{BA} = I_{AB} \tag{21}$$

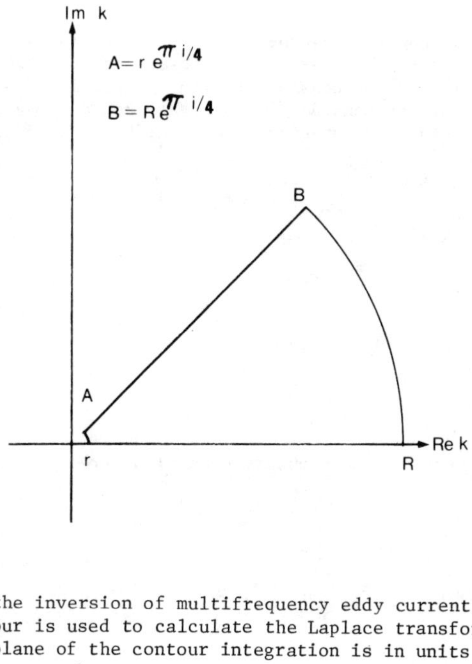

Fig. 5. For the inversion of multifrequency eddy current data the above contour is used to calculate the Laplace transform of equation 14. The plane of the contour integration is in units of the propagation vector k.

Therefore, the explicit integration of equation 14 satisfies the inversion criterions.

It is useful to change the variable of integration to a frequency domain

where $\frac{dk}{k} = \frac{d\omega}{2\omega}$ results in

$$\delta\sigma(x) = \frac{\sigma}{\pi i} \int_0^\infty \frac{b_2(\omega)}{b_1} e^{-2ik(\omega)x} \frac{d\omega}{\omega}, \tag{22}$$

with, ω, being the angular frequency.

The above relation allows us to express the relative deviation in conductivity as a function of the incoming and outgoing wave within the conductor. In order to usefully employ the above integration, we must express b_1 and b_2 in terms of the known and measureable quantities a_1 and a_2.

It is tempting to compute the ratio b_2/b_1 from the boundary conditions in Figure 6 (a). Using these conditions, this results in:

$$\frac{\delta\sigma}{\sigma} = \frac{1}{\pi i} \int_0^\infty \frac{a_1(k-k_o) + a_2(k+k_o)}{a_1(k+k_o) + a_2(k-k_o)} e^{-2ik(\omega)x} \frac{d\omega}{\omega} \tag{23}$$

where k is the mean propagation vector within the material and k_o is the propagation vector in air. However, this imposes the condition that b_2 is a function of b_1, which is not consistent with are original assumptions. Instead, it is necessary to consider two separate boundary values problems, one each for the incoming and outgoing waves, which are shown in Figure 6b.

For the incoming wave, the boundary conditions are:

$$a_1 + a_2 = b_1 \tag{24}$$

$$a_1 - a_2 = \frac{k}{k_o} b_1 \tag{25}$$

Solving for b_1 in terms of a_1

$$b_1 = \frac{2k_o}{k_o + k} a_1 \tag{26}$$

For the outgoing wave, the boundary conditions are:

$$a_2' = b_1' + b_2 \tag{27}$$

$$-\frac{k_o}{k} a_2' = b_1' - b_2 \tag{28}$$

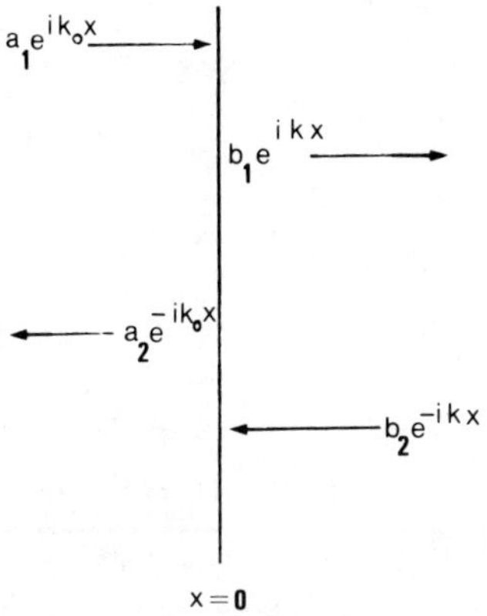

Fig. 6a

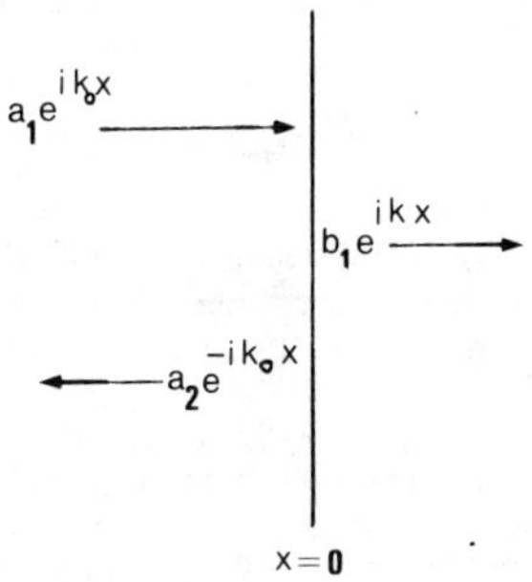

Fig. 6b

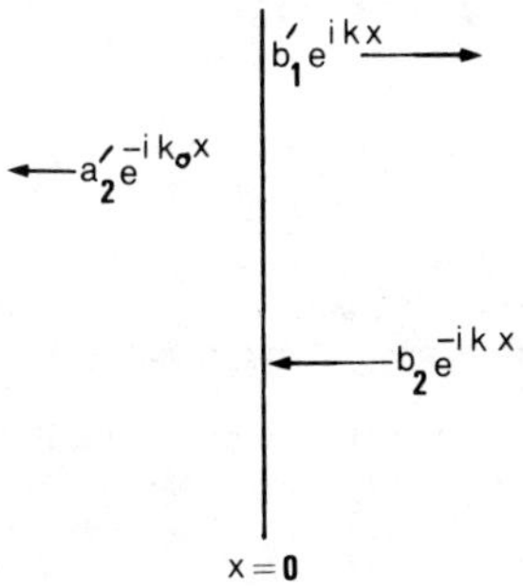

In order to couple the multifrequency inversion process to a detector the solution must be propagated to the external surface of the test volume. In illustration (a), the boundary conditions are too restrictive to allow one to propagate the reflected signal to vacuum $x < o$. However, in illustration (b) there are no restraints placed on the value of b_1, which is consistent with the small signal approximation used to derive the transform.

Fig. 6. Examples of multifrequency inversion process.

Solving for b_2 in terms of a'_2

$$b_2 = \frac{a'_2}{2}\left(\frac{k+k_o}{k}\right) \tag{29}$$

The total measured signal is $a_2^M = a'_2 + a_2$. It is assumed that b_1 in the material is independent of all constraints for small signal reflections.

The integral now becomes:

$$\delta\sigma(x) = \frac{\sigma}{\pi i}\int_o^\infty \frac{a'_2}{a'_1}\,\frac{\left(k+k_o\right)^2}{4k_o k}\, e^{-2ik(\omega)x}\,\frac{d\omega}{\omega} \tag{30}$$

or rewritten in terms of a_2^M

$$\delta\sigma(x) = \frac{\sigma}{\pi i}\int_o^\infty \left\{ \frac{a_2^M}{a_1}\,\frac{\left(k+k_o\right)^2}{4kk_o} - \frac{\left(k_o-k\right)\left(k_o+k\right)}{4k_o\delta k} \right\} \frac{e}{\omega}^{-2ik(\omega)x} d\omega \tag{31}$$

The actual measurement of the quantities a_2^M and a_1 occurs at a particular frequency. In practice, a'_2 would be the most easily measured quantity, if the testing equipment were properly calibrated because any relative change would be solely due to a'_2. This would allow one to compute

$$\frac{\delta\sigma(x)}{\sigma}$$

The quantity a_1 only needs to be calculated, or experimentally measured, once initially, and can be used again during subsequent sampling when needed to compute the transform.

To test the transform technique on multifrequency eddy current data, it is useful to take a set of solutions to a particular boundary value problem that supplies the values $a_2(\omega)$. The first test that is important is the case of a structureless medium with a conductivity of 10^7 mhos/m. Generating $a_2^M(\omega)$for twelve equally spaced frequencies between 25Hz and 300Hz produces the transform using a summation suggested by Brigham (4) shown in Figure 7. Ideally, there should be no change or value in $\delta\sigma(x)$, however, the maximum deviation $\delta\sigma(x)$ to a depth of 3 cm is 100 mhos/m or one part in 10^5. The function $\delta\sigma(x)$ oscillates rapidly beyond this depth with large amplitudes. Similar calculations were performed for a step function and constant changes in conductivity over a finite length. Specific form of the conductivity perturbations and their transforms are shown in figures 8 (a) (b). The transform function breaks up and goes into oscillations at depths greater than where the conductivity fluctuations appear; the transform function therefore bears little resemblance to the conductivity variation. If, however, one plots the position of the transformed peak to the location of the conductivity change one finds a linear relationship. Also, in Figure 9 there is a direct relationship between the product of the conductivity change and the thickness over which the change occured to the amplitude of the transform. These two proportionalities are of value since they directly exhibit the magnitude and position of a conductivity fluctuation with no previous assumptions on that structure

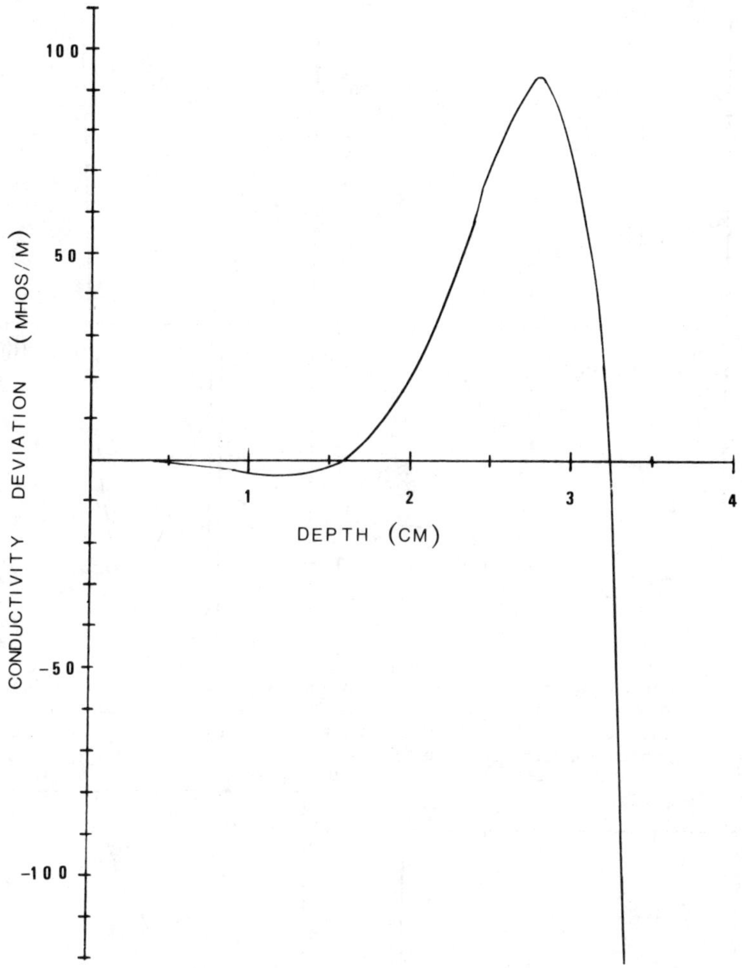

Above is the simulated result of a 12 frequency Laplace inversion of a homogeneous material with a constant conductivity of 10^7 mohs/m. The frequencies were equally spaced from 25 to 250 Hz and the inversion data up to 3 cm shows an error in the true profile of 10^{-5}. The transform begins to oscillate at depths greater than the skin depth of the lower frequency.

Fig. 7. Simulated result of a multifrequency Laplace inversion.

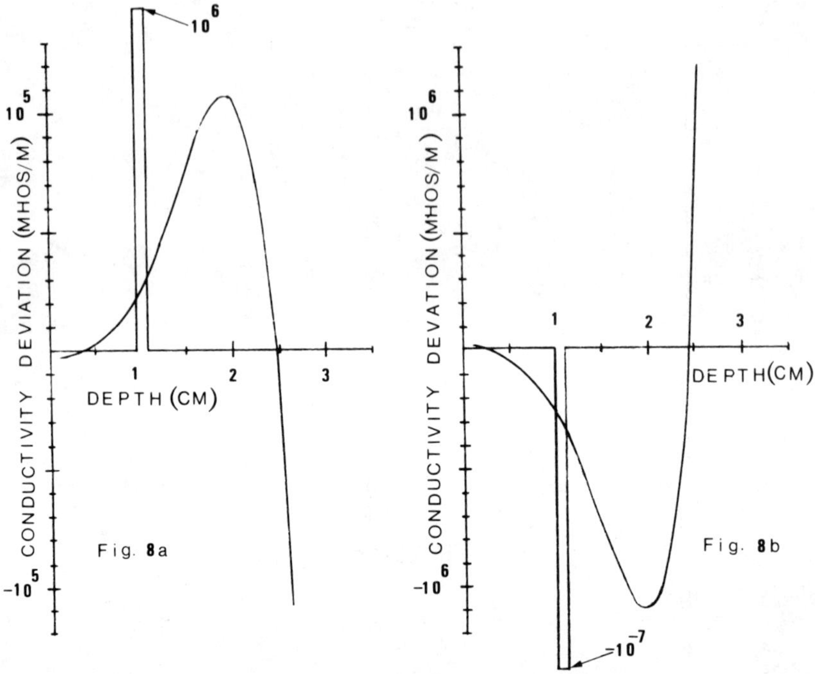

Fig. 8. The figures represent the Laplace inversion of 12 frequency eddy current data for the particular conductivity fluxuation shown. The rectangular profile changes in 8a and 8b are used to produce the raw eddy current signal that is inverted and is graphed as the smooth function in each figure.

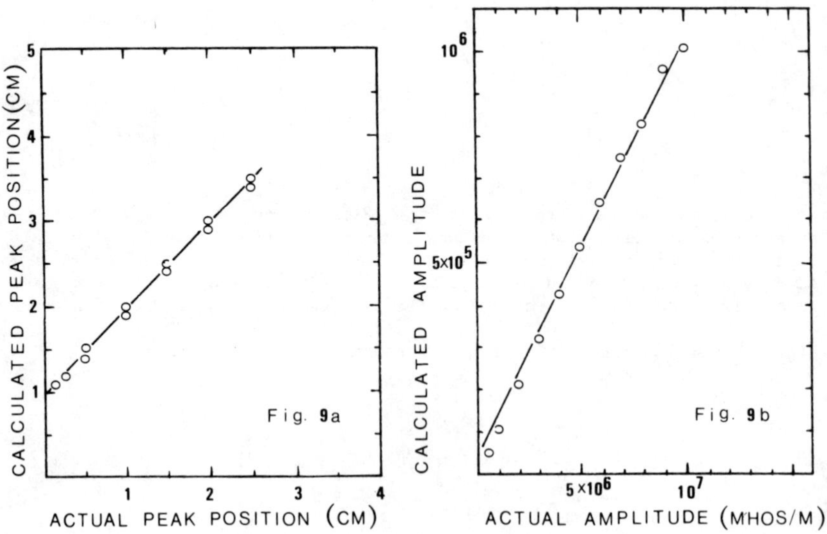

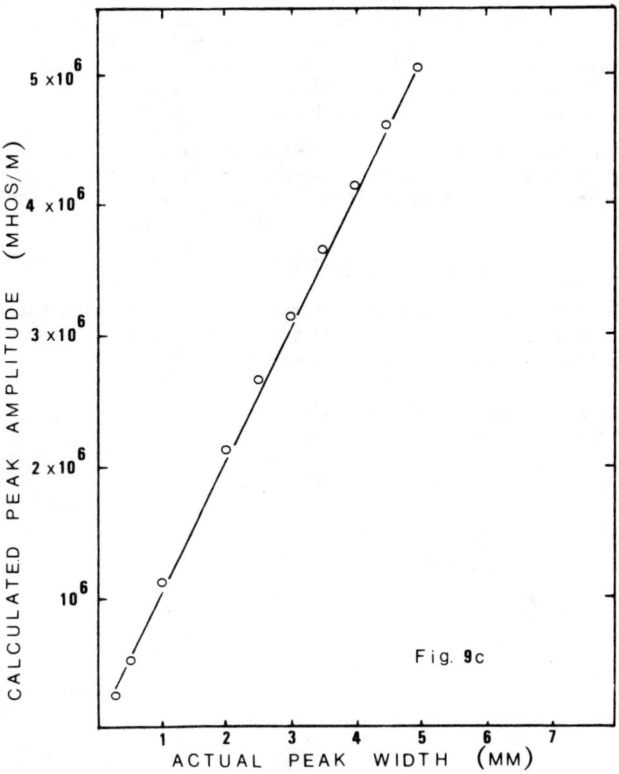

Fig. 9c

The relationship between the inversion peak position and the actual position of the conductivity variation of both signs are plotted in Figure 9(a). There is a systematic offset of 1 cm in the inverted data. In Figure 9(b) and 9(c) the amplitude and width also show a linear relationship between the actual physical structure and the inverted data. This allows spatial and amplitude data on defect position to be deduced without resorting to model comparisons.

Fig. 9. Calculated inversion peak properties versus experimental data.

of the volume of metal being sampled. In this form the data can be machine or operator interpreted without need of direct analysis of raw eddy current data. The form of the transform can be constructed in other ways as shown in the Appendix. For particular experimental situations the transform will have to be extended to appropriate systems and frequencies selected to maximize the useful depth for conductivity measurements.

These results indicate that this integral tranform can be used to detect conductivity changes within a mold and allow accurate determination of the conductivity variation position and magnitude within a sample volume.

Appendix

The inversion of multifrequency data can also be derived through the use of a single set of boundary conditions at x_1 where the propagation vectors on either side of the boundary are k and k'.

$$b_1 e^{ikx} + db_2 e^{-ikx} = Ce^{ik'x} \qquad (1A)$$

$$\frac{k}{k'}(b_2 e^{ikx} - db_2 e^{-ikx}) = Ce^{ik'x} \qquad (2A)$$

where $\delta k = k'-k$

$$db_2 = \frac{\delta k}{\partial k+\delta k} e^{2ikx} \sim \frac{\delta k}{2k} be^{ikx} \qquad (3A)$$

using $\frac{\delta k}{k} = \frac{\delta\sigma}{2\sigma}$

$$db_2 = \frac{\delta\sigma(x)}{4\sigma} b_1 e^{2ikx} \qquad (4A)$$

and $\delta\sigma(x) = \frac{d\sigma(x)}{dx} dx$

$$b_2 = \int db_2 = \int_0^\infty \frac{b_1}{4\sigma} \frac{d\sigma(x)}{dx} e^{2ikx} dx \qquad (5A)$$

Using the Laplace Transform:

$$\frac{d\sigma(x)}{dx} = \frac{4\sigma}{b_1\pi} \int b_2 e^{-2ikx} dx \qquad (6A)$$

$$\sigma(x') = \frac{4\sigma}{b_1\pi} \int_0^{x'} dx \int_0^\infty dk b_2 e^{-2ikx} \qquad (7A)$$

This approach allows one to more easily derive the reflected signal in cylindrical coordinates with the modified Bersel functions $I_1(kr)$ and $k_1(kr)$.

$$b_1 I_1(kr) + db_2 K_1(kr) = cI_1\ (k'r) \tag{8A}$$

By convention $\frac{d}{dr}$ f(kr) = kf'(kr)

$$b_1 kI_1'\ (kr) + db_2 kK_1'(kr) = ck'\ I_1'(k'r) \tag{9A}$$

Solving for db_2

$$db_2 = \frac{k'\ I_1(kr)\ I_1'\ (k'r) - kI_1'\ (kr) I_1(k'r)}{kI_1(k'r)K_1'(kr) - k'I_1'(k'r)K_1(kr)}\ b_1 \tag{10A}$$

where $k' \rightarrow k$, the denomination approaches the Wronskian, which equals $^1/r$ for small dk

$$db_2 = b_1 \left\{ k'I_1(kr)I_1'(k'r) - kI_1'(kr)I_1(k'r) \right\} r \tag{11A}$$

$$db_2 = b_1 rdk \left\{ I_1(kr)\ I_1'(kr) \right\} \tag{12A}$$

$$db_2 = b_1\ r \frac{dk}{dr} \left\{ I_1(kr)I_1'(kr) \right\} dr \tag{13A}$$

$$b_2 = b_1 \int_o^{r'} r \frac{dk}{dr} I_1(kr)I_1'(kr)dr \tag{14A}$$

$$b_2 = \frac{b_1}{2} \int_o^{r'} \frac{dk}{dr} \frac{d}{dr} (I_1(kr)^2)rdr \tag{15A}$$

This result remains to be inverted.

References

1. J. P. Wallace and D. C. Kunerth, Metallurgical Transactions B, 1980, Vol. 11B pp. 267-71.

2. D. C. Kunerth and J. P. Wallace, Metallurgical Transactions B, 1980, Vol. 11B pp. 273-83.

3. C. V. Dodd and W. E. Deeds, Journal of Applied Physics, 1968, Vol. 39 pp. 2829-2838.

4. E. Oran Brigham, The Fast Fourier Transform, Prentice-Hall, Englewood Cliffs, N. J., 1974 p. 8.

5. OFHC Brand Copper/Properties and Applications, p. 72, AMAX, New York, New York, 1974.

6. Aluminum, Vol. I. Kent R. Van Horn, ed., ASM, p. 9-10, Metals Park, Ohio, 1967.

7. The Industrial Graphite Engineering Handbook, p. 5C.01.03, Union Carbide Corporation, New York, New York, 1970.

8. Metals Handbook, Vol. I, Taylor Lyman, ed., ASM, p. 1203-1209, Metals Park, Ohio, 1961.

9. Lead and its Alloys, W. Hoffman, p. 20, Springer Verlag, New York, New York, 1970.

10. Liquid Semiconductors, V. M. Glazov, S. N. Chizhevskaya, N. N. Glagoleva, p. 60, Plenium Press, New York, New York, 1969.

11. Liquid Metals, M. Shimoji, p. 381, Academic Press, London, England, 1977.

RELATIONSHIP BETWEEN PROCESS VARIABLES, MICROSTRUCTURE AND NDE OF A PRECIPITATION-HARDENED ALUMINUM ALLOY

L. J. Swartzendruber, W. J. Boettinger, L. K. Ives, S. R. Coriell, and R. Mehrabian

Metallurgy Division
National Bureau of Standards
Washington, DC 20234

The relationship between process variables used during casting, working, and heat treatment of 2219 aluminum alloy and the resulting microstructures, mechanical properties, and NDE responses were investigated. The behavior of this alloy is dominated by its relatively high copper concentration. The principal age-hardening species in material processed according to T87 and T851 temper requirements is Θ'. The Θ" phase is also present and contributes to strength. The major phases present in as-cast material are α-Al solid solution, Θ-$CuAl_2$, and Cu_2FeAl_7. These phases are also present in the heat treated finished plate product. Macrosegregation, which cannot be completely eliminated from the finished alloy plate, is predictable and due to interdendritic fluid flow. A comprehensive series of C-curves was determined to relate hardness, tensile strength, yield strength and conductivity after aging to various degrees of improper quench from solution heat treatment temperature. These C-curves established a correlation between mechanical properties and electrical conductivity. Such a correlation was found to exist for alloys (of a single lot) given an improper quench from solution heat treatment, other conditions being maintained constant. When more than one condition is varied, the correlation may be destroyed. Finally, these C-curves were used in conjunction with three models of heat transfer to calculate plate mechanical and NDE properties as a function of position in the plate.

Introduction

The compositional homogeneity, microstructure, hardness, electrical conductivity and mechanical properties of 2219 aluminum alloy plates are influenced by the process variables during casting, rolling and thermo-mechanical treatment. The details (1) of these relationships have been investigated for correctly processed 2219 plate as well as for deviations caused by improper quenching after solution heat treatment. A detailed blend of metallurgical examination of the alloy was carried out in conjunction with NDE studies to gain a fundamental knowledge of the features of metallurgical microstructures which cause changes in NDE response. Primary emphasis has been placed on the reliability of eddy current electrical conductivity and hardness as NDE tools to detect variations in mechanical properties.

Specifically, work in this paper is divided into the following areas:

1. Solidification-segregation studies. This is motivated by the realization that the properties vary across plate thickness. These variations can have different effects on various NDE techniques. Part of the variation in properties across finished alloy plate is caused by segregation of the alloy components (Cu, Mn) during ingot casting.

2. Determination of time-temperature transformation diagrams (C-curves) and the relationship between mechanical properties and conductivity. Samples produced during various thermomechanical treatments have a wide range of microstructures and hence of mechanical properties (hardness, yield strength, tensile strength) and conductivity.

3. Predictions of heat flow conditions during malfunctions of the commercial process, specifically slow quenching from the solution heat treatment temperature; and predictions using experimentally determined C-curves of the resultant mechanical property degradations.

Solidification-Segregation Studies

It was expected that across-thickness variations in properties would be most pronounced in the thickest plate. These variations are due to macro-segregation in original DC cast ingots from which plates are rolled or due to the slower cooling rate inherent in thicker plates during quench from the solution heat treatment temperature. Therefore, a commercially processed 12.7 cm. thick plate in the T851 condition was carefully examined for chemical, microstructural and property variation across its thickness*.

The variation in copper concentration, hardness, conductivity and tensile properties across the 12.7 cm thick plate are shown in Figures 1a and b. These data essentially establish the maximum variations in properties due to the macrosegregation remaining in the plate from the original DC cast ingot and the normal variation in cooling rate experienced during quench from the solution heat treatment temperature. The first plot in Figure 1 shows that there is approximately 0.6 wt% variation in copper content across the plate. This was determined by molecular absorption spectrometry (wet chemistry).

---

* Average composition of the 12.7 cm thick plate of 2219 aluminum alloy determined by emission spectroscopy is 6.3 wt% Cu, 0.36 wt% Mn, 0.23 wt% Fe, 0.07 wt% Si, 0.12 wt% Zr, 0.07 wt% V, 0.03 wt% Ti, 0.03 wt% Ni, 0.01 wt% Mg, and 0.03 wt% Zn.

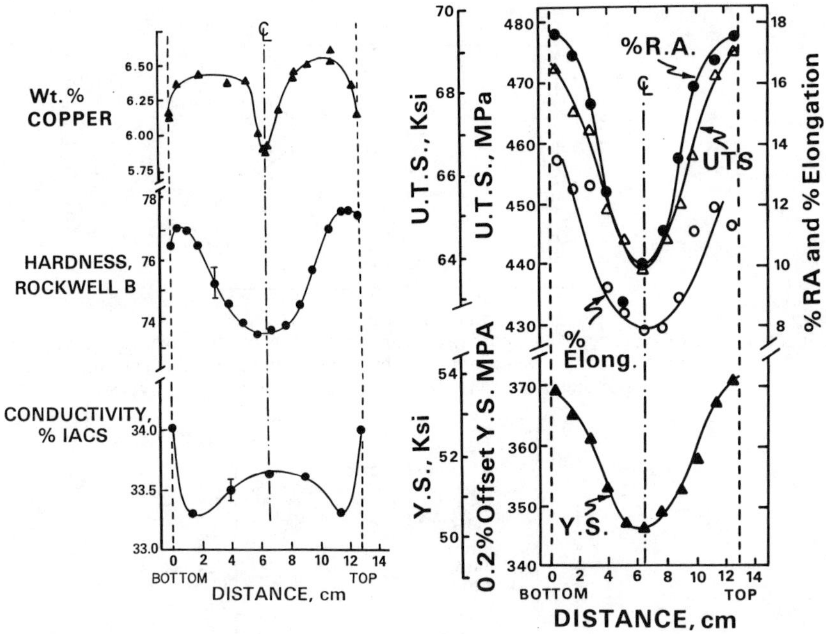

(a) (b)

Figure 1. Measured (a) copper content, hardness and conductivity and (b) yield stress, ultimate tensile stress, percent elongation and reduction in area versus distance through an as-received 12.7 cm (5 inch) thick plate of 2219-T851 aluminum alloy.

This composition profile can be readily correlated to that found in typical DC (direct chill) semi-continuous cast ingots. Figure 2 shows the chill face to centerline composition variation (macrosegregation) in a typical DC cast ingot of 2219 aluminum alloy. This macrosegregation is caused by fluid flow during ingot solidification in conjunction with the rate of heat extraction at the mold metal interface (2). The outermost surface of DC cast ingots are scalped prior to the plate forming operation but some of the negatively segregated region seen, for example, 1 cm from the chill face in Figure 2 often remains at the top and bottom surfaces of the plate. The negative segregation at the ingot centerline is of essentially the same magnitude as that found at the centerline in a processed plate (Figure 1a).

To assess the role of composition variations across finished alloy plate, a special geometry unidirectional ingot was cast with 2219 aluminum alloy to obtain controlled composition samples differing from the nominal. This geometry is shown in the inset of Figure 3 and uses a bottom chill. The variation of copper content determined by atomic absorption spectrometry as a function of distance from the bottom chill in the unidirectional ingot is shown in Figure 3. As expected from previous studies of macrosegregation noted above, high positive segregation (∿7.3 wt% Cu) is observed at the chill face while negative segregation (∿3.6 wt% Cu) occurs in the region of the cross section change. Measurement of electrical conductivity was made on the unidirectional ingot as-cast and is also shown in Figure 3. This data

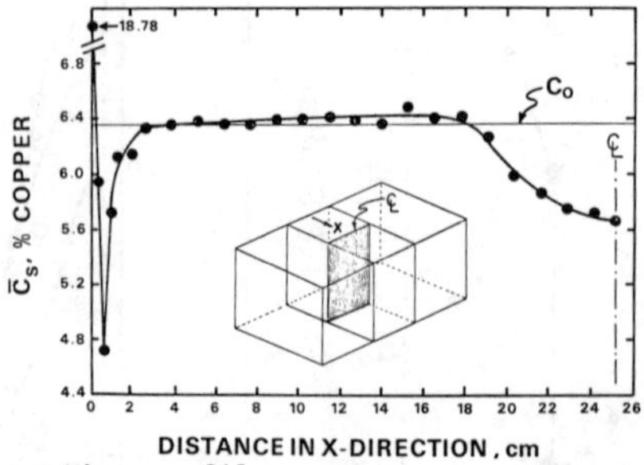

Figure 2. Macrosegregation profile, average copper content versus distance from the chill face, across the short transverse direction of a semi-continuous DC cast ingot of 2219 aluminum alloy.

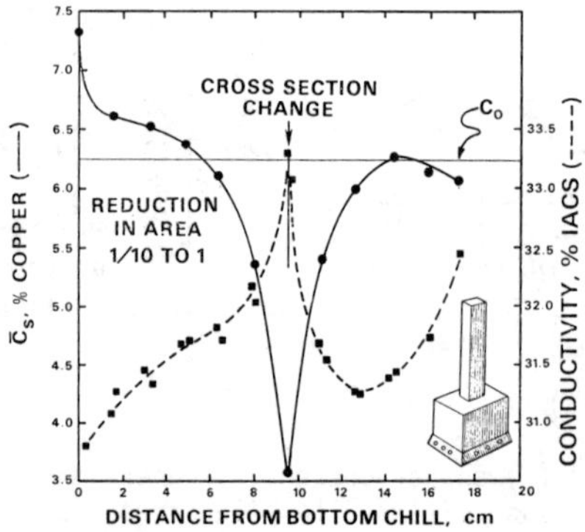

Figure 3. Average copper content and electrical conductivity versus distance from the bottom chill, in an unidirectionally solidified reduced cross section laboratory ingot of 2219 aluminum alloy.

indicates the same trend as observed in the as-received 12.7 cm (5 inch) thick plate; i.e., higher copper content corresponds to lower conductivity.

Samples, ∿3.7 cm square and ∿1.3 cm (∿1/2") thick in the solidification direction, were cut from the ingot at different distances from the bottom chill face to obtain specimens with different compositions. The samples, of course, contained composition gradients through their thickness, but were relatively uniform in perpendicular directions due to the unidirectional solidification. Samples were thermomechanically treated as follows:

(a) homogenization heat treated for 48 hours at 535 °C,
(b) hot rolled at 440 °C to 1/4 of their initial thickness (final thickness 0.32 cm, 1/8"),
(c) solution heat treated at 535 °C for 75 minutes,
(d) stretched 5% and aged at 172 °C for 16 hours.

Step (d) is a commonly used commercial modification of the T87 temper which calls for a 7% stretch rather than a 5% stretch, hence we use here the designation T87*.

Hardness and electrical conductivity measurements were carried out on both surfaces of these 0.3 cm thick samples. In Figures 4a and 4b are plotted hardness and conductivity respectively as a function of the original distance of these surfaces from the bottom chill face along with the copper content. Not until the copper content drops below ∿5.5% wt% does the hardness drop significantly. Hardness data beyond 14 cm from the chill face is probably unreliable due to excessive ingot porosity. Similar results were obtained for the relation of copper content to yield and tensile strength. On the other hand, the variation in electrical conductivity appears to follow the same general trend as that established for the as-cast structure in Figure 3. Note that the variation in copper content shown in Figure 4 is that obtained in the as-cast ingot. Some averaging of variations in composition has occurred during the rolling operation which manifests itself in the conductivity data scatter noted in Figure 4b. With respect to the microstructure typical of the as-cast alloy shown in Figure 5, only three phases appear to be present; the aluminum solid solution (α) plus two phases which occur along with the α-phase in the interdendritic regions. The rounded irregular interdendritic phase is $CuAl_2$ and the bladelike phase is Cu_2FeAl_7 which is sometimes referred to as β-AlCuFe. These phases do not completely dissolve during solution heat treatment and remain in finished alloy plate.

The following conclusions can be drawn at this point regarding the across plate property variations shown in Figures 1a and b. First, the variations in hardness and tensile properties can be ascribed to the changes in cooling rate across the plate during the quench and to the fact that the center of the plate did not experience as much mechanical deformation as the outside. Composition variations due to ingot casting are not the major cause of the mechanical property variations. Second, variations in electrical conductivity, especially the abrupt variation near the surfaces of the plate, are influenced by changes in copper composition in these locations. This raises an important concern regarding the effect of scalping practice on surface alloy composition and consequently eddy current conductivity measurements.

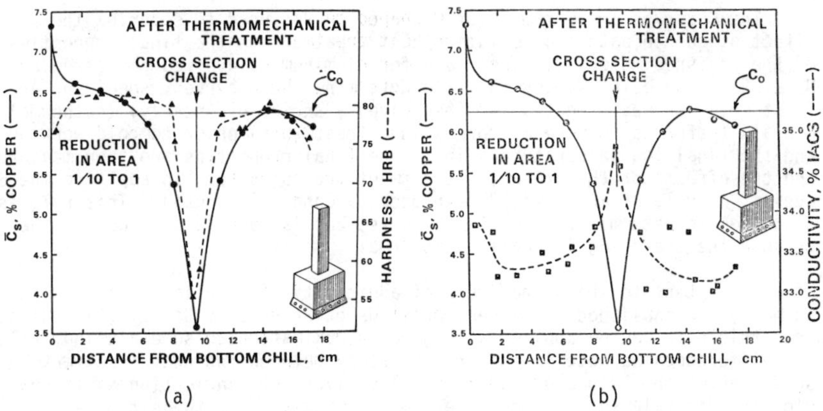

Figure 4. (a) Hardness and (b) conductivity after thermomechanical treatment to the T87* condition and copper content versus distance from the bottom chill in a unidirectionally solidified reduced cross section laboratory ingot.

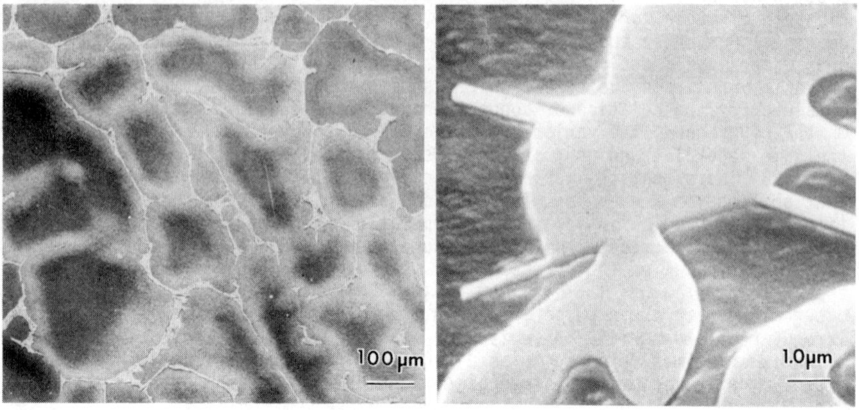

(a) (b)

Figure 5. Typical SEM views of the as-cast microstructure of 2219 aluminum alloy. (a) Dendritic structure. (b) Interdendritic eutectic region showing α-Al (dark), $CuAl_2$ (gray) and Cu_2FeAl_7 (light blades).

C-Curve Determination and NDE Relationships

A comprehensive thermomechanical treatment program was undertaken to provide samples for establishing the effect of process variables on the microstructural, mechanical, and NDE properties of 2219-T87* aluminum alloy and to establish TTT curves (C-curves) that could be used in assessing the effect of various quenching treatments on the final properties of the processed material.

The C-curves are a family of C-shaped curves used to describe the effect of quench path from solution heat treatment on the final properties of the finished material. Their use for aluminum alloys was pioneered by W. L. Fink and L. A. Willey (3). To determine these curves, specimens are first solution heat treated and then given a series of (nearly) isothermal anneals (referred to here as "pre-aging" treatments) prior to cold working and the final low temperature aging. The final properties are then measured and the effects of the pre-aging treatment are assessed. In applying the C-curves, a "rule of additivity" developed by Cahn (4) is used. This rule is also used in determination of the C-curves but is much less important there because the pre-aging is nearly isothermal.

The extent to which the "rule of additivity" is applicable to aluminum is not yet established. However, previous experience by Staley (5) indicates that this rule can be applied with good approximation to some aluminum alloys. The differences we note in the present study between two heat treatment sequences of the 2219 aluminum alloy also gives some indication as to how closely this rule is followed. We use the results from these two sequences in a subsequent section to set limits on what might be obtained under "best" and "worst" quenching conditions for a 2219-T87* aluminum alloy plate.

Because only a 5% stretch was used during the thermomechanical treatment, the temper designation T87* is used here in lieu of T87 which specifies a 7% stretch. A brief study revealed no significant improvement in strength or hardness between a 5 and 7 percent stretch. However, other parameters, especially conductivity, can be expected to vary somewhat.

The treatments were carried out using, as starting material, an 0.625 cm (1/4") thick plate of 2219-T87*. The plate width was 1.22 m (4 ft.) and the plate length was 1.83 m (6 ft.). Three samples of the plate were removed for chemical analysis 0.6 m (2 ft.) from the plate end, one from each edge and one from the middle. These samples were analyzed using emission spectroscopy by the Center for Analytical Chemistry at NBS. Results of this analysis are shown in Table I.

TABLE I

Report of Analysis of three samples from the 0.635 cm thick plate of 2219-T87* aluminum alloy used to establish C-curves. Samples were analyzed by emission spectroscopy. The values labeled 2219 are specifications according to Fed. Spec. QQ-A-250/30.

Sample	Si	Fe	Cu	Mn	Mg	Zn	Ti	V	Zr	Other Total
2219	.2 max	.3 max	5.8-6.8	.2-.4	.02 max	.1 max	.02-.10	.05-.15	.10-.25	.15
LEFT	.08	.04	6.72	.35	.006	.029	.044	.083	.107	.03*
MIDDLE	.08	.05	6.65	.35	.006	.029	.045	.082	.107	.03*
RIGHT	.08	.04	6.79	.35	.006	.029	.045	.084	.108	.03*

* Cr + Ni

The plate was then cut into bars approximately 2.5 x 17 cm. A jig was constructed to hold four bars for simultaneous heat treatment. The samples were instrumented with calibrated thermocouples. Each set of specimens were then given one of two heat treatment sequences which we refer to as <u>sequence A</u> and <u>sequence B</u>. These treatments consisted of:

(i) Solution heat treatment at 535 °C for 75 minutes;
(ii) sequence A alloys; direct transfer to salt bath; sequence B alloys; water quench followed by transfer to salt bath (for both sequences, salt bath temperture was varied between 250 °C and 475 °C; time in salt bath was varied between 2 and 3600 seconds);
(iii) water quench;
(iv) mechanically stretch to 5% permanent strain;
(v) age in air for 16 hours at 172°C.

The time-temperature profiles of these sequences are schematically illustrated in Figure 6. It should be noted that sequence A and B treatments differ only in step (ii) above. The hardness and conductivity of all specimens were measured before the solution heat treatment, after the solution heat treatment and pre-aging treatment, after the 5% stretch, and after the final aging treatment. Two hardness measurements and one conductivity measurement were made on each of 289 samples cut from the 0.635 cm (1/4") thick plate of 2219-T87*. The average hardness in the as-received T87* condition was 78.4 HRB with a standard deviation of 1.0 HRB. The average conductivity obtained was 33.6% IACS with a standard deviation of 0.3% IACS. The longitudinal and transverse yield strengths (0.2% offset) and ultimate tensile strengths were measured on six samples each cut from the same plate.

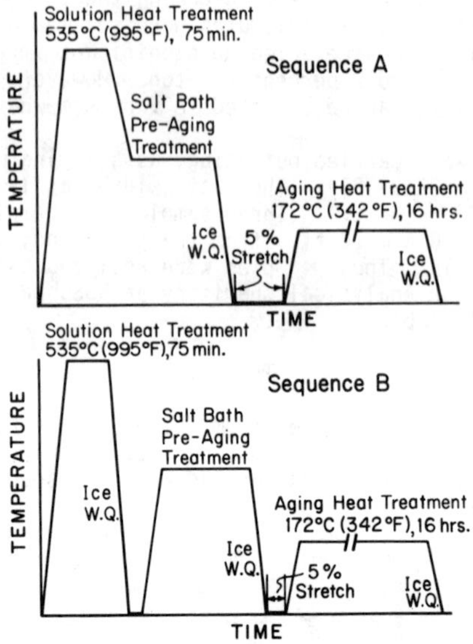

Figure 6. Schematic representation of the thermomechanical treatment sequences given the 2219-T87* aluminum alloys used in this study.

The longitudinal specimens gave a yield strength of 56.4 ± 0.7 ksi and a tensile strength of 70.9 ± 0.7 ksi. All results reported below are for samples with their long axis parallel to the rolling direction.

Twelve specimens from the as-received 2219-T87* were reprocessed with no pre-aging treatments. These reprocessed samples gave an average hardness of 78.1 HRB with a standard deviation of 1.1 HRB, an average conductivity of 33.6% IACS with a standard deviation of 0.4% IACS, a yield strength of 55.4 ± 0.8 ksi and a tensile strength of 69.4 ± 0.8 ksi. These results show that the reprocessed 2219-T87* closely achieves the mechanical and electrical properties of the original plate with perhaps a small degradation (on the order of 1%) in the measured mechanical properties.

Aluminum alloy 2219 as indicated earlier has a nominal concentration of ∿6.3 wt% Cu with other minor constituents (see Table I). The phase transformation characteristics of this alloy are dominated by the presence of Cu and are similar to those exhibited by the binary Al-6.3 wt% Cu alloy. The aluminum rich end of the Al-Cu phase diagram is shown in Figure 7. At equilibrium, the solid consists of a two phase mixture of α-aluminum and the intermetallic compound, Θ-Al_2Cu. This equilibrium, however, is not achieved at low temperatures or with rapid cooling rates. Instead, copper rich G.P. zones, and the metastable Θ'' and Θ' phases that are indicated in Figure 7 may form. The outstanding mechanical properties of Al-Cu alloys are based on the precipitation of a fine dispersion of G.P. zones and the metastable phases. The final microstructural state will depend on the thermomechanical processing history. Although the behavior of the 2219 aluminum alloy is in its essentials similar to the comparable Al-Cu binary alloy, the presence of the minor elemental constituents is important for grain refinement during casting and for other purposes. A typical microstructure after processing to the T87* temper is shown in Figure 8a. A large concentration of Θ'

precipitates (seen edge on) which form during the aging treatment is evident. Also present are Θ" as well as a few large precipitates which may be Θ ($CuAl_2$), present because the copper concentration exceeds the 5% solubility limit at the solution heat treatment temperature of 535 °C or precipitates such as Cu_2FeAl_7 associated with other trace elements in 2219. The yield strength of the sample of Figure 8 was 55 ksi, the conductivity was 33.1 % IACS, and the Rockwell B hardness was 77 HRB.

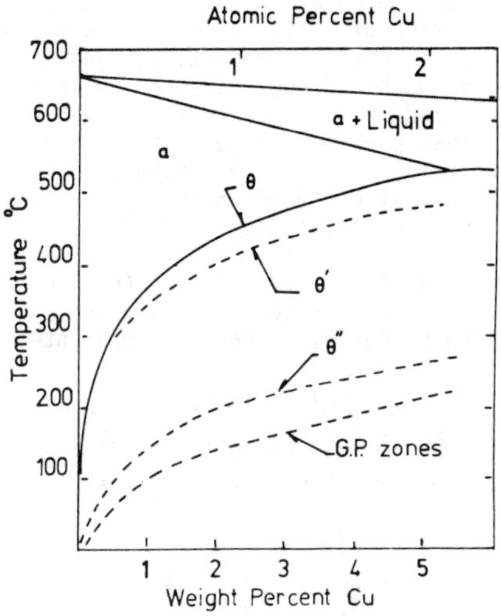

Figure 7. Al-rich portion of the Al-Cu binary equilibrium phase diagram, with G.P. zone, Θ" and Θ' metastable boundaries. From Lorimer (6).

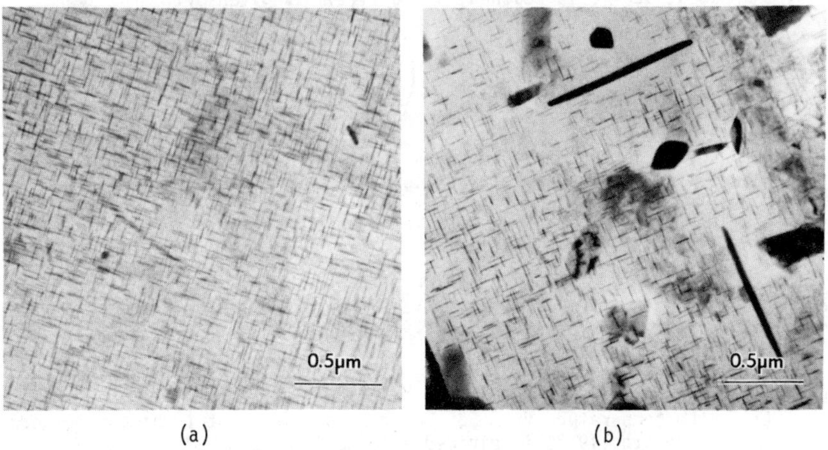

Figure 8. TEM view of the microstructure of 2219 aluminum alloy after processing to (a) the T87* temper. A high concentration of fine Θ' precipitates is evident. (b) Same alloy after a sequence A preaging treatment at 400 °C for 60 seconds. Note the large Θ and Θ' precipitates.

The microstructure of a 2219-T87* sample which was given a sequence A pre-aging treatment at 400 °C for 60 seconds is shown in Figure 8b. Large precipitates of both Θ and Θ' are present. Many of these formed during the quench interrupt period. Again the fine Θ' precipitates which form during the aging treatment are evident. This sample had a yield strength of 44 ksi, a conductivity of 36.2 % IACS, and a hardness of 61 HRB.

The large Θ and Θ' precipitates have little direct effect on hardness or conductivity in either the aged or unaged condition. However, indirectly they lower the concentration of available copper in the aluminum matrix. This yields an increased electrical conductivity and a lower hardness in the unaged material. Since age hardening response is ultimately determined by the concentration of Θ'' and fine Θ' precipitates that form during final aging, the concentration of these precipitates will tend to be lower when the availability of matrix copper is reduced. The increase in conductivity observed in the material with lower strength after aging is caused in this case primarily by the reduced precipitate concentration.

Following Evancho and Staley (7) and Cahn (4), we have used the following parameterization to represent the C-curves: It is assumed that the value of a resulting property, σ, can be represented as

$$\sigma = (\sigma_m - \sigma_o) \exp(-K_{1x}\tau_x) + \sigma_o \tag{1}$$

where σ_m is the maximum achievable property, σ_o is the minimum of "intrinsic" value of the property achieved under the given conditions, τ_x is given by

$$\tau_x = \int_{t_s}^{t_o} \frac{dt}{C_x(T(t))} \tag{2}$$

with t_s being the start time of the quench, t_o the time when a temperature less than about 120 °C is reached, and $C_x(T(t))$ is given by

$$C_x(T(t)) = K_{1x}\, K_2 \exp\left(\frac{K_3 K_4^{\,2}}{RT(K_4 - T(t))^2}\right) \exp\left(\frac{K_5}{RT(t)}\right) \tag{3}$$

where K_2, K_3, K_4, and K_5 are constants to be determined, T is the absolute temperature, and K_{1x} is an arbitrary constant taken to be

$$K_{1x} = \ln\left(\frac{\sigma_x - \sigma_o}{\sigma_m - \sigma_o}\right) \tag{4}$$

K_{1x} is chosen so that for $\tau_x > 1$, σ is less than σ_x so that τ_x becomes a critical parameter for achieving the value σ_x.

To determine the parameters σ_m, σ_o, K_2, K_3, K_4, K_5 the following procedure was used:

(i) the temperature from 535 °C (the solution heat treatment temperature) to 119 °C is divided into 35 intervals,

(ii) the time, t_i, spent in each of the temperature intervals is taken from a chart recorder and used to numerically calculate the integral of Equation (2) according to

$$\tau_x = \sum_{i=1}^{n} \frac{t_i}{C_x(T_i)} \tag{5}$$

where T_i is the average temperature in the interval, and

(iii) an iterative, non-linear, fitting routine was used to obtain values of the parameters which minimize the least squares deviation between measured and calculated properties while requiring K_3, K_4, and K_5 to maintain the same values for a given sequence.

The set of parameters obtained for the C-curves for sequence A and sequence B alloys are given in Table II. The C-curves for yield strength are represented in Figure 9.

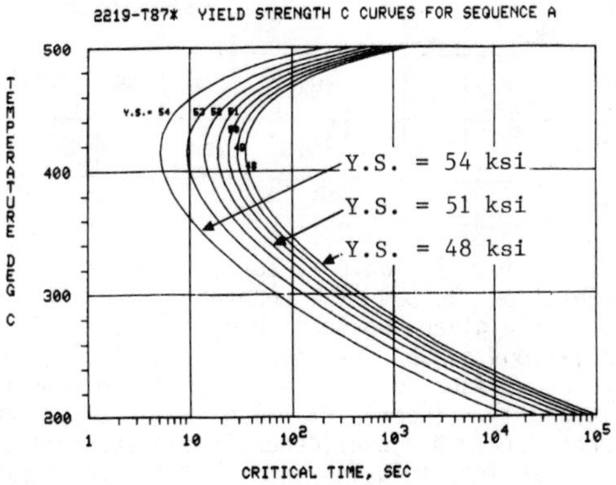

(a)

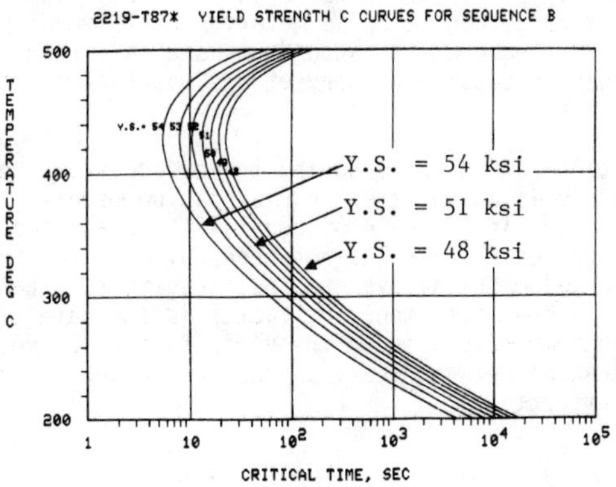

(b)

Figure 9. Yield strength C-curve representation for (a) sequence A and (b) sequence B. The curves give the critical times for obtaining the indicated yield strength (0.2% offset).

TABLE II

C-Curve Parameters for Aluminum Alloy 2219-T87*

Property	Sequence	σ_m	σ_o	K_2 sec	K_3 cal/mol	K_4 °K	K_5 cal/mol	est. std. dev.
Hardness, HRB	A	79.0	46.0	$.86x10^{-10}$	320	900	32,000	2.2
Yield Strength (0.2% offset) ksi	A	55.3	35.4	$.78x10^{-10}$	320	900	32,000	1.5
Tensile Strength ksi	A	69.0	45.1	$.79x10^{-10}$	320	900	32,000	1.7
Conductivity % IACS	A	33.2	37.7	$.59x10^{-10}$	320	900	32,000	0.4
Hardness, HRB	B	80.6	53.6	$.28x10^{-7}$	200	900	25,000	1.9
Yield Strength (0.2% offset) ksi	B	56.5	35.7	$.37x10^{-7}$	200	900	25,000	1.4
Tensile Strength ksi	B	69.5	45.4	$.43x10^{-7}$	200	900	25,000	1.3
Conductivity % IACS	B	32.3	37.4	$.17x10^{-7}$	200	900	25,000	0.4

The set of values given in Table II for the C-curve parameters can be used to obtain correlations between the mechanical and electrical properties. Since the C-curves, for a given sequence, have the same values for K_3, K_4, and K_5, these correlations can be computed parametrically. Two such plots which display the correlations for both sequence A and sequence B alloys are shown in Figures 10 and 11. These plots also show the appropriate data and scatter bands (approximately 95% confidence level) for each correlation. A degree of caution is required in using conductivity measurements to access mechanical properties. Figure 12 shows hardness versus conductivity data for 2219 at intermediate stages in thermomechanical processing to the T87* temper and compares this with the final aged material. It is evident that a single conductivity value can refer to material at three different hardness levels. Conductivity measurements should be used only when the thermomechanical and other processing history of the material is relatively well known.

If the time-temperature curve during the quench from the solution heat treatment temperature is known, the C-curves can be used to predict the final properties of the alloy if the rule of additivity is assumed. The extent to which additivity applies to 2219 has not been fully tested. Previous experience (7) with other aluminum alloys has indicated that it can be applied with good approximation. The most important factor is the rate of temperature decrease while the temperature is in the vicinity of the "nose" of the C-curve. The differences between the C-curves for sequence A and sequence B is small but significant.

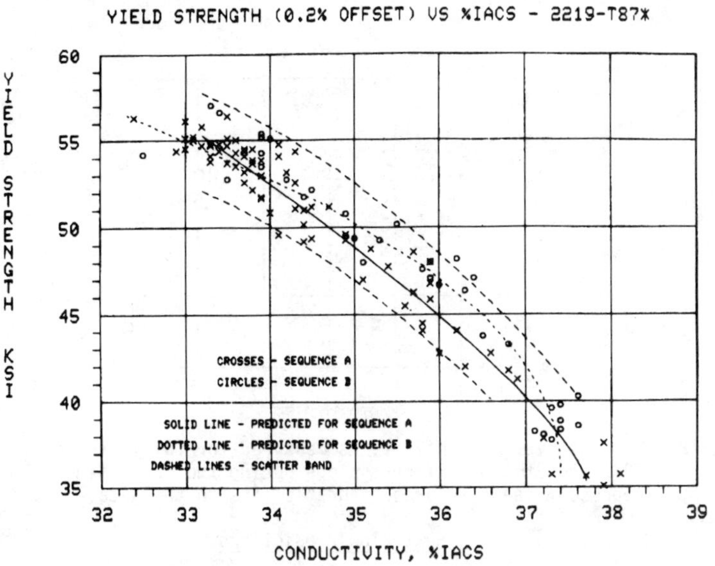

Figure 10. Comparison of yield strength vs. conductivity data with correlations predicted by the C-curves. The dashed lines are the scatter band (~95% confidence level) obtained from a least squares quadratic fit to the data.

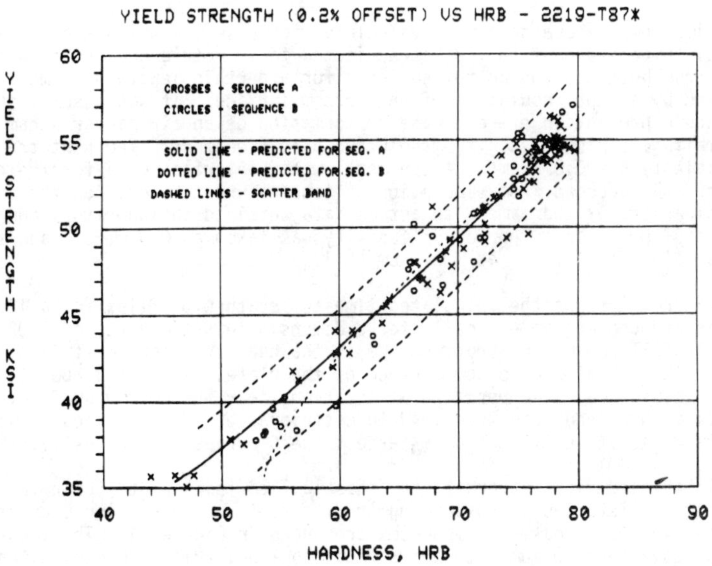

Figure 11. Comparison of yield strength vs. hardness data with the correlations predicted by the C-curves. The dashed lines are scatterband (~95% confidence level) obtained from a least squares quadratic fit to the data.

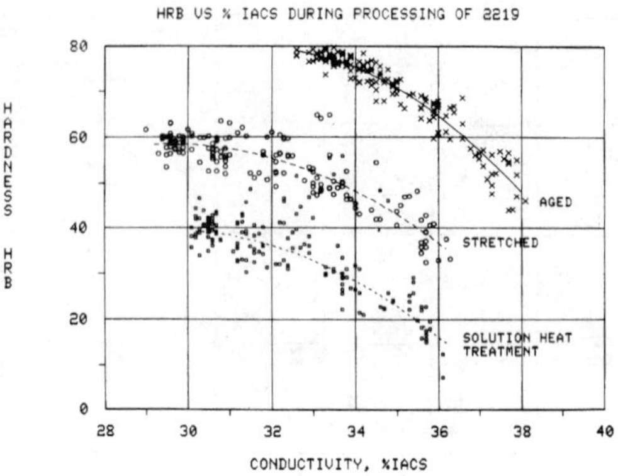

Figure 12. Hardness vs. conductivity as measured for fully processed 2219-T87* and at two different steps during the processing. The lines represent least squares quadratic fits to the data.

Heat Flow Predictions

Heat flow calculations were carried out and combined with C-curves (time-temperature transformation curves) developed for the 2219 aluminum alloy subjected to both sequence A and sequence B heat treatments. The minimum properties in plates of different thicknesses were predicted for the "worst case possible" heat flow condition.

Time-temperature data were calculated for cases in which a plate at an initial temperature of 535 °C was subjected to symmetric cooling by water at 40 °C from both top and bottom surfaces for a certain length of time, followed by abrupt insulation of the bottom surface. It was assumed that this would present an extreme case of formation of an air gap or steam pocket in a malfunctioning quench from solution heat treatment. The heat transfer coefficient, h = 0.8 W/cm^2K at the metal-water interface used for the calculations, was deduced from simulation of temperature-time data on the computer and comparison of the same with actual data obtained in laboratory and proper commercial practice. Thermal diffusivity was taken as 0.5 cm^2s^{-1} and the thermal conductivity, k, as 1.2 W $cm^{-1}K^{-1}$.

Figure 13 shows the calculated time-temperature profiles for a 12.7 cm thick plate cooled symmetrically for a dimensionless time* (τ) of 0.02 (6.45s), followed by an abrupt change in the heat transfer coefficient from 0.8 W/cm^2K to 0 at the bottom surface of the plate. The data shows significant recalescence in temperature at locations near the bottom surface. This reheating phenomenon can then lead to deterioration of properties if it results in additive times that intersect the C-curves for the alloy.

Calculated yield strengths for this 12.7 cm thick plate for a variety of temperature histories corresponding to different τ values using C curve data for the two heat treatment sequences are shown in Figure 14. The value of $\tau=\infty$ is equivalent to complete symmetrical cooling, while $\tau=0$ depicts conditions

* Dimensionless time (F_o) is $\alpha t/L^2$ where α (= 0.5 cm^2/s) is thermal diffusivity, L is plate thickness in cm and t is time in seconds.

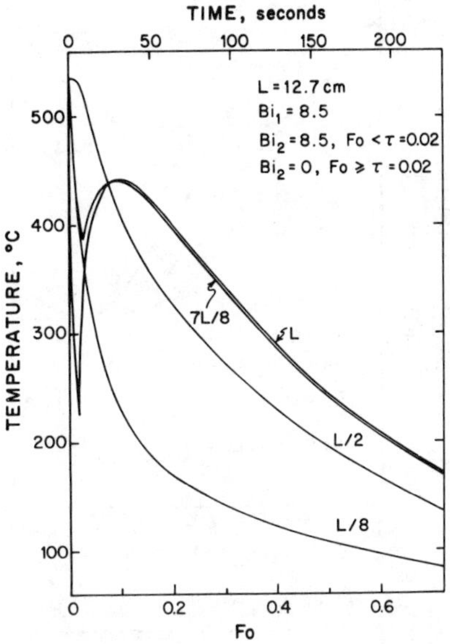

Figure 13. Calculated time-temperature profiles at four locations in a 12.7 cm thick plate of 2219 aluminum alloy subjected to a normal heat extraction (Bi_1 = 8.5) on its top surface and an abrupt discontinuation of heat extraction (Bi_2 = 0 for Fo $\geq$ 0.02) from its bottom surface. Bi = hL/k (Biot Number).

under which the bottom surface is insulated for all times. The predictions of yield strength vs. distance for symmetric cooling ($\tau=\infty$, sequence A) are in good agreement with the yield strength data for the correctly heat treated plate shown in Figure 1b. It is interesting to note that the lowest predicted properties occur for τ=0.02 ($\tau \simeq$ 6.45 seconds). Trial and error showed that this should represent about the "worst" conditions. For example, for sequence A this corresponds to minimum yield strength of 42.6 ksi, at X/L = 0.09 (∿11.4 cm from the top surface). This represents a maximum deterioration in yield strength of 13% below specification. If sequence B values are employed, the minimum predicted yield strength at the same location is 39 ksi. The latter probably represents the "worst" possible case under the assumptions of these calculations.

Similar calculations were carried out for all plate thicknesses up to 15.24 cm (6 inches). The data for the worst properties (lowest yield strength) were established using the computer model and trial and error methodology described above. The predictive "worst possible case" yield strength data are shown in Figure 15. Figure 15 also shows the minimum yield strengths noted in Federal Specification QQ-A-250/30. These predictions show that under the "worst" heat flow conditions, sequence B in Figure 15, plates thinner than ∿4 cm should meet the specifications. On the other hand, a 15.24 cm (6 inch) thick plate could have locations with yield strength of ∿37 ksi.

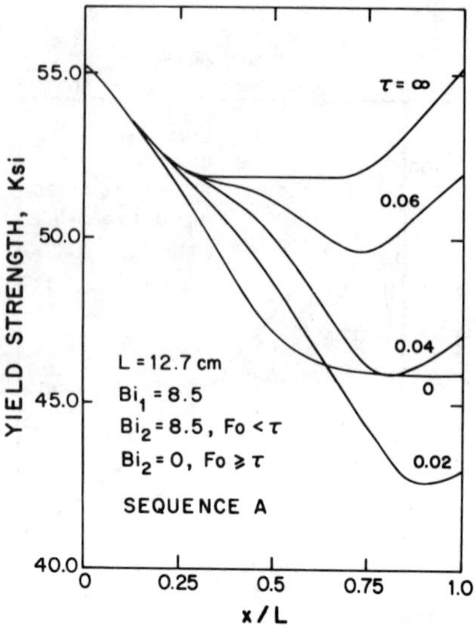

(a)

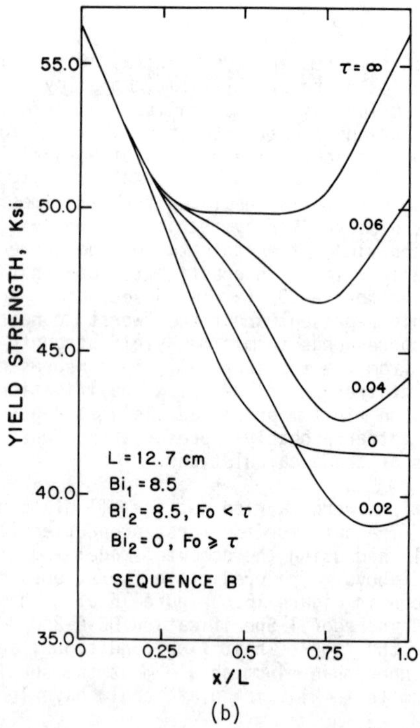

(b)

Figure 14. Predicted yield strength versus distance from the top surface of 12.7 cm thick plates subjected to a variety of heat flow conditions at the bottom surface. Curves are calculated using C-curves previously determined for (a) sequence A and (b) sequence B.

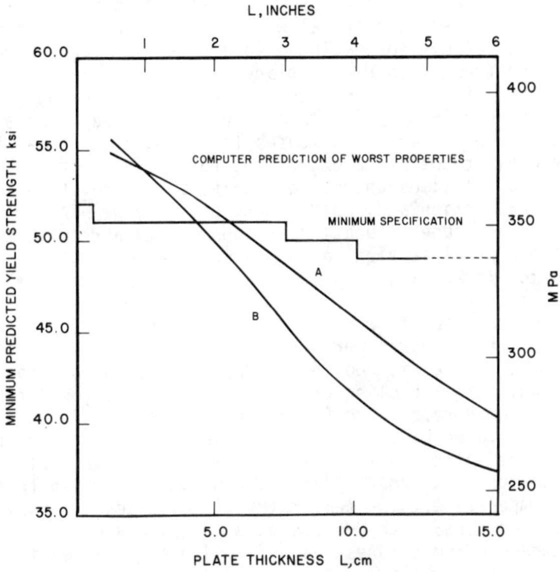

Figure 15. Predicted minimum yield strength in different thickness plates under the "worst possible" heat flow conditions using sequence A or sequence B data.

Summary and Conclusions

Experimental studies were carried out on an industrial size semi-continuous cast ingot and on a smaller laboratory cast ingot of 2219 aluminum alloy. The major phases present in the as-cast condition were α-aluminum solid solution, Θ-$CuAl_2$ and Cu_2FeAl_7. The positive and negative macrosegregation of copper was investigated. This macrosegregation is predictable and is caused by interdendritic fluid flow during casting. It cannot be completely eliminated from finished alloy plate by normal commercial scalping and thermomechanical treatment practice. However, the moderate copper concentration variations found to exist through the thickness of a 12.7 cm thick plate could not account for the measured variation in physical properties. Experiments conducted on correctly heat treated samples from the laboratory ingot indicate that mechanical properties are maintained as long as the copper content is above the maximum solid solubility limit ($\sim$5.5% Cu). Such was the case for the 12.7 cm thick plate examined. Hence property variations across as-received plates are not due to composition variations, but instead is attributed primarily to depth dependent differences in quench rate from the solution treatment temperature.

A comprehensive series of thermomechanical heat treatments on 2219-T87* aluminum alloy was performed. The hardness and electrical conductivity were monitored at each stage of the treatment and the mechanical properties of the finished material were determined. The results were utilized to develop curves (C-curves) that can be used to assess the effect of various quenching treatments on the final mechanical and NDE properties of processed material.

In this study, C-curves were also used to determine correlations between mechanical properties and electrical conductivity. Such a correlation was found to exist for alloys (of a single lot) given an improper quench from solution heat treatment, other conditions being maintained constant. When more than one condition is varied, the correlation may be destroyed, however. The principal age-hardening phase in properly processed material is Θ'. The Θ'' phase is also present and contributes to strength. A pre-aging heat treatment, or other temperature excursion, following solution heat treatment that results in dwell times which are significant in relation to the C-curves, results in the nucleation and rapid growth of large Θ and/or Θ' precipitates, leading to a reduced supersaturation in copper. As a result, the net concentration of the precipitation hardening species, Θ' and Θ'', is decreased and a degradation in mechanical properties is obtained.

One dimensional heat flow calculations for the cooling of aluminum plates of various thicknesses were carried out for three models of heat transfer, viz. (a) symmetric cooling of the plate from both top and bottom surfaces, (b) asymmetric cooling of the plate from the top surface only with an insulated bottom surface, and (c) symmetric cooling of the plates for a specified time followed by an abrupt reduction of the heat treatment coefficient at the bottom surface to zero (insulated surface). The latter situation simulates steam pocket formation during quenching from solution heat treated temperature. The temperature-time curves for various cooling conditions were combined with C-curve data to calculate plate properties, e.g., yield strength and hardness as a function of position in the plate. Such calculations for symmetric cooling are in general agreement with property variations observed in as-received and properly heat treated plate. Assuming proper heat transfer on one side of the plate, the maximum degradation of yield strength was found to occur when temperatures in the plate recalesced due to interrupted quenching.

Acknowledgement

This research was sponsored by the National Aeronautics and Space Administration. We would like to thank R. Clough and P. Blau for mechanical property measurements, D. A. Laughlin for some STEM results, C. H. Brady and D. Ballard for metallography, F. S. Biancaniello for heat treatments and G. Free for measurements of conductivity. The efforts of P. A. Boyer, R. L. Parke, and R. D. Robbins are also greatly appreciated. We also thank J. J. Fillibin for his aid in plotting and analysis of the data. We would also like to thank D. E. Newbury, R. E. Burke, J. Norris and B. I. Diamondstone for chemical analysis, C. J. Bechtoldt for x-ray measurements, and R. F. Boisvert for advice and assistance on the heat flow calculations. Finally, we are indebted to Dr. S. A. Levy and associates of the Metallurgical Research Division of Reynolds Metals Company for frequent consultation and for the ingot and plates made available for this study.

References

1. L. Swartzendruber, L. Ives, W. Boettinger, S. Coriell, D. Ballard, D. Laughlin, R. Clough, F. Biancaniello, P. Blau, J. Cahn, R. Mehrabian, G. Free, H. Berger and L. Mordfin, NBSIR 80-2069 NBS: Nondestructive Evaluation of Nonuniformities in 2219 Aluminum Alloy Plate - Relationship To Processing, 1980.

2. R. Mehrabian and M. C. Flemings, Metallurgical Transactions, 1 (1970) p. 455.

3. W. L. Fink and L. A. Willey, Transactions AIME, 175 (1948) p. 414.

4. J. W. Cahn, Acta Metallurgica, 4 (1956) p. 572; p. 449.

5. J. T. Staley, Industrial Heating, 44 (1977) p. 6.

6. G. W. Lorimer, Precipitation Processes in Solids, Metallurgical Society of AIME, (1976) p. 87.

7. J. W. Evancho and J. T. Staley, Metallurgical Transactions, 5 (1974), p. 43.

ACOUSTIC MICROSCOPY OF CERAMICS*

L. W. Kessler, D. E. Yuhas and C. L. Vorres

Sonoscan Inc.
530 E. Green Street
Bensenville, Illinois

The scanning laser acoustic microscope (SLAM) is applied to the problem of nondestructive testing of ceramic materials. The employment of a very high resolution ultrasonic imaging instrument provides capabilities which supplement ultrasonic pulse-echo testing. In addition, by means of showing a picture of the flaw, the SLAM technique provides its own capabilities which alleviate some of the limitations of other techniques. Flaws of various types are documented in this paper.

* Portions of this work were supported by the ARPA/AFML program on Quantitative NDE administered through Rockwell International Science Center, Thousand Oaks, California.

A desirable performance goal for an ultrasonic inspection method is to be able to confidently and rapidly obtain information on the presence of flaws, on their size, shape and location, and in the case of an inclusion, an identification of the foreign material. With the above information and an a priori knowledge of what constitutes a critical flaw, a decision can be made concerning accepting or rejecting the piece being inspected. With regard to characterizing small flaws (e.g. in ceramic materials) there are obstacles to meeting the above goals by means of "conventional" (pulse echo) ultrasonic testing technique.

In the "pulse-echo" technique the "signal" from the flaw depends upon many factors, for example the flaw characteristics as well as its location within a non-uniform acoustic field produced by the inspection transducer. Since the transducer integrates the acoustic signal over its entire face, a composite signal is generated which depends upon the flaw size relative to the transducer area. The flaw within the material being tested might not be a single-regular-shaped entity such as an idealized spherical inclusion (which would be amenable to mathematical analysis). In fact, a typical flaw may be an irregularly shaped inclusion with pores or microcracks alongside. The sample surface might not be flat as well, thus making the scanning operation difficult in searching for flaws.

No single technique can alleviate all the above problems; however, an imaging approach can serve useful functions in these areas: (1) to provide a basis for interpreting conventional ultrasonic signals, (2) to independently provide diagnostic information on flaws, and (3) to rapidly scan large areas for suspicious structures. Though an imaging system generally requires more sophisticated equipment compared to a pulse-echo type flaw detector, often the information obtained (pictorial image) is simpler to interpret.

With all imaging techniques the acoustic field is probed on a point-by-point basis instead of being integrated into a composite electrical signal. The scanning can be a slow raster generated by a mechanical apparatus, such as a C-scan system or by more rapid electronic scanning with full grey images. If the images are produced rapidly enough, the potential exists for an on-line quality control inspection.

The results obtained in this paper were obtained on a commercially available Scanning Laser Acoustic Microscope (SLAM) operating at 100 MHz. This instrument, by virtue of the high frequencies employed (viz 30-500 MHz) make it capable of achieving very high resolution, and which, by its novel scanning-laser-ultrasonic detection scheme make it capable of very rapid imagery(30 images per second, typical). The principle of operation of the SLAM is described in the literature and will not be discussed here (1-8).

The lower limit on detectable flaw size is set by the wavelength. The resolution, i.e. the smallest defect (of some physical property) which can be identified in terms of lateral dimensions is on the order of 25 microns at 100 MHz, which is half an acoustic wavelength. However, the detectability, i.e. the smallest defect which can be found without focusing, is an order of magnitude better. For example, delaminations that have openings of one micron or less, have been detected as long as their spatial extent is on the order of 25 microns. In the case of microporosity, materials which have sub-resolution porosity can be differentiated on the basis of attenuation characteristics of the sample.

There are two types of acoustic micrographs (images) that will be described. The first is a normal amplitude image in which the brightness of the CRT screen relates to corresponding levels of acoustic energy within the field. The second type of image is an interferogram in which the received acoustic signal is electronically mixed with a phase reference signal so that changes in the velocity of sound propagation through various parts of the sample will be displayed on the screen. Interferograms are characterized by vertical fringes whose spacing will be uniform and straight if the velocity of sound is uniform in a flat specimen. Lateral position shifts of the fringes can be used to calculate change in material properties.

The purpose of this paper is to review the use of SLAM for NDE of ceramics. Results obtained on metals have been reported elsewhere (1,9,10).

Figure 1 is a disc of hot pressed silicon nitride which is shown in both modes, amplitude (top) and interferogram (bottom). The scale is 3 mm horizontally and there appears to be a fair degree of uniformity in this material. The interferogram, whose fringes are spaced by 85 micrometers, reveals slight deviations in the horizontal direction which are due to 3 micron changes in the sample thickness resulting from surface grinding. The grooves also act as acoustic scatterers which are evident as the diagonal structure on the amplitude micrograph.

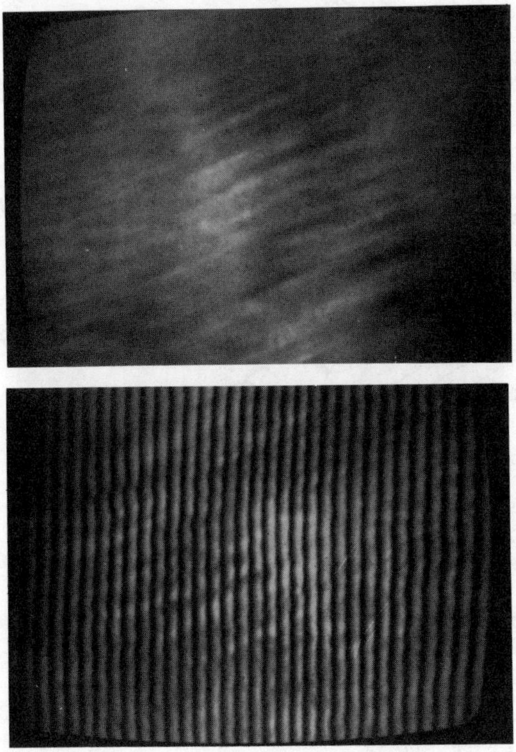

Fig. 1 - Acoustic amplitude micrograph (top) and interferogram (bottom) of a hot pressed silicon nitride disc 7 mm thick. The ultrasonic frequency is 100 MHz and the field of view is 3 mm horizontally.

In order to maximize the information obtained from an acoustic microscope there are several parameters to consider. One would like the highest resolution possible, but, unfortunately, the higher frequencies are also more highly attenuating by the material and there is less signal to noise ratio. Furthermore, as the frequency is increased (and the wavelength gets shorter) surface roughness becomes more critical; here, sample preparation can play an important role. Therefore, one has to compromise resolution, penetration and sample preparation (if it is allowable).

Several types of flaws such as delaminations, cracks, surface cracks and solid or void type inclusions in materials will be described. Figure 2 shows a ceramic chip capacitor which is an electronic component constructed of many layers of barium titanate ceramic material separated by interlayer metal electrodes. Figure 3 is a magnified optical image of a capacitor that has been destructively cross-sectioned. The thickness of each ceramic layer is approximately 25 microns and each electrode layer is about 2 1/2 microns. This image shows typical defects that are found, for example, delamination at the electrode layer and distributed porosity in the ceramic material. When an acoustic micrograph is produced of this component, the acoustic energy is directed up through the stack rather than parallel to the layers.

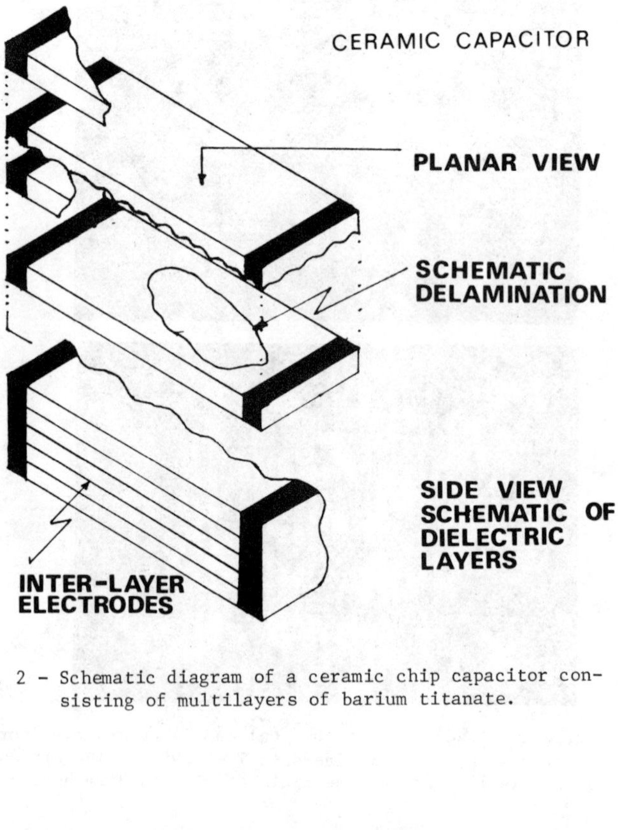

Fig. 2 - Schematic diagram of a ceramic chip capacitor consisting of multilayers of barium titanate.

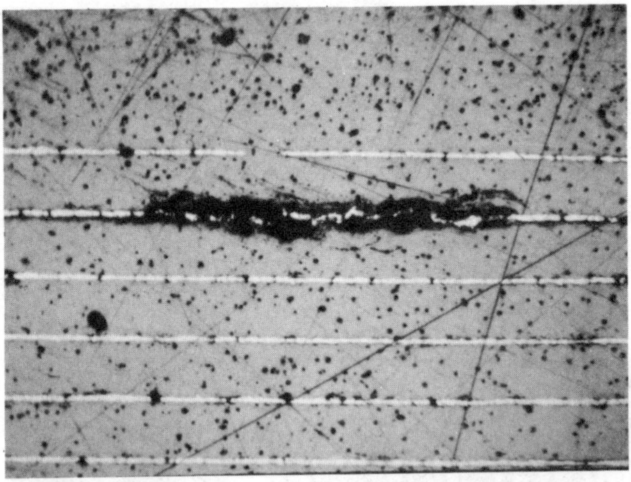

Fig. 3 - Optical view of a cross-sectioned ceramic capacitor revealing delamination at the electrode as well as porosity in the ceramic material.

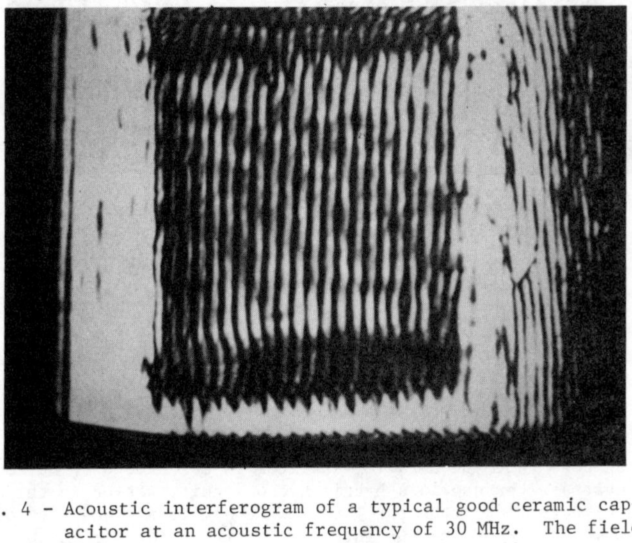

Fig. 4 - Acoustic interferogram of a typical good ceramic capacitor at an acoustic frequency of 30 MHz. The field of view is approximately 10 mm horizontally.

Figure 4 is a 30 MHz acoustic interferogram of a typical good ceramic chip capacitor whose size is approximately 6 x 9 mm. The interferogram fringes are fairly straight and uniform indicating uniform density of the ceramic. It is well established that porosity in sintered materials is associated with increased acoustic attenuation and decreased velocity of sound compared with fully dense material. If there are inhomogeneties evident within the field of view of the microscope, fringe deviations and/or dark areas would be evident. If a delamination exists, a corresponding area of reduced acoustic transmission will be evident as seen in Figure 5. Notice that in a delamination there may still be a slight amount of transmission, but the fringes typically become somewhat scrambled indicating a loss of coherence of the wave. The bright border area, seen in Figs. 4 and 5, is where the acoustic energy is viewed directly, i.e. without passing through the sample. A comparison of the electrical signal levels can be used to ascertain the attentuation characteristics of the sample quantitatively.

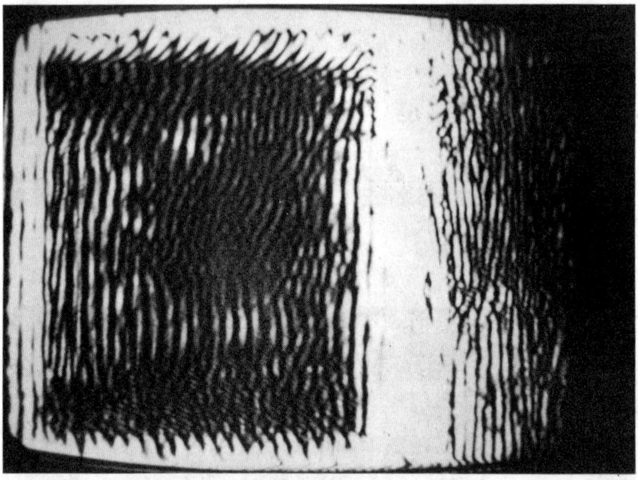

Fig. 5 - Acoustic interferogram of a partially delaminated ceramic chip capacitor.

Now suppose that the ceramic capacitor is soldered onto an alumina (Al_2O_3) substrate for use in a hybrid microcircuit. Prior to the soldering operation we determined that the layers were all intact and free of delamination. Now we can determine the integrity of the solder joints by observing the transmission level through the Al_2O_3-solder-capacitor structure. In Fig. 6, we observe good acoustic transmission in the areas of the bonds, i.e. at either end of the 2 x 1.5 mm capacitor, and therefore, conclude that the bonds are good.

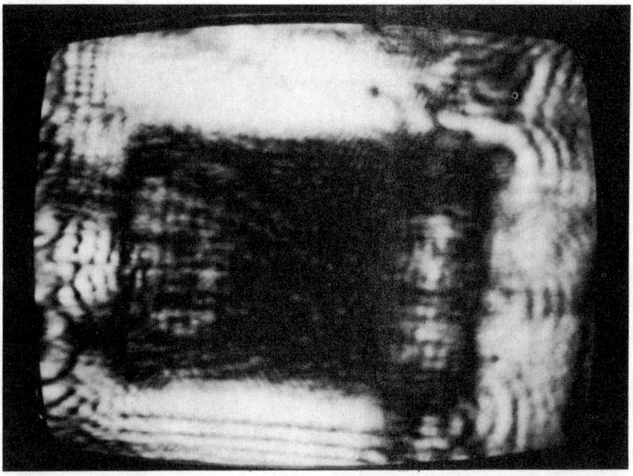

Fig. 6 - Acoustic micrograph (100 MHz) of a capacitor soldered at either end to an alumina substrate. Acoustic transmission through the ends indicates good bonding.

Suppose that a sample has a surface crack. As shown in Fig. 7 acoustic energy incident upon the crack from below will partially convert into propagating modes that travel along and parallel to the surface. The discussion here is particularly relevant to cracks that are shallow, that is, perhaps not more than a few wavelengths of sound deep; at 100 MHz, 25-100 microns would be typical. Referring to the diagram, the area to the far right of the crack at the top surface of the sample has acoustic energy from two sources, a surface mode originating from the crack and a transmitted bulk wave from the bottom. The two waves, which are coherent, will interfere and cause a characteristic cone-shaped pattern in the acoustic micrograph. Fig. 8 shows an acoustic micrograph of a hot pressed silicon nitride sample which has a ground (not polished) surface and a surface crack (arrow) produced by a Vickers hardness tester instrument. The cone-shaped interference pattern is quite evident and extends well beyond the flaw site. This latter observation is useful for easily locating surface flaws since the pattern, not the crack itself, is all that need be detected. Looking at the surface in the SEM, Figure 9, one sees the physical indentation due to the Vickers diamond stylus. Beneath the indent there is a half-penny shaped crack which cannot be seen.

If the crack is deep as illustrated in Fig. 10, and if the incident waves are at an angle with respect to the surface normal, there will be an acoustic shadow, (low transmission area) because the ultrasound will not jump across the air-filled opening. The measured width of the shadow will permit one to determine the depth of the crack provided the angle of insonification is known.

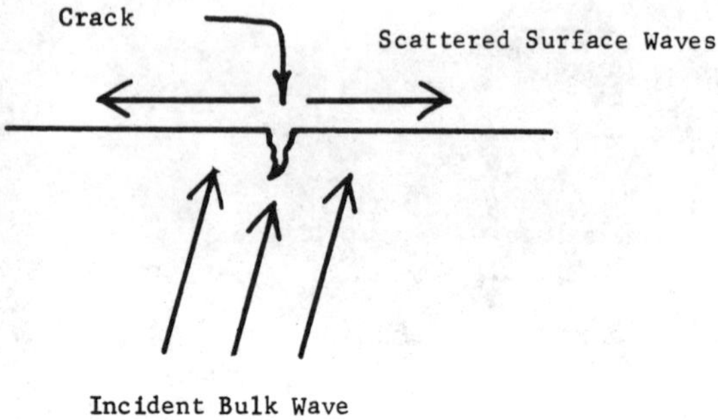

Fig. 7 - Insonification of a sample having a shallow surface crack will result in acoustic energy scattered into modes which propagate along the surface.

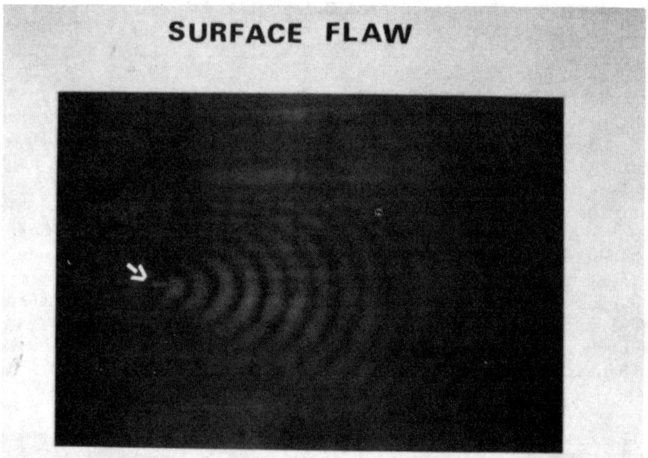

Fig. 8 - Surface flaw in hot pressed silicon nitride. The arrow indicates the site of the crack. The cone shaped interference pattern extending away from the crack is caused by the mixing between the surface skimming wave travelling towards the right and the bulk wave according to Fig. 7.

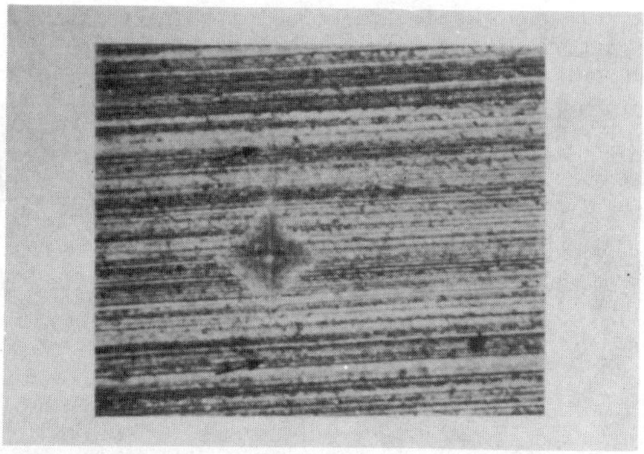

Fig. 9 - Scanning electron microscope view of Vickers indentation which caused the surface flaw on the silicon nitride sample shown in Fig. 8.

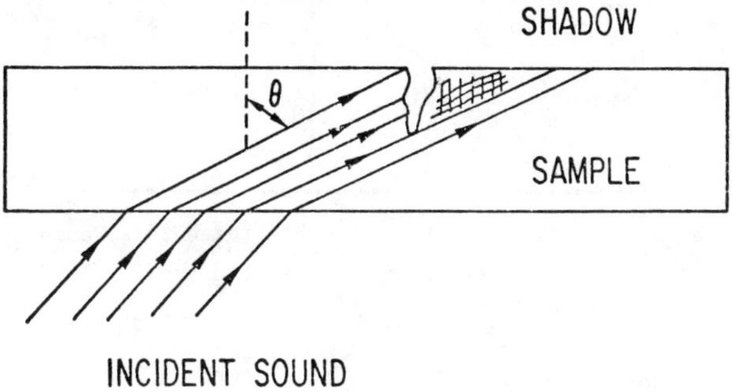

Fig. 10 - Schematic diagram illustrating insonification of a sample having a deep surface crack. A characteristic shadow is produced in the acoustic image whose dimensions are determined by the crack depth and the angle of insonification.

Figure 11 is an acoustic micrograph of a cracked alumina substrate (∿95% density) a few millimeters thick. The sharp onset of the shadow defines the location of the crack at the top surface and the shadow width, about 1 mm, indicates that the crack is also 1 mm deep. (The insonification is at 45°.) The grainy structure in the image is caused by the nonuniform microstructure of the alumina itself. In contrast, high density alumina (>99%) has a very clean acoustic transmission characteristic. The nature of the diffraction pattern will depend on the size and shape of the flaw, and also upon whether the flaw is opaque or transmissive to the acoustical wave.

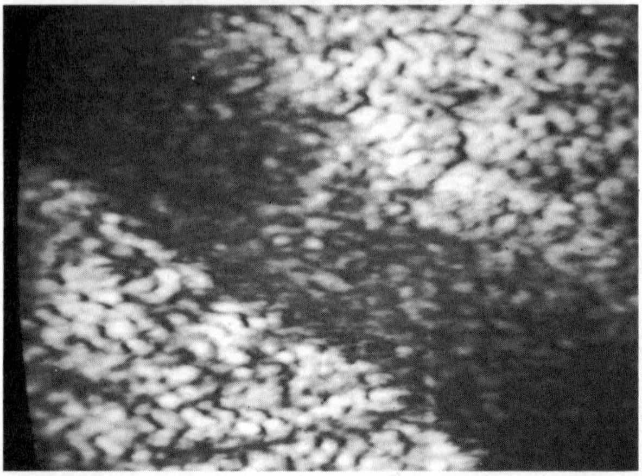

Fig. 11 - Acoustic micrograph of a 1 mm deep crack in an alumina ceramic slab.

If an inclusion or a void is buried deep beneath the surface of a material and viewed in the SLAM there will be scattered and transmitted energy which will interfere with each other giving rise to a diffraction pattern. This is illustrated in Fig. 12.

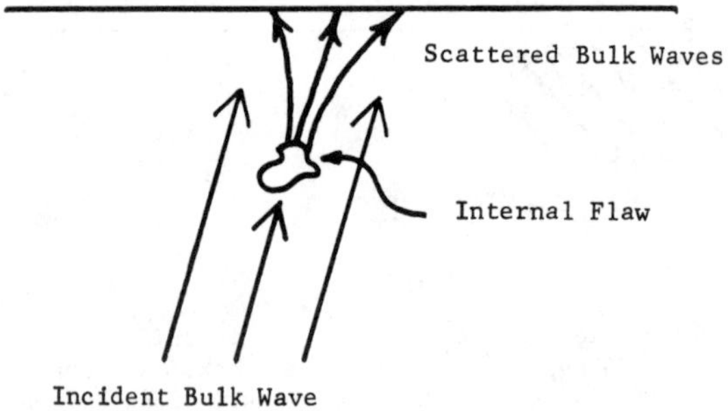

Fig. 12 - Insonification of sample having a void or inclusion deep beneath the surface.

Figure 13 is the acoustic micrograph of a 400 micron iron inclusion implanted within a silicon nitride disc. It is located about 3 mm below the surface which is beyond the near field zone of the flaw. There is a characteristic bright diffraction ring in the center and some higher order rings surrounding it. As the flaw gets closer to the surface the image characteristic changes in a well defined manner. A series of acoustic micrographs was generated by sequentially grinding the surface thereby moving it closer towards the flaw. This sequence is shown in Figure 14. Note that as the flaw gets closer to the image plane (the surface) the diffraction effects are reduced. These effects are an inevitable consequence of any imaging system. In order to determine the flaw size from the image we measure the image of the flaw to the first ring, then the flaw depth by stereometry. With these two data points the flaw size is determined. In the near field of the flaw we don't see diffraction rings and the flaw size is the same as its image.

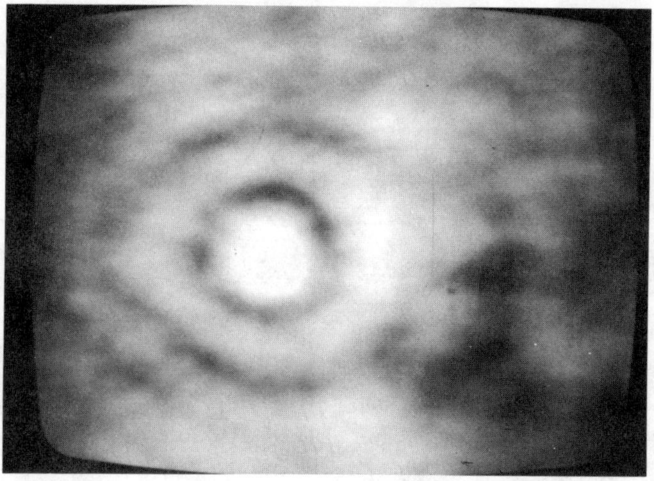

Fig. 13 - An iron inclusion, 400 microns in diameter, is located 3 mm below the surface of a hot pressed silicon nitride sample giving rise to this diffraction pattern.

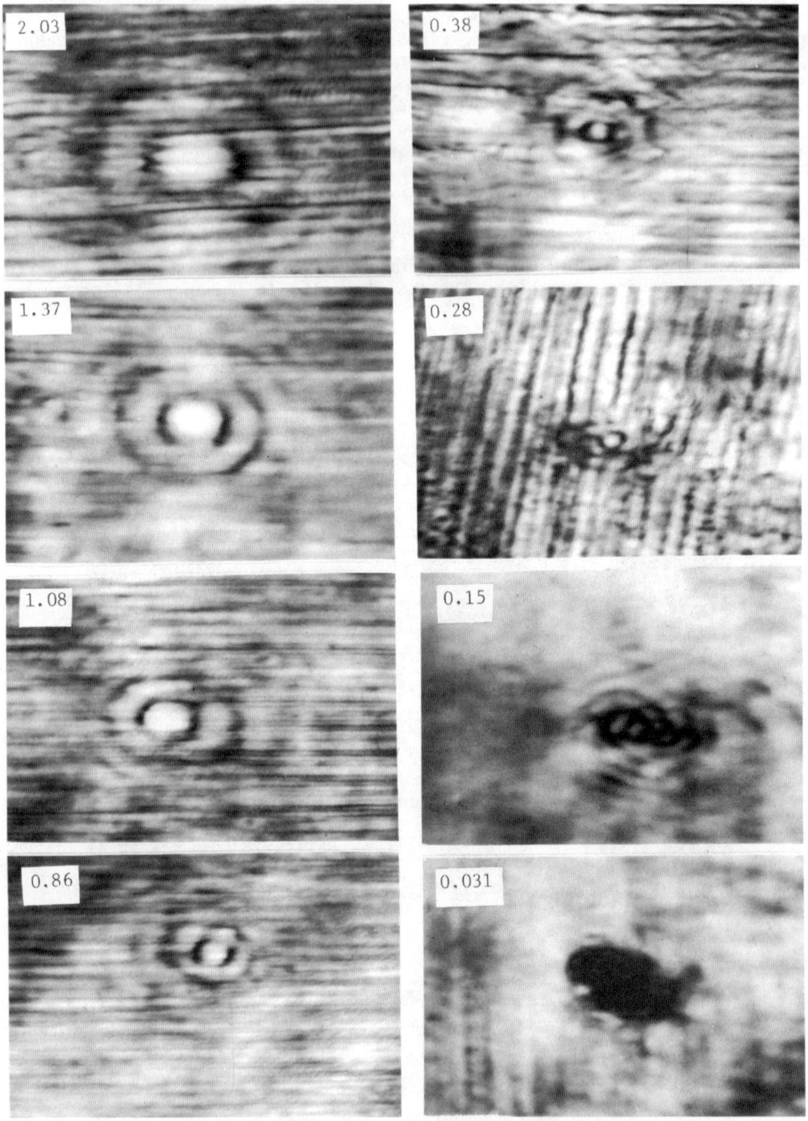

Fig. 14 - Acoustic micrographs (100 MHz) of the same iron inclusion as Fig. 12 as the thickness of the sample was reduced by grinding to bring the flaw closer to the surface. The grooves in each image are due to the rough grinding operation. The depths were measured by a simple stereoscopy technique with the acoustic microscope. The numbers in the flaw location are in millimeters below surface.

To complicate life a little more, Fig. 15 indicates a common problem: flaws are not necessarily singular entities. This is a silicon inclusion located in silicon nitride and there is also some concentrated porosity (dark zones) in the same area of the sample. The SLAM image clearly differentiates the two effects. Suppose we were to try conventional high frequency ultrasonic testing. In this particular image, the horizontal field of view is 3 mm. As a side note, typical pulse-echo transducers are about this size or larger. Thus we could expect some difficulty distinguishing the two flaw types from echo signals.

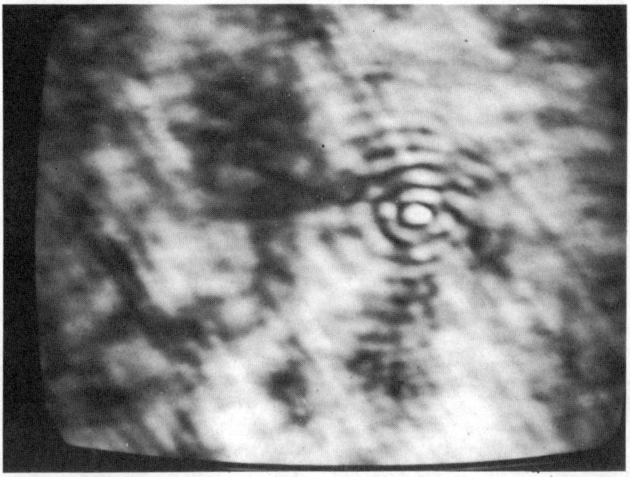

Fig. 15 - Acoustic micrograph of a silicon inclusion and porosity within a silicon nitride sample. The porosity, which is attenuating to the acoustic wave, is located in such a position relative to the inclusion that for certain orientations of the sample one flaw masks the other.

In addition to flat and test samples, the SLAM can be adapted for use on complex geometries. Work has already been performed on turbine blades, tubes and other odd shaped components. In each of these cases, it is up to the investigator to select and optimize an entrance and exit port for the ultrasonic energy. Curved samples will necessarily cause corresponding curvatures to the interference fringes. For example, preliminary experiments have been performed on cylindrical ceramic test bars that are going to be subjected to tensile strength testing. The material is silicon nitride and a photograph of the sample is shown in Fig. 16. An interferogram of a good zone as compared with a suspicious zone is shown in Fig. 17 and it is evident that the differences are clearly distinguishable even though the fringes are curved.

Our approach for further work involves 2 parallel tasks. First, analytical procedures for diffraction phenomenon which have been developed will be brought to bear on SLAM images in order to get as much quantative information about the flaw as is possible. The second task involves establishing a data-flaw library with destructive correlative analysis on naturally occurring flaws. This will guide the analysis work and can be employed to interpret flaw images on actual components which are too complicated for mathematical analysis.

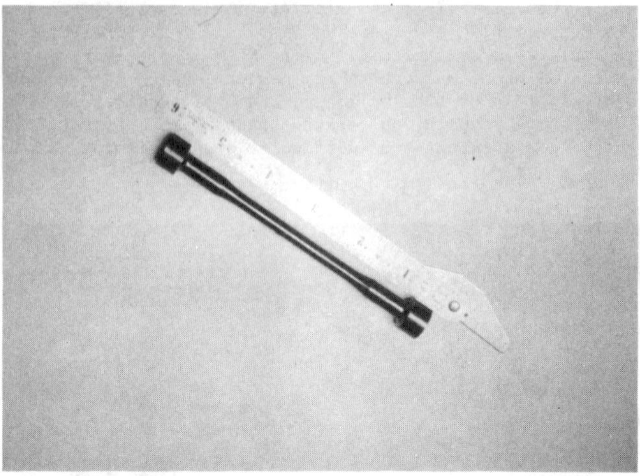

Fig. 16 - Photograph of a cylincrical tensile test bar which is made of silicon nitride.

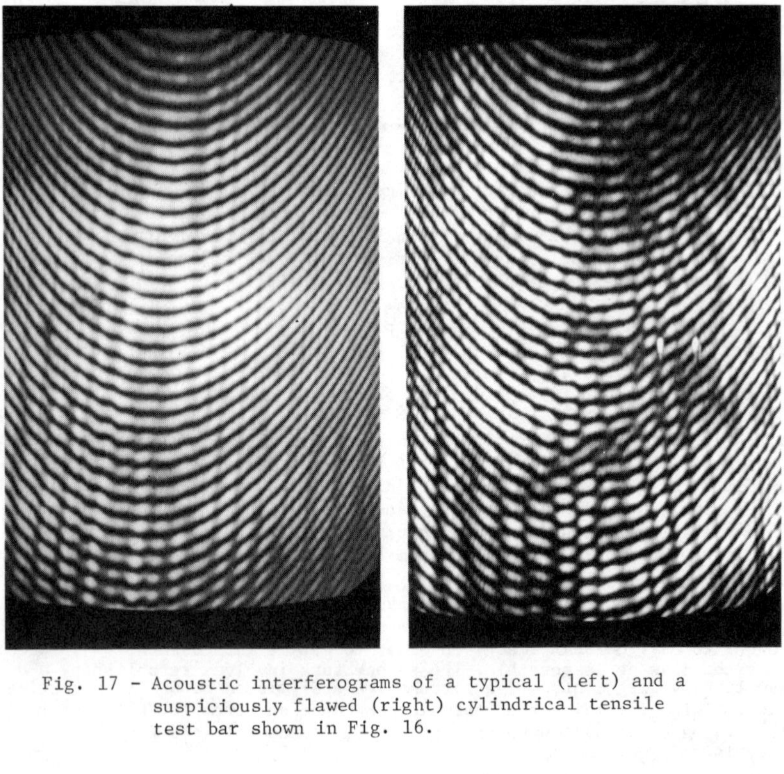

Fig. 17 - Acoustic interferograms of a typical (left) and a suspiciously flawed (right) cylindrical tensile test bar shown in Fig. 16.

References

1. L. W. Kessler and D. E. Yuhas, "Acoustic Microscopy 1979," Proc. IEEE, 67, (4) pp. 526-536 (1979).

2. L. W. Kessler and D. E. Yuhas, "Principles and Analytical Capabilities of the Scanning Laser Acoustic Microscope," Proc. Scanning Electron Microscopy, 1978; Vol. 1, pp. 555-560, AFM O'Hare, IL. 60666.

3. D. S. Kupperman, et al., "Acoustic Microscopy Techniques for Structural Ceramics," Bulletin of the American Ceramic Society, 58, 814 (1980).

4. D. E. Yuhas and L. W. Kessler, "Defect Characterization by Means of the Scanning Laser Acoustic Microscope (SLAM)," Proc. Scanning Electron Microscopy, 1980; Vol. 1, pp. 385-391, SEM Inc., AFM O'Hare, Chicago, IL. 60666.

5. D. E. Yuhas and T. E. McGraw, "Acoustic Microscopy, SEM and Optical Microscopy: Correlative Investigations in Ceramics," Proc. Scanning Electron Microscopy, 1979; Vol. 1, pp. 103-110, SEM Inc., AMF O'Hare, IL. 60666.

6. D. S. Kupperman, et al., "Evaluation of Ceramic Turbine Blade with an Acoustic Microscope," Proc. First Int'l. Symp. on Ultrasonic Materials Characterization, June 7-9, 1978, National Bureau of Standards, Gaithersburg, MD., H. Berger, ed. (in press).

7. D. E. Yuhas and L. W. Kessler, "Scanning Laser Acoustic Microscope Applied to Failure Analysis," Proc. ATFA - 78; pp. 25-29, IEEE Inc., New York, NY., Catalog No. 78CH1407-6 REG 6 (1978).

8. D. E. Yuhas and L. W. Kessler, "SLAM the Door on Material Failure from Internal Flaws," Industrial Research and Development, July, pp. 108-112 (1980).

9. A. Madeyski and L. W. Kessler, "Initial Experiments in the Application of Acoustic Microscopy to the Characterization of Steel and to the Study of Fracture Phenomena," IEEE Transactions of Sonics and Ultrasonics, SU-23 (5), 363-369 (1976).

10. D. E. Yuhas, P. E. McGraw and L. W. Kessler, "Nondestructive Evaluation of Stainless Steel Bellows and Poco-Graphites by Means of the Scanning Laser Acoustic Microscope," Sandia Technical Report, SAND 79-7047, Sandia Laboratories, Albuquerque, NM (1979).

ACOUSTIC MICROSCOPY OF HYBRID MICROELECTRONIC JOINTS*

Chin C. Lee, Jung K. Wang, and Chen S. Tsai
School of Engineering
University of California
Irvine, CA 92717

Recent progress on diagnosis and characterization of defects in hybrid microelectronic components using a transmission scanning acoustic microscope (SAM) operating at 150 MHz is summarized. Acoustic imaging of defects in thick-film circuits and multilayer chip capacitors was carried out. Several simple methods have been established to determine both the defect location (in depth) and the defect type. Acoustic transmission resonance spectra have been observed in liquid- and air-filled cavities in host materials such as glass and steel. A methodology which is comprised of resonance spectra and differential phase measurements has been devised for determining simultaneously the longitudinal dimensions and the acoustic velocity of the defects. This methodology has been verified by the experiments with the liquid- and air-filled cavities.

* This work was supported in part by the MRL Section of NSF Grant No. DMR 72-03297-A03.

Introduction

Acoustic microscopy is concerned with the study of a specimen based on its material parameters such as the elastic modulus, mass density, and acoustic absorption. One of the instruments that has evolved recently is the Scanning Acoustic Microscope (SAM) (1). The SAM does not require destructive sample preparation and, unlike the optical microscope, can examine details underneath the surface of an optically opaque specimen. A reflection version of this instrument (2) has demonstrated a spatial resolution comparable to that of the optical microscope and the capability of viewing the subsurface regions of integrated circuits. This paper is concerned with the application of a transmission version of this instrument (1,3) for nondestructive diagnosis of hybrid microelectronic components and specifically made cavities filled with air and liquid.

Defects in hybrid microelectronic components may be categorized into two general types, the "inclusion" and the "void." Examples of the former are foreign substances and the host substance of different phase. Examples of the latter are air gaps, cracks, and delamination. Since both types of defects are accompanied with large local variations in the aforementioned material parameters the SAM should be highly useful for nondestructive diagnosis of such defects. In fact, we had previously demonstrated that a transmission version of this instrument operating at a relatively low frequency (3) is capable of detecting such defects in thick production-line hybrid microelectronic components (3, 4). In this paper, some of the advances we have made most recently in connection with this particular application are reported. First, a brief review on the basic components, working principles and modes of operation of the transmission SAM is made. Imaging and detection of defects in production-line thick-film circuits and multilayer chip capacitators are then detailed. Characterization of specially made liquid inclusions and air cavities by use of transmission resonance technique is also discussed.

Basic Components and Capability of a Transmission Scanning Acoustic Microscope

The block diagram of a transmission SAM is shown in Fig. 1(5). The acoustic wave is generated by a planar piezoelectric transducer on one end

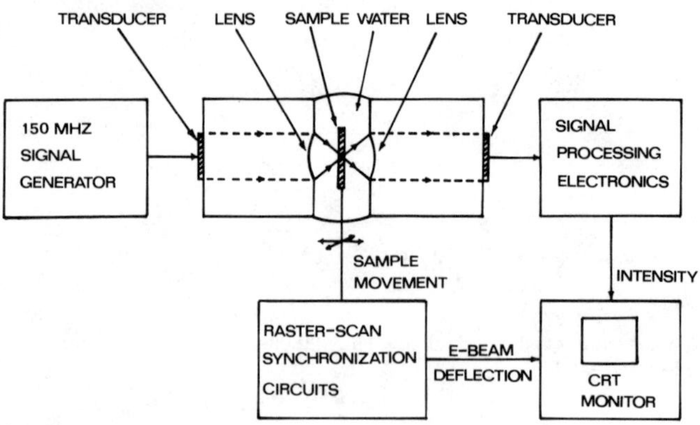

Fig. 1
Block diagram of a transmission scanning acoustic microscope

face of either a sapphire or a quartz rod. A spherical concave surface at the other end of this rod acts as an acoustic lens and focuses the acoustic wave into a diffraction-limited spot. The focused acoustic beam then illuminates the specimen suspended in liquid, e.g. water. The acoustic beam encounters reflection, refraction, scattering, and attenuation at the regions in the specimen where variations in the elastic parameters occur. The transmitted acoustic signal is collected and collimated by a similar acoustic lens and reconverted into an electrical signal by a second piezo-electric transducer. The electrical signal is processed by electronics and used to modulate the intensity of a CRT display. A two-dimensional image is then produced by raster-scanning the specimen across the focused acoustic beam. As the specimen is scanned, the transmitted acoustic signal is both amplitude- and phase-modulated. By electronically separating the amplitude and phase information, both amplitude and phase images are obtained. The amplitude image is a measure of the acoustic opacity of the specimen and the phase image is related to the local variations in elastic modulus and/or mass density of the specimen. By examining amplitude information together with phase information, the defects can be identified more easily. For example, the correlation between the phase and the amplitude information was used not only to detect "alloy spikes" in high power transistors but also to determine their sizes (6). The spatial resolution of the transmission SAM is on the order of the acoustic wavelength in the specimen. The depth of focus can be varied by adjusting the separation between the transmitting lens and the receiving lens(7). For the acoustic microscope employed in this study the operating frequency is 150MHz and the spatial resolution ranges from about 15μm to 40 μm in commonly used semi-conductors and metals. At this relatively low operating frequency the microscope is capable of examining thick specimens, e.g., up to a few millimeters thick with the two classes of materials just mentioned. In addition to the amplitude and phase modes of operation, we have most recently employed swept-frequency mode of operation to obtain the trans-mission resonance spectra of liquid- and air-filled cavities. From the resonance spectra and the phase information, the longitudinal dimension of the cavities and the acoustic velocity of the liquid and air in the cavities can be determined simultaneously.

Characterization of Hybrid Microelectronic Components, Liquid Inclusions and Air Cavities

We had earlier employed the transmission SAM to image and detect the irregularities in simple material joints (3, 7). Recently, we have also employed the same microscope to examine the interfacial regions of thick production-line microelectronic components (4, 5). Some defects in these specimens were detected. Characterization of some of these defects was also made. We have since made further progress on the diagnosis of defects in hybrid microelectronic components and specially made specimens. Some of the results obtained are now described.

(A) Thick-Film Circuits

In thick-film circuits the defects which result from contaminants and blisters introduced during the fabrication process greatly affect the resistance, adhesion, and solderability of the finished products. It is thus desirable to detect such defects nondestructively using the transmis-sion SAM for failure analysis and quality control. For example, Figs. 2(a) and 2(b) show, respectively, the optical micrographs and acoustic micro-

Fig. 2 Imaging of thick-film circuits (thickness: 760 μm)

(a) Optical micrographs for three resistors (100 Ω, 1.3 KΩ, and 500 KΩ from left to right)

(b) The corresponding acoustic micrographs depicting high degree of contrast

graphs of a portion of a production-line thick-film resistor† circuit fabricated on an alumina substrate of 760 μm in thickness (10, 11). Note that all three acoustic micrographs were recorded using the same level of power. The dark and gray areas in the optical micrographs correspond, respectively, to three resistors (100Ω, 1.3kΩ and 500kΩ) and the conductors. Contrary to the optical micrographs, the acoustic micrographs show clearly large nonuniformity and some defects in the resistor region. In addition, the acoustic micrographs show a high degree of contrast as a function of the resistance values. Measurement of the amplitude of the transmitted acoustic signals has indicated that the smaller the resistance the larger the acoustic attenuation. Specially, a differential attenuation of 43dB was measured between the 100Ω and 500kΩ resistors.

†Proprietory composition

(B) Multilayer Chip Capacitors

Both voids and inclusions are potential causes for a leaky capacitor, and end metalization of poor quality is often sufficient to cause open-circuit failure (9). We have employed the transmission SAM to detect such types of defects (10,11). We were able to determine not only the type (inclusion or void) of the defects but also the location (in three dimensions) of the defects by displacing the capacitor along the lens axis at small increments and recording the corresponding amplitudes and phases

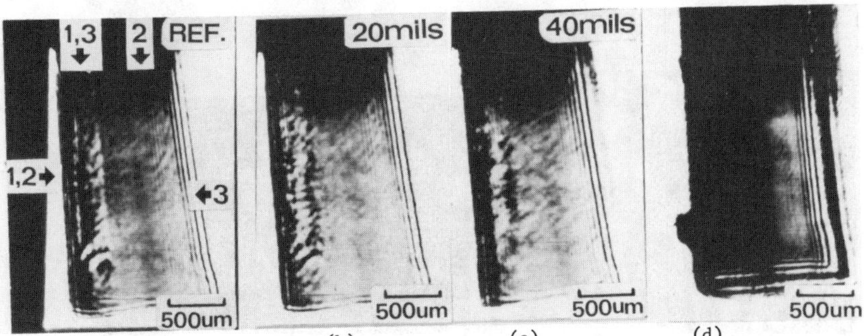

(a) (b) (c) (d)

Fig. 3 Acoustic micrographs of a production-line chip capacitor (thickness: 690 μm)

(b) and (c) Obtained by displacing the capacitor along the lens axis by 500 μm (20 mils) and 1000 μm (40 mils), respectively, from the capacitor position of (a).

(d) Obtained after the capacitor was heated at 100° for 80 minutes

Fig. 4 Photograph of a production-line chip capacitor (thickness: 690 μm)

of the transmitted signal from a particular defect. For example, Figs. 3(a) to 3(c) show a series of the acoustic micrographs obtained for the capacitor shown in Fig. 4. We have found that the image area designated by coordinates (2.2) is defect-free, the image area (1.1) has an inclusion-type of defect and the image area (3.3) has two void-type of defects (10, 11). Furthermore, the two defects in area (3.3) are located at different depths which are about 100 μm apart. Fig. 3(d) shows the detrimental effect on the capacitor created by heating it for an extended period.

(C) Liquid- and Air-Filled Cavities

We have most recently employed the swept-frequency mode of operation to observe the transmission resonance spectra of a cavity (8). The resonance spectra were obtained by focusing the acoustic beam at a fixed position in the cavity and sweeping the acoustic frequency over a frequency band centering at 90MHz. We were able to determine simultaneously the longitudinal dimension and the acoustic velocity of the cavity by using the resonance spectra together with the differential phase (8). For example, Fig 5(b) shows the transmission resonance spectrum of an air-cavity in glass. The size of the cavity is 9.14 x 6.60 mm^2 and its actual longitudinal dimension is 280 μm. We see that the resonant peaks are separated by

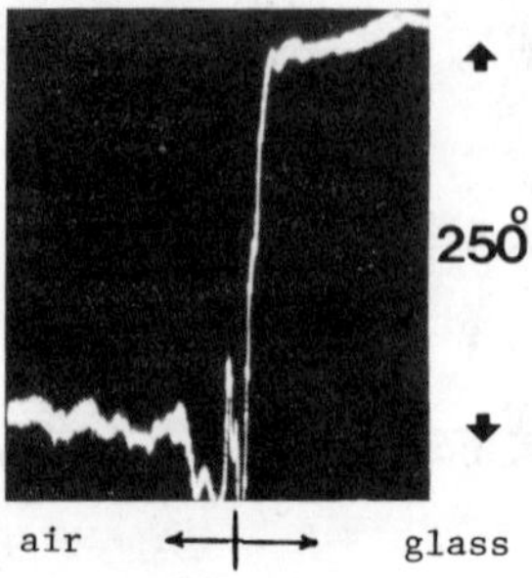

Fig. 5(a) Differential phase between transmitted acoustic signals

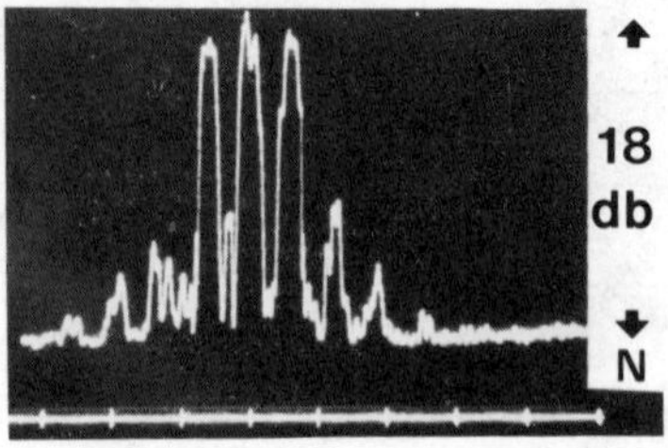

Fig. 5(b) Transmission resonance spectrum for air-cavity in glass N: noise

.77 MHz. Fig 5(a) shows the differential phase as the acoustic beam is scanned across the cavity boundary at a direction perpendicular to the lens axis. From the differential phase and the separation of the resonance peaks, we deduce that the longitudinal dimension of the cavity is 250 μm which is very close to the actual dimension of 280 μm, and the acoustic velocity of the air in the cavity is .38 X 10^5 cm/sec, again close to the published value of .332 X 10^5 cm/sec. Other cavities filled with liquid such as water, trichloroethylene, motor oil and toluene have also been characterized by the same technique and the results obtained are in close agreement with the published values (8).

Conclusion

We have demonstrated that the transmission SAM is capable of detecting not only the location but also the type of defects in hybrid microelectronic components. We have also shown that a combination of transmission resonance spectra and differential phase measurements can be employed to simultaneously determine the longitudinal dimension and acoustic velocity of the defect region in specially made thick specimens. This technique should be highly useful for diagnosis and characterization of defects in hybrid components.

References

1. R. A. Lemons and C. F. Quate, "Acoustic Microscope- Scanning Version," Appl. Phys. Lett., 24, (Feb. 1974), p. 163.

2. V. Jipson and C. F. Quate, "Acoustic Microscopy at Optical Wavelengths," Appl. Phys. Lett., 32, (June 1978) p. 789.

3. C. S. Tsai, S. K, Wang and C. C. Lee, "Visualization of Solid Material Joints Using a Transmission-Type Scanning Acoustic Microscope,"

Appl Phys. Lett., 31, (Sept. 1977) p. 317.

4. J. K. Wang, C. C. Lee and C. S. Tsai, "Nondestructive Diagnosis of Thick Production-Line Microelectronic Components Using Transmission Acoustic Microscope," 1978 International Electron Devices Meeting, Washington, D. C., Dec. 4-6, 1978, Technical Digest, pp. 449-451.

5. C. C. Lee, "Integrated Optics for RF Spectrum Analysis and Acoustic Microscopy for the Study of Material Joints," Ph. D. Thesis, Carnegie Mellon University, Pittsburgh, PA, Aug. 1979.

6. C. C. Lee, J. K. Wang, S. K. Wang, P. Hower, and C. S. Tsai, "Detection and Characterization of Alloy Spikes in Power Transistors Using Transmission Acoustic Microscopy," First International Symposium on Ultrasonic Materials Characterization, National Bureau of Standards, Maryland, June 7-9, 1979, Ultrasonic Materials Characterization, NBS Special Publication 596, pp. 387-391.

7. S. K. Wang, C. C. Lee and C. S. Tsai, "Nondestructive Visualization and Characterization of Material Joints Using A Scanning Acoustic Microscope," 1977 Ultrasonic Symposium Proceedings, pp. 171-175.

8. J. K. Wang, C. S. Tsai and C. C. Lee, "Spectroscopic Study of Defects in Thick Specimen Using Transmission Scanning Acoustic Microscopy," To Appear in the Proceeding of International Symposium on "Scanned Image Microscopy" at the Royal Society, London. Academic Press In Print (1980).

9. T. F. Brennen, "Ceramic Capacitor Insulation Resistance Failures Accelerated by Low Voltage, "IEEE Trans. on Electron Devices, ED-26, (Jan. 1979), pp. 102-108.

10. C. S. Tsai, C. C. Lee and J. K. Wang, "Diagnosis of Hybrid Microelec-Tronics Using Transmission Acoustic Microscopy," Proc. of 1979 International Reliability Phys. Symp., San Francisco, CA April 24-26, 1979, pp. 178-182.

11. J. K. Wang, C. C. Lee and C. S. Tsai, "Further Progress on Nondestructive Diagnosis of Hybrid Microelectronic Components Using Transmission Acoustic Microscopy," ARPA/AFML Review of Progress in Quantative NDE, La Jolla, CA, July 8-13, 1979.

III. NDE Characterization of Deformation & Fracture Processes

CHARACTERIZATION OF CRACK TIP PLASTIC ZONE PARAMETERS

AND THEIR INTERRELATIONSHIP WITH NDE TECHNIQUES

David L. Davidson and James Lankford
Southwest Research Institute
San Antonio, Texas

This paper summarizes the known data on plastic zone size, maximum stress and strain magnitude at the crack tip, distribution of stress and strain within the plastic zone, crack tip opening displacement, and energy expended in forming new crack surface, and it compares these values to analytical efforts to determine each of these parameters. Also, laboratory techniques used for experimentally determining each of these parameters are briefly discussed. Application of current NDE technology to the characterization of crack tip plasticity is considered, and needed research is suggested.

Acknowledgement

This paper was prepared at the request of the symposium organizers, to whom we are grateful for this opportunity. Our own research which is reported here has been funded by AFOSR, ONR and Southwest Research Institute; we gratefully acknowledge the support of these organizations.

Introduction

The crack tip plastic zone is important because it is indicative of both how the crack grew as well as how the crack is likely to grow; thus, if it were possible to detect a crack and characterize its plastic zone, assessment of structural lifetime could be greatly enhanced. This paper will examine how crack tip plastic zones are characterized, discuss briefly some of the methods used to determine the relevant parameters, and consider how these might be measured nondestructively.

The relevant parameters for describing a crack are its size, shape and location within the body (1). For assessing the danger to the structure caused by the crack, knowledge of the concentration of stress and strain at the crack tip is important, together with the fracture characteristics of the material. Thus, for a complete assessment of the importance of the crack, a continuum mechanics description of structural loading and crack geometry must be combined with a description of material behavior; it is these factors which govern formation of the plastic zone and are responsible for its characteristics. The environment at the crack tip also may be an important factor in the formation of the plastic zone (2), and temperature and loading rates are certainly important factors in material plasticity. This paper will concentrate on describing crack tip plasticity in structural materials at ambient temperature, within ambient and inert atmospheres, because most of the knowledge to date is limited to these conditions.

An important objective of crack tip plasticity studies has been to compute the parameters describing the plastic zone based on descriptions of the crack and the material. A longer range goal is to describe growth of the crack from the plastic zone and material characteristics.

Characterization of the Plastic Zone

The size and shape of the plastic zone, the stress and strain magnitude and distributions within it, together with the work dissipated during its formation--these factors constitute the useful information found within the plastic zone.* The crack itself also may be described by its opening (the crack tip opening displacement, or CTOD) and by the mode (the combination of tensile and shear) of that opening. These factors are, of course, related to the plastic zone characteristics, and if this relationship could be established and shown to be unique, it might be possible to measure CTOD nondestructively and deduce the plastic zone parameters. Thus, the crack tip opening mode and magnitude will also be considered as crack tip plasticity parameters.

Plastic Zone Size

Analytical

There have been numerous analytical procedures for computing the plastic zone dimensions, but most of these are for unidirectional loading conditions. There have been essentially three approaches: 1) the material

* Inhomogeneities such as voids and cavities or inclusions or second phase particles may also exist within the crack tip plastic zone, so knowledge of these and how they affect plastic zone characteristics is also important.

is treated as a continuum, 2) the material is considered as being a series of finite elements, and 3) the plastic zone and crack are modeled by continuously distributed dislocations. These methods were reviewed in 1977 by the authors (3); therefore, only summary information will be given here. The continuum approach also is thoroughly reviewed by Rice (4) and the finite element approach has recently been summarized by Rice, McMeeking, Parks and Sorensen (5). The continuous dislocation distribution approach is that of Bilby, Cottrell and Swinden (6) (BCS theory).

Table I presents the various analytical and computational estimates (excluding the BCS theory) for plastic zone size, r_p, in the form

$$r_p = \alpha\left(\frac{K}{\sigma_y}\right)^2 \tag{1}$$

where K = stress intensity factor and σ_y = tensile yield strength.

Table I. Plastic Zone Size Computations**

No.	$\theta(°)^*$	α Plane Stress	α Plane Strain Continuum Mechanics	Ref. No.
1	0	$\frac{1}{2\pi}$(=0.159)	$\frac{(1-2\nu)^2}{2\pi}$(=0.017)	7
2	90	$\frac{5}{8\pi}$(=0.2)	$\frac{2(1-2\nu)^2+3}{8\pi}$(=0.128)	7
3	0	.318		4
	90	.392		4
4	0	.382		8
5	0		$\frac{1}{6\pi}$ (=0.05)- $\frac{1}{3\pi}$ (=0.106)	9
			Finite Element	
6	70		.175	4
7	0		.036	10
8	70		.157	10
9	90		.138	10
10	0		.041	11
11	71		.152	11
12	0		.035	12
13	70		.14	12

*θ = 0° is parallel to the crack plane.

**For unidirectionally loaded cracks.

Note that all these estimates are for unidirectionally loaded cracks. For fatigue cracks, Rice has postulated a cyclic yield zone within the main yield zone; by assuming that under reverse loading the yield stress would be replaced by two times its value, a cyclic yield zone one-fourth the plastic zone dimension given in Table I would be estimated.

Average values for α given in Table I are ($\theta = 0°$):

$$\text{Plane stress: } \alpha_{ave} = .29$$

$$\text{Plane strain: } \alpha_{ave} = .036$$

In the BCS theory, plastic zone size depends on the applied stress σ, as follows:

$$r_p = \frac{K^2}{\pi\sigma^2} \sin\left(\frac{\pi}{4}\frac{\sigma}{\sigma_y}\right) \tag{2}$$

Experimental

High spatial resolution techniques are required to measure the aforementioned parameters in the laboratory because of the relative small size of the plastic zone and the need to obtain information at numerous places within it. Table II lists many of the techniques which have been used, together with the spatial resolution and type of measurement which can be made.

Table II. Laboratory Techniques for Plastic Zone Measurements

No.	Technique	Type of Measurement	Spatial Resolution (μm)	Resolution (%)	Ref. No.
1	Oblique Lighting	Size	30	0.5-3	13
2	Etching	Size, strain	10	0.3	14
3	Transformation	Size	5	1	15
4	Microhardness	Size	20	2-5	16
5	Recrystallization	Size, strain	10	1	17
6	Interferometry	Size	10	0.1	18
7	Moiré Fringes	Size, strain	15	0.1	19
8	Strain Gages	Size, strain	200	0.1	20
9	Stereoimaging	Size, strain	0.5	1	21
	Electron Channeling				
10	Patterns	Size, strain	10	0.3	22
11	Contrast	Size, stress	1	20 MPa	23
12	Microfocus X-ray	Size, stress	100	30-60 MPa	24
13	TEM - Dislocations	Substructure	50Å	-	25
14	Magnetic Perturbation	Size	100-1000	-	26
15	Exoelectron Emission	Size	3-5	0.5-3	27
16	Holography	Size, strain	100	-	28

As will be noted from the Table, many of the techniques can only measure size of the plastic zone, while others can measure both the stress or strain distribution and the size. The major differences between the techniques are in spatial resolution, measurement sensitivity, and the quantity actually being measured. This point can be illustrated by considering three of the most used techniques: etching, interferometry and microhardness. Etching causes a pit to form at each dislocation, or group of dislocations, intersecting the surface. Size of the plastic zone is defined by the excess density of dislocations, and strain magnitude is inferred from the degree of excess density. Thus, size may be measured accurately, and strain less so. Interferometry measures displacements perpendicular to the plane of observation, and these must be transformed into in-plane strains; therefore, accuracy of the strains is dependent on this transformation being done in an elasto-plastic region, which is difficult. Plastic zone size is usually determined by extrapolation of displacements to the elastic value. Microhardness is an indirect measurement of dislocation density. Plastic zone size is determined best for those materials that either work harden or work soften to a large extent, but strain sensitivity of this technique is usually a problem in estimating both strain magnitude and plastic zone size.

It is also desirable to know the characteristics of the plastic zone deep within the material as well as on the surface; this has been an especially difficult problem, and only a few of the techniques can be used to get this information--none of them can do it nondestructively.

Only two of the techniques for measuring crack tip plasticity parameters will be further discussed; these are techniques with which we have the most knowledge and experience, and, therefore, confidence. Electron channeling is a technique of obtaining contrast due to crystalline orientation in scanning electron beam instruments. It cannot be used on all materials, such as martensitic steels, because of their highly faulted structures, but channeling does have the capability of obtaining information deep within a specimen, after it has been sectioned, and from the fracture surface. The phenomenon, first discovered in 1967 (29), has been used to study overload and fatigue fractures. By the use of channeling patterns, crystallographic orientation may be obtained within the plastic zone, and indirect assessment of plasticity may be made from channeling pattern changes. Practical resolution limits in current commercial scanning electron microscopes are 5-10 μm. There is also an upper limit of 10-15% on the magnitudes of strain which may be measured, but the technique is relatively sensitive on the lower end--plastic strains of 0.2-0.4% may be detected. Strains are determined by comparison of channeling patterns of unknown deformation with channeling patterns made from calibration specimens in which strains are known. This has been done by using tensile specimens (30) and expressing the observed changes in channeling patterns within the plastic zone as "equivalent tensile strains." Channeling contrast in micrographs may be used to observe the formation of subgrains within the plastic zone. Since subgrain size is related to stress, mapping of the stresses within the plastic zone may be determined from the observed subgrain distribution. Again, not all materials are suitable for analysis by this technique either because of a highly defected microstructure, e.g., martensitic steels, or due to the non-formation of equilibrium subgrains (except just adjacent to the crack plane, e.g., high strength aluminum alloys). When subgrains do form and are distributed within the plastic zone, their size at the crack plane is usually 0.5-1 μm; therefore, quantitative information may be obtained very close to the crack tip. Conversely, mapping of the plastic zone boundary is difficult because the approach of subgrain size to the grain size is gradual and very sensitive in this low stress region to the nominally random distribution of grain orientation. Both channeling techniques may be used equally as well on the interior of a specimen, after sectioning, as on the outer surface.

A summary of values of α in Eq. (1), as derived by the various techniques which have been discussed, is given in Table III. This Table streamlines data previously given in Tables 3 and 4 of Ref. 3, and includes new data for aluminum alloys.

Table III. Comparison of Measured Plastic Zone Size with Theory

	Measured α at θ = 0°	Location	Theoretical α (Average) Plane Stress	Theoretical α (Average) Plane Strain	Ref. No.
Aluminum Alloys					
2024	0.15	Surface	0.29		31
6061	0.22	Surface			31,33
7075	0.62	Surface			31
7075	0.30	Surface			32
Steels					
Fe-3Si	0.17	Surface	0.29		14
Low-C/-3Si	0.10	Interior		0.036	3,14
Maraging	0.06-0.1	Interior		0.036	16
Stainless & Fe-34Ni	0.053	Interior		0.036	16
Nickel Alloys					
Inco 718	0.08	Surface	0.29		15
Inco 718	0.04	Interior		0.036	15
Incoloy 901	0.6	Surface	0.29		34
Copper Alloys					
Single Crystal	0.08	Interior		0.036	35
Polycrystal	0.056	Interior		0.036	36

Data in the Table are referenced to the site where the measurements were performed (surface or interior), but this does not simply imply plane stress or plane strain conditions; rather, it is necessary to consider the maximum stress intensity and specimen thickness when the measurements were made--conditions not known for all the data available. Data reported for the aluminum alloys was derived at $K \doteq 10\ MN/m^{3/2}$ for specimens 6mm thick; these parameters are sufficient to ensure the specimen is about thirty times the maximum plastic zone size, which suggests that a nearly plane strain condition exists across the whole specimen thickness. Yet, the plastic zone dimensions in Table III correlate better with theoretical analyses for plane stress than for plane strain. This contradiction is apparently unique to the aluminum alloys; for all others, theoretical plane strain analysis correlates reasonably well with plastic zone dimensions measured in the specimen interior.

Stress and Strain Magnitude and Distribution

There exists considerably less data on stress and strain distributions within the plastic zone than on plastic zone size, principally because few techniques 1) have sufficient spatial resolution to make useful measurements, 2) measure in-plane displacements from which strains may be derived, and 3) are capable of measuring the large displacements which exist for many engineering alloys near the crack tip. Therefore, data, where available, are sketchy and limited to the region nearer the plastic zone boundary than the crack tip.

Analytical

All of the analytical stress and strain distribution functions which have been derived are for unidirectionally loaded cracks with no prior history of deformation. The various distribution functions are listed in Table IV. All result in a singularity at the crack tip.

Table IV. Analytically Derived Stress and Strain Distributions*

No.	Stress	Strain	Reference
1	$r^{-\frac{N}{1+N}}$	$r^{-\frac{1}{1+N}}$	Rice & Rosengren (37) N = work hardening coefficient
2	$r^{-\frac{1}{2}}$	$r^{-\frac{1}{2}}$	Linear elastic fracture mechanics (LEFM)
3		r^{-1}	Perfectly Plastic, Hutchinson (38)
4	$\sigma_o(\frac{\varepsilon}{\varepsilon_y})^N$	$\varepsilon_y(1 + \ln \frac{R}{r})$	Rice, moving crack (39)** R = plastic zone size ε_y = strain at elastic limit

*All solutions are for a unidirectionally loaded crack.

**As interpreted by Vermilyea and Diegle (40).

Experimental

Distribution. Of the many laboratory techniques listed in Table II, only eight or nine potentially have sufficient resolution for obtaining stress or strain distributions within the plastic zone, and of these, etching and recrystallization are quantitative only in a sketchy way, and the strain gage, holographic and microfocus X-ray techniques have marginal resolution. The Moiré fringe and electron channeling pattern techniques are most suited for measurements near the plastic zone boundary.

The stereoimaging technique is capable of determining the in-plane differences in material displacements within the plastic zone with a high degree of spatial resolution. Three elements of the symmetric strain tensor may be directly derived from those displacements (21). The technique is capable of determining quite large strains, but lacks sensitivity at the lower end (minimum detectable values are about 1% at present). Accuracy and spatial resolution are greatly dependent upon both resolution of the photographs being analyzed and on the accuracy of the measuring equipment (21). Plastic zone size must be determined by extrapolation of the strains measured near the crack tip to the far field. If electron channeling or Moiré fringe and stereoimaging were to be used together on one material, then it should be possible to accurately map the size and shape of the plastic zone and determine the magnitude and distribution of the stresses and strains within it.

Measured distributions of strain are shown, somewhat schematically, in Figure 1. The two types of functional dependences which have been measured are given both on the Figure and in Table V, along with other relevant data. On the basis of currently available data, it appears that a crack loaded in tension has a Type A strain distribution, while fatigue cracks have a Type B strain distribution, although further experimental confirmation is required.

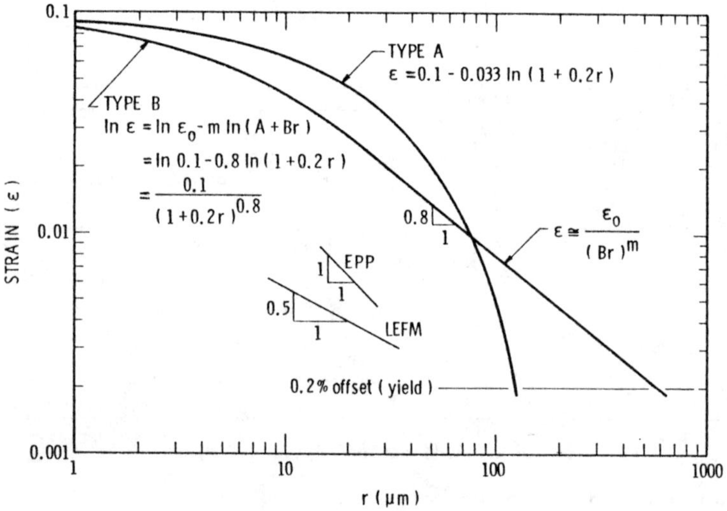

Figure 1. Schematic representation of the two strain distribution functions discussed in Tables IV and V. Curvature of both functions in the near crack tip region ($1<r<50$) makes separation of the functional form difficult; data near the plastic zone boundary is essential.

<u>Magnitude</u>. Even though it now appears that the distribution functions for stress and strain are reasonably well known, maximum values of these variables at the crack tip are known with less confidence. Relatively few analytical efforts have attempted estimates of the magnitudes of these parameters, but there have been several experimental determinations. Both are given in Table VI; it should be noted that most of the experimental values are of recent origin, and many more will be measured in the near future now that the techniques are available. Although not clearly the

case, it appears as though analytically derived maximum values of stress and strain consistently exceed those measured, and preliminary results also strongly imply that the maximum strains for unidirectionally loaded cracks are larger than those at cyclicly loaded cracks.

Table V. Experimental Distribution Functions for Stress and Strain

Distribution Function* (see Fig. 1)	Variable	Measuring Technique (Ref No.)	Range of Validity (r/r_p)	Material	Theoretical Analysis (Table IV)
A	Strain	Moiré fringe (19) (Note 1)	0.04-1	Brass 2024-0 Zr-Nb	4
A	Strain	Channeling Patterns (31) (Note 2)	0.06-1	Al alloys Fe-3Si	4
A	Strain	Replication (33)	0.01-0.2	6061-T6 304SS	4
B	Stress	Channeling Contrast (41)	0-0.8	Fe-.05C	1
B	Strain	Moiré fringe (19)	0.1-1	2024-T351	1
B	Strain	Replication (33)	0.001-0.01	6061-T6, 304SS	1
B	Strain	Stereo-imaging (21) (Note 3)	0-0.6	Fe-.05C Al Alloys	1
B	Strain	Strain gages (42)	0.15-1	Fe-.05C	1
B	Stress	X-ray (43)	0-1	1020, 1025 steel	1

*Distribution Function: Type A $\varepsilon = \varepsilon_y[1 + m \ln \frac{r_p}{(r+Z)}]$

Type B $\ln \varepsilon = \ln \varepsilon_o - m \ln(A+Br)$

Type B $\sigma = \frac{\sigma_o}{(A+Br)^n}$

Notes: (1) This is the only distribution which has been derived for tensile loading; all others are for cyclic loading.

(2) Channeling patterns calibrated with "equivalent tensile strain."

(3) This is based on a limited amount of data; it is not yet completely clear that the Type B distribution function fits better than the Type A.

Table VI. Maximum Values of Crack Tip Stress and Strain
UNI = Unidirectional Loading FAT = Fatigue Loading

Loading	Stress	Strain	Material	Method	Ref.
UNI	$3\sigma_y$			Analytical	10
UNI	$5\sigma_y$		N = 0.3	Analytical	12
UNI	$2\text{-}3\sigma_y$	0.8		Analytical	38
UNI		0.05	7075-T6	Replica	44
FAT	$1.4\sigma_y$*		1020 Steel	X-Ray	43
FAT	$2.25\sigma_y$		Fe-.05C	Subgrains	41
FAT		0.20	Fe-.05C	Stereoimaging	41
FAT		0.06	2024-T4	ECP#	31
FAT		0.05	6061-T6	ECP#	31
FAT		0.10	7075-T6	ECP#	31
FAT		~0.50	6061-T6	Replica	34
FAT		~1	304SS	Replica	34
FAT		0.09	2024-T4	Stereoimaging##	
FAT		0.05	6061-T6	"	
FAT		0.03	7075-T6	"	
UNI**		0.12	2024-T4	Stereoimaging##	
UNI**		0.10	6061-T6	"	
UNI**		0.36	7075-T6	"	

*Value is ΔK dependent, which may be due to large beam size (625 μm). Magnitude shown is comparable with that obtained from channeling contrast observation of subgrains.

**Value is that determined after an overload of 2ΔK for a crack previously cycled at ΔK = 10 $MN/m^{3/2}$.

#Obtained by extrapolating Type A strain function to r = 0.

##Preliminary results recently obtained within 2 μm of crack tip.

Crack Tip Opening Displacement (CTOD)

Although the magnitude of the crack tip opening is not a characteristic of the plastic zone, it is obviously related to the magnitude and extent of crack tip plasticity. It is also of interest because CTOD is one of the most discussed of the crack tip plasticity related parameters.

Analytical

Magnitude of the CTOD may be computed using the equations of Rice (4) for unidirectionally loaded cracks:

$$\text{CTOD} \cong 0.5 \frac{K^2}{E\sigma_y} \text{ for plane stress} \tag{3}$$

$$\text{CTOD} \cong 0.225 \frac{K^2}{E\sigma_y} \text{ for plane strain} \tag{4}$$

The numerical factor for $\frac{K^2}{E\sigma_y}$ in Eq. (4) is for $\theta = 90°$ (the angle of inclination of the slipline from the crack plane); the factor is 0.253 for $\theta = 60°$ and 0.192 for $\theta = 45°$. For cyclicly loaded cracks, Rice (4) gives no numerical value, so that

$$\text{CTOD} = \beta \frac{K^2}{E\sigma_y} \tag{5}$$

Experimental

Only a few direct measurements (about 15) of CTOD exist for cracks with small plastic zones in relatively thick specimens. An analysis of these results suggests that for cyclic loading $\beta = 0.13 \pm .05$, and for fatigue cracks which have experienced a single cycle of overload, $\beta = 0.5 \pm .1$, where β is defined by Eq. (5). These values correlate relatively well with numerical estimates made using Eqs. (3) and (4).

Energy Expended in Fatigue Crack Growth

The energy expended in the plastic zone as the fatigue crack grows has also been derived for a number of aluminum alloys and steels. Fine (45) has tabulated most of the results to date, which vary considerably according to alloy type and cyclic stress intensity factor. For high strength aluminum alloys, the values are $0.5\text{-}3.2 \times 10^5$ J/m^2 and for the steels $2.2\text{-}53 \times 10^5$ J/m^2. Most of these values have been determined using strain gages, which cannot measure closer than about 200 μm to the crack tip; for Fe-.05C steel, an estimate of the energy using subgrains, as detected by electron channeling contrast (42), which allows measurement up to the crack tip, gave magnitudes approximating the value determined by the strain gage technique. Strains, as determined by the stereoimaging technique, may be used for any single cycle to obtain the work density distribution in the region very close to the crack tip. For Fe-.05C, the density of work done has the value of approximately 40 MJ/m^3 in the region 0-10 μm from the crack tip, decreasing radially approximately as $\frac{1}{r}$.

Nondestructive Detection of Crack Tip Plasticity

The ability to characterize a crack as a geometric entity, having a particular size, shape, orientation, and location, has increased markedly in recent years. Progress in the characterization of crack tip plastic zones in terms of NDE physics has been much slower.

As shown in Table II, a wide variety of techniques can be used within the laboratory to study and define plastic zones attending crack tips. However, many of these techniques are clearly inapplicable to technological NDE applications in which instrument portability, the need to detect buried (inaccessible) cracks, and specimen size, for example, are important considerations.

NDE techniques which, in principle, are capable of characterizing crack tip plastic zones under such conditions, are based upon either the modification or generation of energetic fields. Only a few of the approaches listed in Table II seem to be suitable for practical NDE. A few other techniques also appear applicable under operational NDE conditions; summarized in Table VII are five approaches which, in our judgement, may provide such plastic zone characterization.

Table VII. NDE Techniques for Plastic Zone Characterization

Technique	Field Effect	Crack Location	Characteristic Measured	Theory	Ref.
Acoustic Emission	Sound Wave Generation	Surface, Subsurface	Change in Plastic Zone Size	Yes	46-48
			Size of Plastic Zone	Yes	46-48
			Stress Intensity Factor	Yes	46-48
Exoelectron Emission	Ejected Electron Generation	Surface	Size of Plastic Zone	No	
Eddy Current	Electric Field Modification	Surface	Unknown	No	
Magnetic Field	Magnetic Field Modification	Surface	Unknown	No	
Acoustoelastic Measurement	Sound Wave Velocity Modification	Surface, Subsurface	Crack Tip Stress Field	Yes	55
			J-Integral	Yes	55
			Stress Intensity Factor	Yes	55

Acoustic Emission

Techniques for measuring the acoustic emission from cracked specimens have developed over a number of years. Associated with this work has been progress in the development of a theoretical basis for the relationship between acoustic emission and plastic zone size and growth. Several workers (46,47) have argued that acoustic emission counts (N) emitted by a cracked structure should be linearly related to the volume of the crack tip plastic zone V_p, i.e.,

$$N = kV_p$$

where k is a constant of proportionality that depends on material and instrumentation sensitivity. V_p for a through crack can be derived from Eq. (1) as

$$V_p \cong \frac{t}{4\pi} \left(\frac{K}{\sigma_y}\right)^4$$

where t is the specimen thickness and Pollack (48) has shown that this relationship is obeyed for A516 steel. However, it should be noted that for a buried penny-shaped crack,

$$V_p = \frac{a}{16} \left(\frac{K}{\sigma_y}\right)^4$$

where a is the radius of the crack, hence

$$N = \frac{ka}{16} \left(\frac{K}{\sigma_y}\right)^4$$

In this case, which is the one most relevant to NDE practice (it would hold for a small semi-elliptical surface crack of length 2a as well), it is necessary to determine a independently in order to interpret N in terms of either K or V_p. This requires the use of some other NDE technique which is able to locate and characterize the geometry and size of the acoustic emission source.

The acoustic emission generated by plastic zone growth is of low magnitude, in comparison to that produced by larger, discrete events such as microfracture and twinning. In fact, the plasticity-related emission level is so low, that in practical situations, primary acoustic emissions derived from plastic zone growth may actually represent (49) the brittle fracture of small intermetallic particles within the plastic zone. Secondary contributions would derive from brittle fracture of matrix material. This problem is currently being investigated for 7075-T6 by McBride, et al. (50).

Further work on small cracks is clearly required in order to assess the ultimate feasibility of acoustic emission as an NDE technique; nearly all studies to date have involved large through cracks. In particular, surface cracks should be studied so that crack size is not a question, and plastic zone dimensions should be established independently, using a technique such as electron channeling. The role of intermetallic particles and discrete microfracture processes should be established by studying materials of varying constituent particle makeup and plastic flow character.

Exoelectron Emission

This technique may be used to directly reveal the presence of the plastic zone of a surface crack. Moreover, the physics of the general effect has been understood for many years (51); NDE practice usually involves photostimulated (enhanced) exoelectron emission, through one of two possible surface mechanisms (49). One effect involves electron emission in vacuo, during oxide breakup, and has been used to monitor fatigue crack tip deformation (27). The other approach, involving surface plasmons produced by the surface roughness generated by fatigue, can be used in air, although certain other problems must be addressed (49).

Unfortunately, knowledge of the physics of exoelectron emission has yet to be coupled with knowledge of the nature and extent of deformation associated with the crack tip plastic zone. Probably more unfortunate, however, is the fact that no experiments have yet been carried out using specimens with well characterized plastic zones. Thus, the portions of the plastic zone volume, or strain levels, which are mapped out by emitted exoelectrons are not known. Exoelectron emission experiments are required in which small surface cracks of known stress intensities and plastic zone characteristics are studied, and the results correlated with the strain distributions within the plastic zones. The minimum stress intensities for consistent plastic zone delineation via exoelectron emission has yet to be established for various materials.

Eddy Current and Magnetic Field Techniques

These two approaches are discussed together, since they are based upon a similar concept, i.e., the perturbation of an impressed electromagnetic field by a material inhomogeneity. Historically, the inhomogeneity which usually has been addressed is that of the crack as a geometrical entity, and both theoretical and experimental efforts have been successful in predicting and demonstrating the ability of small coil probes to detect surface cracks.

There is reason to believe that both techniques should be sensitive to crack tip plastic zones as well; the question is, how sensitive? It has already been shown (52) that electrical techniques can detect surface dislocation damage similar to that produced by fatigue, although thermal noise is a serious problem. Similarly, flux leakage around surface fatigue cracks in 4340 steel has produced Hall probe signatures in which the primary signal, caused by the crack interfaces, is attended by smaller "satellite" signals (53), thought to be due to the plastic zone of the crack. If this is correct, the reason for the appearance of these signals would be that the permeability of magnetic materials is heavily affected by deformation.

The degree to which either of these techniques might be exploited for NDE characterization of plastic zones requires the development of theories suitable for relating the plastic zone signals to the field, probe, and signal processing parameters, so that instrumentation may be optimized for detection. The effects of macroscopic stress and bulk deformation substructure upon resistivity and permeability already have been analyzed theoretically (49,54). However, theories combining this information into a microscopic flux leakage context do not exist. It also should be noted that none of the eddy current and magnetic leakage experiments to date have utilized specimens with well-characterized plastic zones; thus, minimum damage zone sizes required for electromagnetic flux leakage detection are unknown.

Acoustoelastic Measurements

This effect, which has been developed only recently (55), is based upon the relative velocity shift of a longitudinal plane wave ΔV traveling through a deformed solid, as compared to the velocity V_o of the same wave propagating through deformed material. The relation is

$$\frac{\Delta V}{V_o} = B(\sigma_1 + \sigma_2)$$

where σ_1 and σ_2 are the principal stresses in the plane normal to the propagation direction, and B is a constant depending on the state of stress (plane strain/plane stress), the mass density, and the second and third order elastic constants. Once B is determined by acoustic measurements on uniaxially deformed samples, measurement of $\Delta V/V_o$ at a point in a sample can provide an estimate of the sum of the in-plane principal stresses at that point. Moreover, if a crack is present, acoustic data taken from the crack tip field serves to provide the data required for J-integral analysis (55), which yields the crack tip stress intensity directly.

This is clearly a very attractive technique with potential for translating crack tip plasticity characteristics into useful information by NDE. However, considerable research will be required for its development. For example, the effect of deformation upon $\Delta V/V_o$ through the elastic constants is not as large in some materials as in others. Most of the work carried out to date has involved large, stressed through-notches in panels, so that the effect of the yielded zone about the notch tip upon $\Delta V/V_o$ is integrated across the width of the panel. It will be important to apply the technique to small flaws, such as surface cracks, whereby the percentage of total path length of the interrogating beam intercepted by the plastic zone will be only a fraction of the distance traversed through undeformed material. For a given specimen thickness the minimum ratio of plastic zone dimension to specimen thickness for which velocity changes can be measured will determine the minimum crack size/stress intensity for which the technique can be used to calculate J.

Relation Between Crack Tip Opening Displacement Measurement and NDE Physics

Buck and Alers (49) have reviewed the state-of-the-art regarding NDE capabilities for detecting crack tip opening; the most promising approaches are based upon ultrasonic attenuation. The two factors which characterize the crack tip are 1) the CTOD magnitude at maximum load, and 2) the load at which the crack tip opens. If NDE is to play a realistic role, it will probably be in measurement of the latter.

When a semielliptical surface fatigue crack is gradually loaded, the crack opens first at the center. With higher loads, the crack opens closer and closer to its tips, until, at an often rather high fraction of the maximum cyclic load, the tips open and blunt. If it is possible to determine the opening load, as well as the crack length and stress field, one may then determine the effective stress intensity factor ΔK_{eff}, i.e.,

$$\Delta K_{eff} = A(\sigma - \sigma_{open})\sqrt{a}$$

where A is a geometrical constant, σ and σ_{open} the maximum and opening load stress levels, and 2a the crack length.

The problem in determining the crack tip opening load is one of resolution. Recent work has shown that even in the laboratory, the crack tip opening load is difficult to measure. Techniques such as direct observation of crack opening using an SEM loading stage (56), optical interferometry (57), and microstrain gages located near the crack tip (58), have been used. The problem with most current NDE techniques is that they lack the necessary resolution to detect the last increments of crack opening which must be sensed in order to specify σ_{open}. Development of such resolution will require the correlation of NDE measurements with one of the sensitive laboratory techniques.

Discussion

Plastic Zone Characterization

Crack tip plastic zone parameters are slowly being measured by laboratory techniques, so that some idea of their magnitude for various engineering materials is beginning to emerge. The descriptive parameter which best fits analytical determination is the CTOD, and the ones about which least is known are stress and strain magnitudes at the crack tip, and energy expended in forming new crack surface and energy density within the plastic zone. Determination of the size of the plastic zone is still not on a firm basis either experimentally or analytically, as may be seen from Table III. It appears as though the distribution of stress and strain within the plastic zone may be described by one of two functions, Table V, based on available evidence, but it is not yet certain which function applies to which loading condition.

Complicating the interpretation of the laboratory measurements are the problems of stress state; surface observations on a thick specimen with a small plastic zone are not the same as on a relatively thinner specimen with large plastic zone. For specimens of both types, surface observations are likely to be nearly like interior observations, but there is a region of stress intensities for thick specimens where surface and interior observations are not likely to be the same; the surface is probably best described by a condition of plane stress, while the interior is still in a condition most nearly described as plane strain. This complicating factor is particularly a problem in interpreting plastic zone size data.

Nondestructive Techniques

Although much useful information can be extracted from nondestructive evaluation of crack tip plastic zones, this aspect of NDE currently is not nearly so advanced as is characterization of crack size, location, and geometry. In order to make progress, more use must be made of conventional techniques of plastic zone characterization in conjunction with advanced nondestructive evaluation, so that the latter can be physically interpreted in terms of crack tip damage or strain, for example, rather than merely the change in an undefined "signal." To this end, more emphasis also should be placed upon theoretical analysis of the interaction between various NDE fields (ultrasonic, electromagnetic, etc.) and perturbing crack tip defect structures (dislocation density gradient, strains, stresses). Techniques should be emphasized which hold promise of characterizing actual, discrete flaws such as small surface half-penny, or buried penny-shaped cracks as might be found in real structures,and not just large through cracks.

References

1. R. W. Hertzberg, Deformation and Fracture Mechanics of Engineering Materials, p. 255, Wiley, New York, 1976.

2. D. L. Davidson and J. Lankford, "The Influence of Water Vapor on Fatigue Crack Plasticity in Low Carbon Steel," pp. 897-904 in Fracture 1977, D. M. R. Taplin, ed., U. of Waterloo Press, Waterloo Ontario, Canada, 1977.

3. J. Lankford, D. L. Davidson and T. S. Cook, "Fatigue Crack Tip Plasticity," pp. 36-55, in Fatigue Crack Growth, ASTM STP 637, ASTM, Philadelphia, 1977.

4. J. R. Rice, "Mechanics of Crack Tip Deformation and Extension by Fatigue," pp. 247-311 in Fatigue Crack Propagation, ASTM STP 415, ASTM, Philadelphia, 1967.

5. J. R. Rice, R. M. McMeeking, D. M. Parks, and E. P. Sorensen, "Recent Finite Element Studies in Plasticity and Fracture Mechanics," pp. 411-442, in Computer Methods in Applied Mechanics and Engineering 17/18, 1979.

6. B. A. Bilby, A. H. Cottrell and K. H. Swinden, "The Spread of Plastic Yield from a Notch," Proc. Royal Soc. A272 (1963) pp. 304-310.

7. D. Broek, Elementary Engineering Fracture Mechanics, Noordhoff Inter. Publ., The Netherlands, 1974.

8. D. S. Dugdale, "Yielding of Ship Steel Sheets Containing Slits," J. Mech. Phys. Solids 8 (1960) pp. 100-104.

9. F. A. McClintock and G. R. Irwin, "Plasticity Effects in Fracture Mechanics," pp. 84-113 in Fracture Toughness Testing and Its Applications, ASTM STP 381, Philadelphia, 1965.

10. N. Levy, P. V. Marcal, W. J. Ostergren, and J. R. Rice, "Small Scale Yielding Near a Crack in Plane Strain: A Finite Element Analysis," Inter. J. of Fracture Mech. 7 (1971) pp. 143-156.

11. J. R. Rice and D. M. Tracey, "Computational Fracture Mechanics," pp. 585-623 in Numerical and Computer Methods in Struct. Mech., ed. S. J. Fenves, Academic Press, N.Y., 1973.

12. D. M. Tracey, "Finite Element Solution for Crack Tip Behavior in Small Scale Yielding," J. Engineering Materials and Technology (ASME) 98 (1976) pp. 146-151.

13. C. Bathias and M. Vancon, "Mechanisms of Overload Effect on Fatigue Crack Growth in Aluminum Alloys," Eng. Fract. Mech. 10 (1978) pp. 409-424.

14. G. T. Hahn, R. G. Hoagland and A. R. Rosenfield, "Local Yielding Attending Fatigue Crack Growth," Met. Trans. 3 (1972) pp. 1189-1202.

15. M. Clavel, D. Fournier and A. Pineau, "Plastic Zone Sizes in Fatigued Specimens of INCO 718," Met. Trans. 6A (1975) pp. 2305-2309.

16. C. Bathias and R. M. Pelloux, "Fatigue Crack Propagation in Martensitic and Austenitic Steels," Met. Trans. 4 (1973) pp. 1265-1273.

17. Y. Iino, "The Recrystallization Technique for Local Plastic Zone Observation in 304SS in the Range -196 to 950°C," Exp. Mech. 20 (1980) pp. 316-319.

18. G. Marci, "The Effect of the Active Plastic Zone on Fatigue Crack Growth Rates," pp. 168-175 in Fracture Mechanics, ASTM STP 677, ASTM, Philadelphia, 1979.

19. H. W. Liu and J. S. Ke, "Moire Method," pp. 111-165 in Experimental Techniques in Fracture Mechanics 2, A. S. Kobayashi, ed., Iowa State University Press, Ames, 1975.

20. S. Ikeda, Y. Izumi and M. E. Fine, "Plastic Work During Fatigue Crack Propagation in a High Strength Low Alloy Steel and 7050 Al-Alloy," Eng. Fract. Mech. 9 (1977) pp. 123-129.

21. D. R. Williams, D. L. Davidson and J. Lankford, "Fatigue-Crack-Tip Plastic Strains by the Stereoimaging Technique," Experimental Mechanics 20 (1980) pp. 134-139.

22. D. L. Davidson, "The Study of Fatigue Mechanisms with Electron Channeling," pp. 254-275 in Fatigue Mechanisms, ASTM STP 675, ASTM, Philadelphia, 1979.

23. D. L. Davidson, "The Use of Subgrains as Imaged in the SEM from Bulk Specimens in the Study of Localized Plasticity," Scanning 2 (1979) pp. 225-229.

24. S. Taira and K. Tanaka, "Study of Fatigue Crack Propagation by X-Ray Diffraction Approach," Eng. Fract. Mech. 4 (1972) pp. 925-938.

25. M. A. Wilkins and G. C. Smith, "Dislocation Structures Near a Propagating Fatigue Crack in an Aluminum Alloy," Acta. Met. 18 (1970) pp. 1035-1043.

26. B-Z. Weiss and M. R. Meyerson, "Plastic Zone Formation and Fatigue Crack Extension During High Cyclic Bending of Steels," Eng. Fract. Mech. 3 (1971) pp. 475-491.

27. W. J. Baxter and S. R. Rouze, "Photoelectron Microscopy of Plastic Zone Around a Fatigue Crack," Met. Trans. A, 7A, (1976) pp. 647-654.

28. T. D. Dudderas, "Applications of Holography to Fracture Mechanics," Experimental Mechanics 9 (1969) pp. 281-285.

29. D. G. Coates, "Kikuchi-Like Reflection Patterns Obtained with Scanning Electron Microscope," Phil. Mag. 16 (1967) pp. 1179-1184.

30. D. L. Davidson, "A Method for Quantifying Electron Channeling Pattern Degradation Due to Material Deformation," pp. 927-934 SEM 1974, Om Johari, ed., IITRI, Chicago, 1974.

31. D. L. Davidson and J. Lankford, "Fatigue Crack Tip Plasticity in High Strength Aluminum Alloys," Fatigue of Eng. Materials and Structs. (1980), in press.

32. G. R. Chanani, "Determination of Plastic-Zone Sizes at Fatigue-Cracks by Optical Interference Techniques," Inter. J. of Fracture 13 (1977) pp. 394-399.

33. D. L. Davidson and J. Lankford, "Plastic Strain Distribution at the Tips of Propagating Fatigue Cracks," J. Eng. Mater. and Tech. ASME Series H 98 (1976) pp. 24-29.

34. J. M. Larson, "An Investigation of the Shape and Extent of Crack Tip Plastic Zones in Fatigue Crack Propagation and the Role of the Overload Plastic Zone in Retardation," Master of Science Thesis, Vanderbilt University, Nashville, TN, 1977.

35. A. H. Purcell and J. Weertman, "Crack Tip Area in Fatigued Copper Single Crystals," Met. Trans. 5 (1974) pp. 1805-1809.

36. A. Saxena and S. D. Antolovitch, "Low Cycle Fatigue, Fatigue Crack Propagation and Substructures in a Series of Polycrystalline Cu-Al Alloys," Met. Trans. A, 6A (1975) pp. 1809-1828.

37. J. R. Rice and G. F. Rosengren, "Plane Strain Deformation Near a Crack Tip in a Power-Law Hardening Material," J. Mech. Phys. Solids 16 (1968) pp. 1-12.

38. J. W. Hutchinson, "Plastic Stress and Strain Fields at a Crack Tip," J. Mech. Phys. Solids 16 (1968) pp. 337-347.

39. J. R. Rice, "Mathematical Analysis in the Mechanics of Fracture," pp. 191-311, eq. 289, p. 287 in Fracture V II, H. Liebowitz, ed., Academic Press, New York, 1968.

40. D. A. Vermilyea and P. R. Diegle, "Concerning Strain-Enhanced-Corrosion Mechanisms in SCC," Corrosion 32 (1976) pp. 26-29.

41. D. L. Davidson and J. Lankford, "The Effect of Water Vapor on Fatigue Crack Tip Stress and Strain Range Distributions and the Energy Required for Crack Propagation in Low-Carbon Steel," In. J. of Fracture (1980) in press.

42. P. K. Liaw, M. E. Fine and D. L. Davidson, "Comparison of Plastic Work of Fatigue Crack Propagation in Low Carbon Steel Measured by Strain-Gages and Electron Channeling," Fatigue of Eng. Mater. and Struct. 3 (1980) pp. 59-74.

43. J. E. Allison, "The Measurement of Crack Tip Stresses by X-Ray Diffraction," pp. 550-562 in Fracture Mechanics, ASTM STP 677, C. W. Smith, ed., ASTM, Philadelphia, 1978.

44. D. P. Rooke and F. J. Bradshaw, "A Study of Crack Tip Deformation and a Derivation of Fracture Energy," pp. 46-57 in Fracture 1969, P. L. Pratt, ed., Chapman and Hall, London, 1969.

45. M. E. Fine, "Fatigue Resistance of Metals," Met. Trans. 11A (1980) pp. 365-379.

46. H. L. Dunegan, D. O. Harris, A. S. Tetelman, "Fracture Analysis by Use of Acoustic Emission," Eng. Frac. Mech. 1 (1968) pp. 105-122.

47. I. G. Palmer and P. T. Heald, "The Application of Acoustic Emission Measurements to Fracture Mechanics," Met. Sci. and Eng. 11 (1973) pp. 181-184.

48. A. A. Pollock, "Quantitative Evaluation of Acoustic Emission From Plastic Zone Growth," Int. Adv. NDT 6 (1979) pp. 239-262.

49. O. Buck and G. A. Alers, "New Techniques for Detection and Monitoring of Fatigue Damage," Fatigue and Microstructure, American Society for Metals, Metals Park, Ohio (1979) pp. 101-147.

50. S. L. McBride, D. H. Harold and B. P. Paradis, "Acoustic Emission Due to Fatigue Crack Growth in 7075-T6 Aluminum," ARPA/AF Review of Progress in Quantitative NDE, 1979.

51. J. Kramer, Der Metallische Zustand, Vanhoeck and Ruprecht, Goettingen, 1950.

52. O. Buck, D. Schumacher and A. Seeger, Phys. Stat. Sol. (b) 60 (1973) pp. 707-720.

53. J. Lankford and P. H. Francis, "Magnetic Field Perturbation Due to Metallurgical Defects," Advanced Experimental Techniques in the Mechanics of Materials, ed. P. H. Francis and U. S. Lindholm, Gordon and Breach, N.Y. (1973) pp. 385-402.

54. H. Kronmuller, "Magnetic Techniques for the Study of Dislocations in Ferromagnetic Materials," ibid, pp. 33-68.

55. G. S. Kino, J. H. Shaw, "Acoustic Techniques for Measuring Stress Regions in Materials," EPRI Report NP-1043, April, 1979.

56. J. Lankford and D. L. Davidson, "The Effect of Overloads Upon Fatigue Crack Tip Opening Displacement and Crack Tip Opening/Closing Loads in Aluminum Alloys," Fracture 1981, in press.

57. D. E. Macha, D. M. Corbly, and J. W. Jones, "On the Variation of Fatigue-Crack-Opening Load with Measurement Location," Exp. Mech. 19 (1979) pp. 207-213.

58. A. Ohta, M. Kosuge and E. Sasaki, "Change of Fatigue Crack Closure Level with Gauge Location Along Crack Line," Int. Jour. Fract. 15 (1979) pp. R53-R57.

A REVIEW OF ACOUSTIC EMISSION FROM SOURCES*

CONTROLLED BY GRAIN SIZE AND PARTICLE FRACTURE

W.W. Gerberich and K. Jatavallabhula

University of Minnesota
Minneapolis, Minnesota 55455

Some recent developments on using both amplitude and frequency analysis to understand acoustic emission sources are reviewed. The theoretical background uses the proposed model of Rouby and Fleischmann to describe discontinuous dynamical events such as free dislocation flight stopping at a grain boundary. A large variety of models for dislocation induced acoustic emission controlled by grain size are then discussed. The emphasis, however, is on relatively isotropic, multi-phase systems with a uniform array of second phase particles, specific examples being given on high-strength aluminum alloys, steels and super alloys. In terms of either linear or log normal particle distributions and a particle fracture criteria, it is shown that a prediction of acoustic emission is straightforward. Either mean-free path dislocation or particle fracture models are capable of explaining why acoustic emission rate experiences a maximum with increasing strain. It is projected that acoustic emission techniques should be able to assist in the analysis of

i) strains at which large non-coherent particles fracture;

ii) determination of critical strains for two overlapping particle distributions to undergo particle fracture and/or decohesion;

iii) grain size and particle interactions in dislocation dynamics.

Such applications should lead to a clearer understanding of how microstructure influences work hardening and ductile fracture.

* This work was supported by the U.S. Department of Energy under Grants EY-76-S-02-2212 and DOE AC02-79-ER10433.

Introduction

For some time, it has bothered many investigators, both in and out of the NDE field, that acoustic emission count rate or acoustic power should exhibit a maxima at some plastic strain well short of failure. Since this phenomenon was exhibited in both ultra-pure single crystals as well as complex alloy polycrystals, there was an early tendency to interpret it in terms of fractional mobile dislocation densities. It is becoming apparent that such approaches do not address the entire spectrum of events (e.g. dislocation phenomena versus particle fracture) nor do they qualitatively explain the observed acoustic emission (AE) response to various deformation phenomena (e.g. AE maxima as a function of strain and grain size (1, 2) or tension versus compression testing. (3, 4)) This present paper will attempt to clarify several contributions with particular emphasis on grain size and particle fracture effects. In that regard, this effort is indebted to the recent review of Wadley, et al. (5) which made the task if not infinitely at least considerably easier.

This study will not address macroscopic crack propagation phenomena except where it provides additional insight into microscopic fracture phenomena. The aim will be an understanding of acoustic emission emanating from a uniaxially loaded bar throughout its strain history. In doing this, special effects peculiar to specific materials such as phase transformations (6, 7, 8) and magnetic domains (9) will be ignored. This interpretation will be attempted by viewing AE as a result of phonon emission, i.e. from the start and stop motions of dislocations or cracks as others have recently proposed. (5, 10, 11, 12, 13) This generalization of acoustic emission is in terms of either an incremental area swept out by a dislocation segment of length, L, moving a distance, λ, or a particle of diameter, d, fracturing. As is indicated in Figure 1, the resulting areas are $\Delta A = L\lambda$ and $\Delta A \geq \pi d^2/4$. The $\geq$ in the later case is due to different particle orientations since ΔA may grow into the matrix where the macroscopic crack plane is coplanar with the particle fracture plane but may be restricted to $\pi d^2/4$ if it is at a large angle ϕ with the particle fracture plane.

The approach will be first to present a general format as taken from a model by Fleischmann, et al.(10, 12) apply this to the grain size effect in high purity metals, and then analyze complex polycrystalline materials containing second phase particles. In doing this some rigor will necessarily be sacrificed to obtain semi-quantitative relationships.

Acoustic Emission as Accelerating or Decelerating Waves

The interpretation of both dislocation and cracking phenomena as mechanical instabilities which may be interpreted in terms of radiating sound waves has been recently quantified. (10, 11, 12) Both Fleischmann, et al. (10, 12) and Kuribayashi and Kishi (11) have started out with the same premise, that the contribution from a longitudinal sound source is given by

$$d\sigma(t,r) = \frac{\rho_o}{2\pi r}\left(\frac{\partial \dot{u}}{\partial t}\right) d\mathcal{S} \tag{1}$$

The intensity of the acoustic wave is given in terms of the density, ρ_o, the distance r from the source, the displacement, u, at the source, the surface, $\mathcal{S}$, of the displacement, and the acceleration vector, $\partial\dot{u}/\partial t$. Considering the analogy between a moving line dislocation and a moving sound wave, using Laplace transforms and integrating over a dislocation segment of length, L, Fleischmann, et al.(10) find

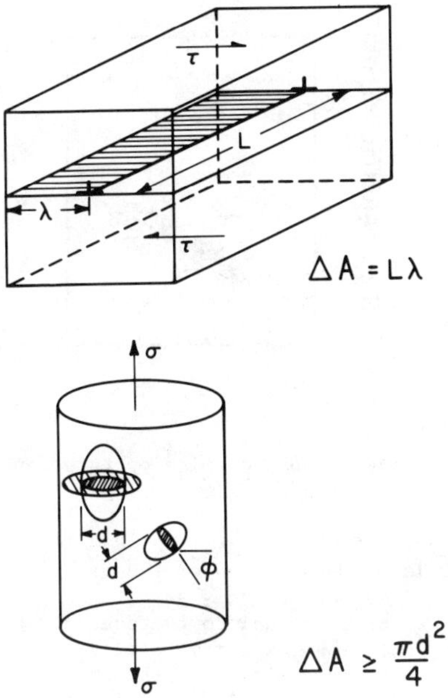

Fig. 1 Schematic of two AE sources producing discontinuous phonon emission.

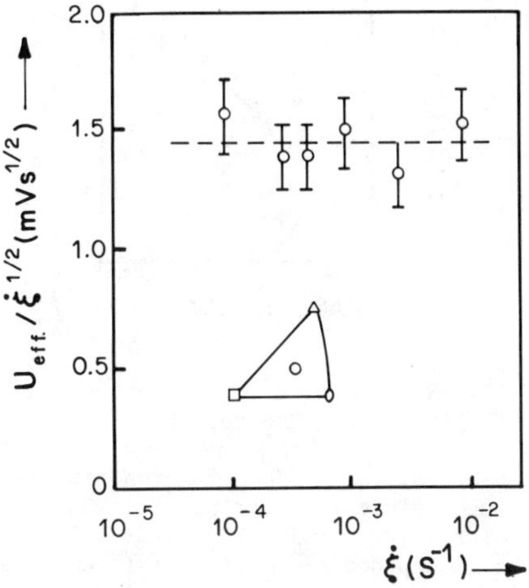

Fig. 2 Strain rate dependence of the acoustic emission of Al 99.99% (transducer 140 kHz, τ=22 MPa) After Kiesesetter and Schiller. (1)

$$|\sigma(\nu,r_o)| = \frac{\rho_o\, u_o\, L\, v_o}{\pi r_o} \left|\sin \frac{\pi\nu\lambda}{v_o}\right| \tag{2}$$

where r_o is the initial distance from the source, λ is the glide distance, v_o is the mean dislocation velocity, and ν is the frequency. Since the spectral energy density, S, is given in terms of the number of sources, N, and a frequency factor, $K'(\nu)$, one can write

$$S(\nu) = K'(\nu)N|\sigma(\nu)|^2 \tag{3}$$

$$\text{with } N = \frac{\dot{\varepsilon}\Delta t V}{L\lambda b} \tag{3a}$$

Here, $\dot{\varepsilon}$ is the strain rate, Δt is the sampling time, V is sample volume and b is the Burgers vector. This analysis may be applied to a tensile test by considering how $S(\nu)$ might change with plastic deformation. Incorporating the first part of Eq. (2) into $K(\nu)$ and writing the glide distance and velocity as a function of strain, one may approximate the spectral energy density as (12)

$$S(\nu,\varepsilon) \approx K(\nu)N\, L^2 v_o^{\,2}(\varepsilon)\, \sin^2\left(\frac{\pi\nu\lambda(\varepsilon)}{v_o(\varepsilon)}\right) \tag{4}$$

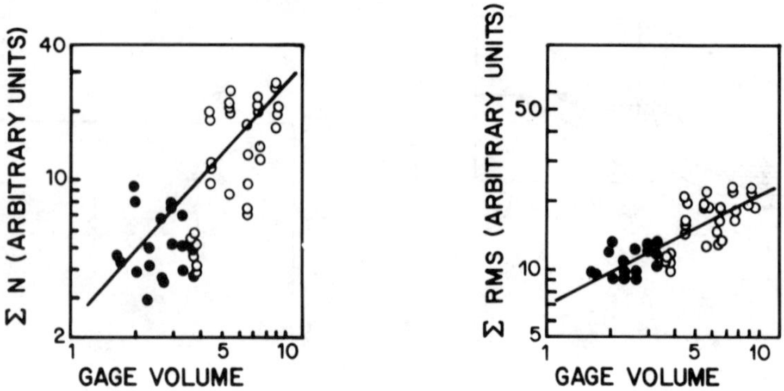

Fig. 3 Effect of gage volume on (a) integrated acoustic emission rate, and (b) integrated RMS voltage. After James and Carpenter. (15)

Since $S(\nu_1\varepsilon) = AE^2_{rms}$ is proportional to both strain rate (1) and sample gage volume (15) through Eq. (3a), a partial demonstration of the validity of Eq. (4) is given in Figures 2 and 3. These respectively show $AE_{rms}/\dot{\varepsilon}^{\frac{1}{2}}$ as independent of strain rate and ΣAE_{rms} as proportional to the square root of the gage volume. Furthermore, one may note the maximum of $S(\nu,\varepsilon)$ occurs at $\nu\lambda(\varepsilon)/v_o(\varepsilon) = 1/2$ where $\sin^2(\pi/2)$ is unity, assuming that the pre-trigonometric term increases throughout a given test. In fact, with this approach Rouby and Fleischmann (12) have convincingly demonstrated that the peak of $S(\nu,\varepsilon)$ occurs at larger strains as the transducer frequency increases. This is shown in Figure 4 where single crystal aluminum has been tested with a broad band piezoelectric transducer evaluated between 0.3 and 3 MH_Z. As the sampling frequency increased, the strain at which $S(\nu,\varepsilon)$ peaks increases, qualitatively in agreement with Eq. (4).

Quantification of this acoustic emission maximum may be approximately accomplished with some recent results of Kim and Kishi (14) realizing that their aluminum was 99.9% purity compared to the 99.99% purity of Rouby and Fleischmann. (12) In their study of grain size effects, Kim and Kishi (14) demonstrated that the flow stress, τ_{flow}, at room temperature could be given in terms of an athermal strain dependent term, $\tau_{i(\varepsilon)}$, and the grain size term, giving

$$\tau_{flow} = \tau_{i(\varepsilon)} + k_{(\varepsilon)}D^{-\frac{1}{2}} \tag{5}$$

Kim and Kishi (14) found that the Hall-Petch slope was strain dependent which will be approximated by $k(\varepsilon) \approx \alpha\varepsilon^{\frac{1}{2}}$. Since the deformation process in aluminum is nearly athermal at room temperature, one may write velocity in terms of some viscous phonon or electron drag mechanism, giving

$$v_o(\varepsilon) \approx \frac{\left\{\tau_{i(\varepsilon)} + \alpha\varepsilon^{\frac{1}{2}} D^{-\frac{1}{2}}\right\} b}{B} \tag{6}$$

where b is the Burgers vector and B is the viscous damping coefficient. If the glide distance is interpreted in terms of the mean free path of dislocations with a density, $\rho \approx M\varepsilon$, where M is the multiplication coefficient, then

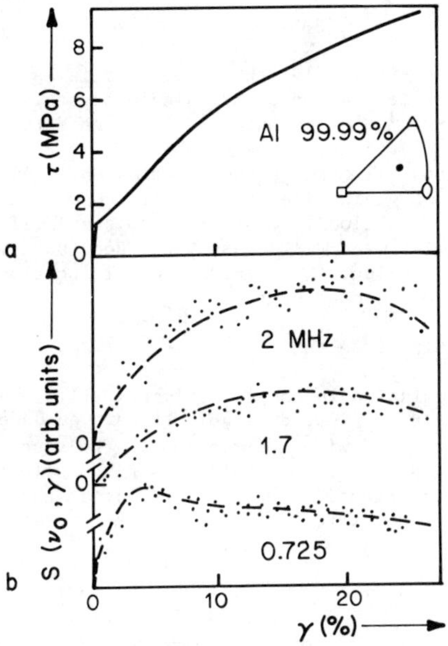

Fig. 4 Change in AE output as a function of sampling frequency for high purity aluminum. After Rouby and Fleischmann. (12)

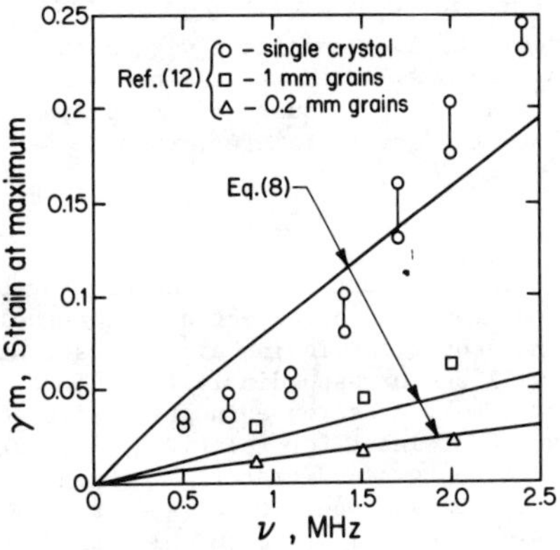

Fig. 5 Prediction of strain for peak acoustic emission rate as a function of sampling frequency.

$$\lambda(\varepsilon) \simeq 1/\rho^{\frac{1}{2}} \simeq 1/M^{\frac{1}{2}}\varepsilon^{\frac{1}{2}} \tag{7}$$

Combining the last term of Eq. (4) with Eqs. (6) and (7) leads to an estimate of the maximum in the spectral energy density at $\nu\lambda(\varepsilon)/v_o(\varepsilon) = 1/2$ which gives

$$\nu\Big|_{S(\nu,\varepsilon)_{max}} \simeq \frac{M^{\frac{1}{2}}\varepsilon^{\frac{1}{2}}\tau_{i(\varepsilon)}b}{2B} + \frac{\alpha\varepsilon M^{\frac{1}{2}}b}{2BD^{\frac{1}{2}}} \tag{8}$$

From the flow stress data of Kim and Kishi, (14) the value of α is approximately 1.5 MPa$\cdot$m$^{\frac{1}{2}}$. From Rouby and Fleischmann's (12) single crystal data, a value of $M^{\frac{1}{2}}b/2B$ equal to 0.76 MPa^{-1} s^{-1} is obtained from one data point since the second term in Eq. (8) is zero. As can be seen in Figure 5, this fix on one parameter predicts all other acoustic emission maxima reasonably well even as a function of grain size. The severe knee in the single crystal data is due to the transition from stage II to stage III slip as postulated by Rouby and Fleischmann. (12) Assuming then that this is a reasonable model of acoustic emission as generated from dislocation events, grain size effects will be considered next.

Grain Size Effects

In order to analyze acoustic emission in complex polycrystalline systems containing brittle particles, it is first desirable to have an understanding of dislocation events associated with deformation in the absence of second phases. Unfortunately, grain size apparently has mixed effects on acoustic emission resulting during plastic strain. Thus, it is useful to present a brief overview, survey one system in depth and then present a possible theoretical interpretation.

Brief Overview

Two of the leading investigators (2, 14) of grain size relationships have developed different interpretations as to how grain size affects acoustic emission. Frederick and Felbeck, et al. (2, 16, 17) have concluded that both the width of grain boundary sources, f, and the slip area associated with a source, Δs. are important. They estimated that the number of activated sources was proportional to 1/f and that the incremental displacement, Δx, associated with a source sweeping across a grain was $b\Delta s/2A_o$ where A_o is the specimen cross sectional area. Since Δs should be proportional to the square of the grain diameter, and f should be proportional to the grain diameter, then

$$\Delta AE \approx \frac{1}{f}\cdot\frac{\Delta x}{\Delta x_o} \approx \frac{1}{D}\cdot\frac{ZbD^2}{2A_o} \approx \frac{bD}{2A_o} \tag{9}$$

where Z is a proportionality factor near unity and Δx_o is the minimum displacement that can be detected at threshold. If Δx_o is reasonably large, then the acoustic emission events associated with slip across the entire grain will be correspondingly low. Thus, they anticipate that at the same trigger level, a decrease in grain size would produce fewer detectable events and at least for relatively fine grains, that ΔAE would be proportional to the grain diameter, D. Siegel (18) has analysed acoustic emission during the Bauschinger effect and came to the same conclusion that decreasing grain size will decrease emission rate.

On the other hand, Kim and Kishi (14) have interpreted their results solely in terms of grain boundary sources for dislocation emission and obtained

$$\Sigma AE_{rms}^{2} = H(\varepsilon) \exp (\beta D^{-1}) \tag{10}$$

where β is nearly constant and H(ε) is a mildly increasing function with increasing strain. Thus, there is the possibility of emission strongly increasing with grain size as in Eq. (9) or strongly decreasing with grain size as in Eq. (10).

There are ample data in the literature to support both of these approaches. First, there are the data of Mintzer, et al. (19) shown in Figure 6 which approximately indicate ΣAE to be proportional to the grain diameter for Cd and Cu, thus supporting Eq. (9). They use the detection threshold and mean free path argument to explain these observations, in agreement with Frederick and Felbeck, et al.(2, 16, 17, 18) In the opposite sense, Tandon and Tangri (20) demonstrated that AE from deformed Fe-3%Si drastically decreased with grain size approximately in agreement with Eq. (10). As is seen in Figure 7, a factor of 30 decrease in ΣAE occurred as the grain size increased by a factor of 30. Arguments were given in terms of a decrease in grain size producing a greater number of dislocation breakaways because of a finer substructure. One could just as easily interpret it in terms of a greater amount of dislocation emission from the greater grain boundary area as suggested by the thin film transmission investigation of Murr and Hecker (21) and espoused by Kim and Kishi. (14)

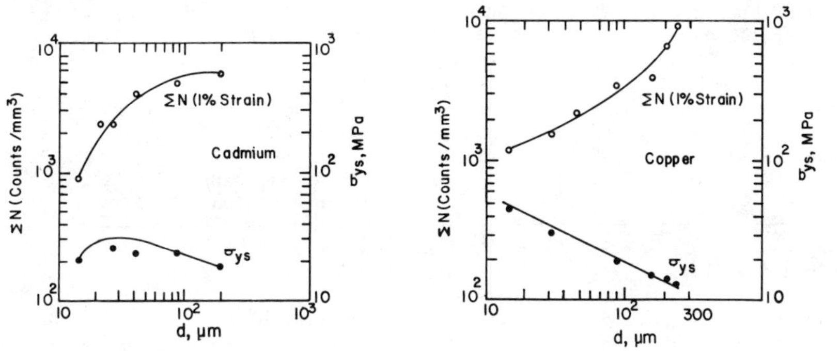

Fig. 6 Total AE to a strain of 0.01 for various grain sizes of Cu and Cd. After Mintzer, et al. (19)

Considering the approach represented by Eq. (4), both of these analyses may represent an oversimplification. In an attempt to rationalize such varying results, it was considered necessary to analyze one system in depth where a large amount of multiple-investigator data were available.

High-Purity Polycrystalline Aluminum

There have been a number of investigations on single and polycrystalline aluminum as indicated in Table I. Of particular note is the fact that the frequency range has both increased and broadened as recording and analysis equipment has become more sophisticated. A second point of importance is that several investigators have reported AE in terms of stress as the independent variable rather than strain. These two factors, the instrumentation and reporting variations, have led to a number of apparent discrepancies in the literature.

Table I: Polycrystalline or Single Crystal Aluminum AE Data

Investigators	Purity	Grain Size Range	Transducer & Instrumentation	Parameter Reported	Independent Variables
Sankar, et al (1970) and Siegel (1977)	99.99%	——	PZT-5 6-12kHz	ΣAE (load & unload)	ε $\dot{\varepsilon}=2\times10^{-5}s^{-1}$
Bill, et al (1970) (1979)	99.99%	10-1000 μm	120-180kHz, 80dB	ΣAE	σ $\dot{\varepsilon}=2\times10^{-5}s^{-1}$
Imanaka, et al (1973)	99.6%	350 μm	PZT (140kHz resonance) 100-300kHz, 80dB	ΣAE	ε $\dot{\varepsilon}=5\times10^{-5}s^{-1}$
Kiesewetter and Schiller (1976)	99.5%-99.99%	50 μm-single crystal	PZT-5 140 or 900kHz, 0.1s integration	AE^2_{rms}	τ $\dot{\varepsilon}=1.4\times10^{-4}s^{-1}$
Kuribayashi and Kishi (1977)	97.0-99.9% Al-Mg	single crystal	PZT (140kHz resonance) 20-200kHz, 70dB 0.1s integration	AE^2_{rms}	ε $\dot{\varepsilon}=2.7\times10^{-3}s^{-1}$
Rouby and Fleischmann (1978)	99.99%	200 μm-single crystal	Broad Band 0.3-3 MHz 25 μs integration	AE^2_{rms}	γ $\dot{\varepsilon}=4.77\times10^{-3}s^{-1}$
Kim and Kishi (1979)	99.9%	26-153 μm	PZT, 140kHz, 70dB 0.25s integration	ΣAE^2_{rms}	ε or σ $\dot{\varepsilon}=8.3\times10^{-4}s^{-1}$
Fleischmann, et al (1977)	99.99%	200-3600 μm	Broad Band 0.6-3.3MHz 25 μs integration	AE^2_{rms}	ε $\dot{\varepsilon}=1.4\times10^{-2}$ 2.8×10^{-3}

For example, Wadley, et al. (5) reviewed the polycrystalline data for aluminum (2, 1, 14) and found disagreement. The first investigation reported ΣAE to increase and then decrease as grain size increased. The second reported AE^2_{rms} to increase continuously as grain size increased. Finally, the third demonstrated that ΣAE^2_{rms} continuously decreased as grain size increased. Wadley, et al. (5) considered that the later discrepancy

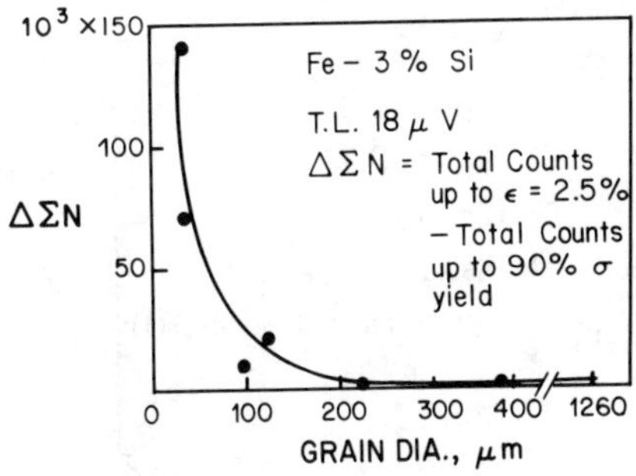

Fig. 7 Total AE to a strain of 0.025 for various grain sizes of Fe-3%Si. After Tandon and Tangri. (20)

could be due to a consequence of higher impurities in Kim and Kishi's (14) material (99.9%) versus the others (99.99%). Although clearly a factor, we do not believe this to be the major cause of this apparent discrepancy. In examining Table 1, the "discrepancy" cannot be due to the instrumentation since they were comparable in the three investigations. One may note, however, that both the parameter reported and the independent variables used were different. For example, in the first two acoustic emission was reported as a function of stress while in the later investigation it was reported as a function of strain. In the comparison of grain size effects, it does not seem appropriate to report data as a function of applied stress if dislocation events are the primary source of acoustic emission. To illustrate the extreme, Bill, et al's data (2, 22) reported at a stress level of 8.27 MPa (their Fig. 9(b)) indicate a clear ΣAE peak at a grain size of about 400 μm. However, at that stress level, the coarsest grain samples had undergone a plastic strain magnitude about two orders of magnitude greater than the finest grain samples and the finest grain sample exhibited almost wholly elastic behavior. When plotted at a constant stress in this manner, it is not surprising that ΣAE would increase at some larger grain size as gross yielding proceeded.

For a clear picture of the true grain size effect in the context of summed dynamical dislocation events, a plot of total acoustic emission to a given strain level is probably the most representative. Using established stress-strain data for high-purity polycrystalline aluminum, the data were reconstructed as best as possible so that the same parameter could be compared under the same given strain level. We chose ΣAE or ΣAE^2_{rms} and plotted this as a function of a given strain (either 0.002 or 0.01). As is seen in Figure 8, such a plot always produced a decrease in acoustic emission as grain size increased.† The differences in slope may be partly due to small instrumentation differences, strain differences and/or crystal purity but the fact is that this removes any apparent inconsistency in the literature.

Such a demonstration does not, however, explain the gross differences in Figs. 6 and 7 since similar transducers, instrumentation settings, measured parameters, and independent variables were used. For that reason, it is highly probable that different grain size dependencies do exist for different materials and a theoretical analysis of the various possibilities is in order.

Theoretical Interpretations

A theoretical interpretation of ΣAE^2_{rms} will be made using Eq. (4) from

---

† We do not mean to suggest that acoustic emissions cannot increase at very large grain sizes. Considering the gross deviation of the summed acoustic energy at the three largest grain sizes, it is quite possible that the emission goes through a minimum or a plateau. The reason for this will be discussed in the next section.

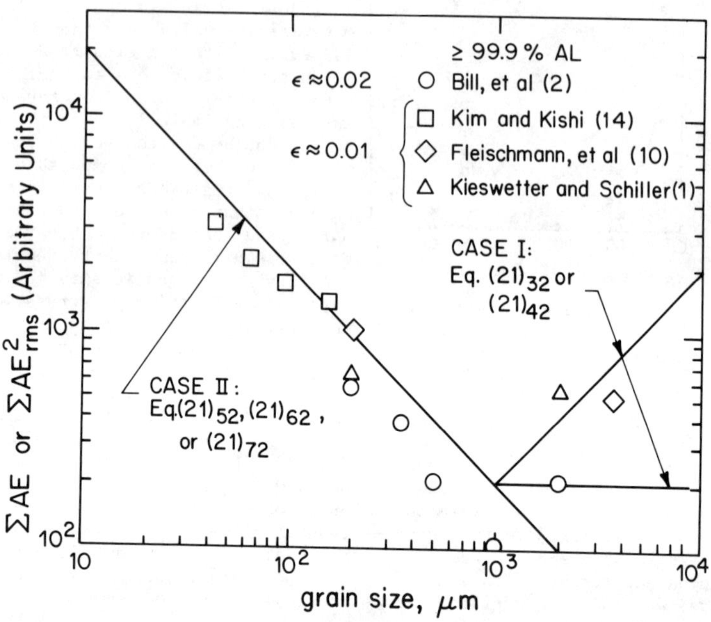

Fig. 8 Effect of grain size on total AE to a given strain in aluminum.

Rouby and Fleischmann. (12) Two general cases for the number of acoustic emission sources, N, will be evaluated for two general types of dislocation mechanisms, athermal and thermal. The first case will consider the number of sources as others have, (10, 12) e.g. Eq. 3(a), giving

Case I: $$N_I = \frac{\dot{\varepsilon}\Delta t V}{Ab} = \frac{V d\varepsilon}{\lambda L b} \tag{11}$$

The second equality follows since the area swept out by the emitting dislocation is λL and the incremental strain, $d\varepsilon$, is given in terms of the strain rate and sampling time. The second case will consider that the initial dislocation content is relatively immobile and a new dislocation density is created as a response to plastic strain in terms of displacement accomodation at grain boundaries, i.e. Ashby's (24) geometrically necessary dislocation model. In terms of the geometrically necessary density, $\rho^G = \varepsilon/4bD$, the number of sources becomes

Case II: $$N_{II} = \frac{\dot{\varepsilon}\Delta t V}{\lambda L b} = \frac{V d\rho^G}{L} = \frac{V d\varepsilon}{4bDL} \tag{12}$$

Each of these cases may be considered in materials which are relatively strain rate independent or dependent. For a given temperature and stress regime, a material may behave athermally, so that dislocation velocity is given by

athermal: $$v_o = \frac{\sigma b}{B} \tag{13}$$

For a thermally-activated mechanism a power or exponential empirical equation is often used. Rather than the more common $\exp(-A/\sigma)$ term used to represent dislocation velocities, a similar exponential relationship is picked, but one which yields simpler analytical relationships. This is given by

$$\text{thermal:} \qquad v_o = B'\exp(\sigma/\sigma_o) \qquad ; \quad v_o \ll c \qquad (14)$$

Note that the validity of this relationship breaks down as velocities become a significant fraction of the sound velocity, c; however, even before this limit an athermal mechanism intercedes and this is not a reason to reject Eq. (14).

A third set of possibilities arises from the detection capability of the acoustic emission instrumentation and transducer, and from the frequency content of the emitted signal. If one considers a relatively low frequency pick-up, e.g. 100 kHz, it is quite possible that $\nu\lambda/v_o$ in Eq. (4) is small and therefore,

$$\text{"low" frequency:} \qquad \sin^2\left(\frac{\pi\ \nu\ \lambda}{v_o}\right) = \sin^2\psi \approx \frac{\pi^2\nu^2\lambda^2}{v_o^{\,2}} \qquad (15)$$

On the other hand, for broad band frequencies in the MHz range, it is likely that $\nu\lambda/v_o$ is sufficiently large so that

$$\text{"high" frequency:} \qquad \sin^2 \frac{\pi\ \nu\ \lambda}{v_o} = \sin^2\psi \approx 1 \qquad (16)$$

The reader is cautioned here that what might be a low frequency transducer setting for one material could be high frequency for another since it depends on the ratio of $\nu\lambda/v_o$ with both the glide distance and dislocation velocity involved.

Within each one of these eight possible combinations, e.g. Case II-thermal-high frequency, there are a subset of possible assumptions with regards to the average glide distances and dislocation segment lengths. One example will make the point. Consider Case II with testing at low temperatures such that Eq. (14) is appropriate for dislocation velocities. In addition, the velocities are low enough so that Eq. (16) is applicable. With Eqs. (12) and (16), integration of Eq. (4) to represent the sum of the root-mean-square acoustic emission gives

$$\Sigma\ S(\nu,\varepsilon) \simeq \int_o^{\varepsilon} K(\nu)\ \frac{V}{4bDL}\ L^2B'^2\exp\left(\frac{2\sigma(\varepsilon)}{\sigma_o}\right)d\varepsilon \qquad (17)$$

Some decision with regards to the dislocation segment length and the functional dependence of stress on strain must be made. One choice for the dislocation segment size is a dependency on the inverse square root of the geometrically necessary dislocations, giving

$$L \simeq 1/(\rho^G)^{\frac{1}{2}} \simeq \left(\frac{4bD}{\varepsilon}\right)^{\frac{1}{2}} \qquad (18)$$

For the stress dependence on deformation, a slight oversimplification of Eq. (5) is

$$\sigma_{(\varepsilon)} \simeq \alpha\ \varepsilon^{\frac{1}{2}}\ D^{-\frac{1}{2}} \qquad (19)$$

which gives a first order estimate of Kim and Kishi's data (14), with $\alpha \simeq$ 1.5 MPa $m^{\frac{1}{2}}$. Combining Eqs. (17) through (19) gives

$$\Sigma\ S(\nu,\varepsilon) \simeq \int_0^{\varepsilon} K(\nu)\ \frac{V}{(4bD)^{\frac{1}{2}}}\ B'^2 \exp\left(\frac{2\alpha\varepsilon^{\frac{1}{2}}}{\sigma_o D^{\frac{1}{2}}}\right) d\varepsilon \tag{20}$$

Integration gives

$$\Sigma\, S(\nu,\varepsilon) \simeq \frac{VK(\nu)B'^2}{4b^{\frac{1}{2}}}\ \frac{\sigma_o}{\alpha}\left[\exp\left(\frac{\alpha\varepsilon^{\frac{1}{2}}}{\sigma_o D^{\frac{1}{2}}}\right) - 1\right] \qquad (21)_{83}$$

which is the equation in the lower right hand corner of Table 2.

Table II: Possible functional dependencies of acoustic emission on grain size and strain. (Eqs. $(21)_{11}$ through $(21)_{83}$)

$\Sigma S(\nu,\varepsilon)$	Segment and Glide Distance	Athermal, $v_o \simeq \frac{\sigma b}{B}$ High ν: $\sin^2\psi \simeq 1$	Athermal, $v_o \simeq \frac{\sigma b}{B}$ Low ν: $\sin^2\psi \simeq \frac{\pi^2\nu^2\lambda^2}{v_o^2}$	Thermal, $v_o \simeq B'\exp\left(\frac{\sigma}{\sigma_o}\right)$ High ν: $\sin^2\psi \simeq 1$
CASE I $N = \frac{V}{\lambda Lb}\, d\varepsilon$	L~const. λ~const.	$\frac{VLb}{2\lambda B^2 D}\ K(\nu)\alpha^2\varepsilon^2$	$\frac{VL\lambda}{b}\pi^2 K(\nu)\nu^2\varepsilon$	$\frac{VL}{\lambda b}\ K(\nu)B'^2\ f(\varepsilon,D)$ †
	λ~L~ const.	$\frac{Vb}{2B^2 D}\ K(\nu)\alpha^2\varepsilon^2$	$\frac{V\lambda^2}{b}\ \pi^2 K(\nu)\nu^2\varepsilon$	$\frac{V}{b}\ K(\nu)B'^2\ f(\varepsilon,D)$ †
	λ~L~ $1/M^{\frac{1}{2}}\varepsilon^{\frac{1}{2}}$	"	$\frac{V}{bM} K(\nu)\nu^2 \ln\left(\frac{\varepsilon}{\varepsilon_o}\right)$	"
	λ~L~ $\left[\frac{4bD}{\varepsilon}\right]^{\frac{1}{2}}$	"	$V4\pi^2 K(\nu)\nu^2 D\cdot \ln\left(\frac{\varepsilon}{\varepsilon_o}\right)$	"
CASE II $N = \frac{V}{4bDL}\ d\varepsilon$	L~const. λ~const.	$\frac{VLb}{8B^2D^2}\ K(\nu)\sigma^2\varepsilon^2$	$\frac{V\lambda^2 L}{4bD}\pi^2 K(\nu)\nu^2\varepsilon$	$\frac{VL}{4bD} K(\nu)B'^2\ f(\varepsilon,D)$ †
	λ~L~ const.	"	$\frac{V\lambda^3}{4bD}\pi^2 K(\nu)\nu^2\varepsilon$	"
	λ~L~ $1/M^{\frac{1}{2}}\varepsilon^{\frac{1}{2}}$	$\frac{Vb}{6M^{\frac{1}{2}}B^2D^2} K(\nu)\sigma^2\varepsilon^{3/2}$	$\frac{V\pi^2 K(\nu)\nu^2}{2bM^{3/2}D}\left[\frac{1}{\varepsilon_o^{\frac{1}{2}}} - \frac{1}{\varepsilon^{\frac{1}{2}}}\right]$	$\frac{VK(\nu)B'^2}{8bM^{\frac{1}{2}}D^{\frac{1}{2}}}\ \frac{\sigma_o}{\alpha}\left[\exp\left(\frac{\alpha\varepsilon^{\frac{1}{2}}}{\sigma_o D^{\frac{1}{2}}}\right) - 1\right]$
	λ~L~ $\left[\frac{4bD}{\varepsilon}\right]^{\frac{1}{2}}$	$\frac{Vb^{3/2}}{3B^2D^{3/2}}\ K(\nu)\alpha^2\varepsilon^{3/2}$	$8Vb^2D^2K(\nu)\nu^2 \times \left[\frac{1}{\varepsilon_o^{2}} - \frac{1}{\varepsilon^2}\right]$	$\frac{VK(\nu)B'^2}{4b^{\frac{1}{2}}}\ \frac{\sigma_o}{\alpha}\left[\exp\left(\frac{\alpha\varepsilon^{\frac{1}{2}}}{\sigma_o D^{\frac{1}{2}}}\right) - 1\right]$

$$\dagger \qquad f(\varepsilon,D) = \left[\frac{\sigma_o{}^2 D\ \exp\left(\frac{2\alpha\varepsilon^{\frac{1}{2}}}{\sigma_o D^{\frac{1}{2}}}\right)}{2\alpha^2}\left(\frac{2\alpha\varepsilon^{\frac{1}{2}}}{\sigma_o D^{\frac{1}{2}}} - 1\right) + \frac{\sigma_o{}^2 D}{2\alpha^2}\right]$$

The subscript in Eq. $(21)_{83}$ indicates that the equation is in row eight, column three of Table II. Here, the other possibilities are also presented using the various cases along with further assumptions on segment length, L, or glide distance, λ. In the column under thermal activation, low ν equations are missing since v_o^2 cancels in the solution giving rise to identical equations as are presented for the athermal model. The only additional assumption that was made for Table II was in those solutions which would predict unbounded acoustic power at zero strain. There, a threshold strain, ε_o, was invoked for the start of substantial acoustic emission, which is the observation in many instances. In fact, for the other equations, one could replace ε by $\varepsilon-\varepsilon_o$ (or $\varepsilon^2-\varepsilon_o^2$, etc.) but this does not substantially alter the result for modest strain levels where $\varepsilon>>\varepsilon_o$. The important point is that a large number of strain and grain size dependencies are possible for acoustic emission including ε, ε/D, $(\varepsilon/D)^{3/2}$, $(\varepsilon/D)^2$ and $D \ln(\varepsilon/\varepsilon_o)$.

For the aluminum case, the dislocation mechanism is largely athermal and since v_o is fairly large, one might consider the low frequency analysis to apply giving a grain size independence (CASE I) or a 1/D relationship (CASE II). This latter case assumes that grain boundaries are the principal dislocation sources as has been demonstrated by the work of Bill, et al. (2, 22). As is seen in Figure 8, this agrees very well with the trend of published data for $\Sigma S(\nu)$ keeping in mind that the actual magnitudes could not be assessed since either data were reported in arbitrary units or there was no way to independently determine $K(\nu)$. The three samples with the largest grains in Figure 8 seem to deviate from this relation. The following reason is proposed. If one examines the three investigations, it evolves that the sample dimensions are marginally larger than the grain size. This is documented as follows:

Ref.	Grain Size, mm	Sample Dimensions, mm	No. of Grains in Cross Section
(1)	2	3 x 6	~5
(2)	2	4mm round	~4
(10)	3.6	6mm round	~3

Even this is an optimistic number of grains in the cross section since the grain size reported is an average value. Thus, in many instances, the cross section might be a single grain, and if oriented properly for slip, the results for small strains would approach single crystal behavior. This would negate using CASE II to model the number of acoustic emission sources. CASE I would now become dominant with the mean free path either being limited by dislocation or grain size considerations. Thus, it might be expected that Eqs. $(21)_{32}$ or $(21)_{42}$ are more appropriate for very large grains where the crossover occurs when the average grain size approaches about one-fourth the sample diameter.

The next comparison was made to the data of Mintzer, et al. (19) on 99.9% Cd and on Zn-0.4%Al made from high purity metals. Both of these represent materials with rapid dislocation velocities and high homologous temperatures at room temperature. For example, it is known (25) that the dislocation process in Zn is athermal with modest stresses producing velocities of 20m/s. Considering the mean free path in coarse grains to be controlled by the dislocation intersection process, then one can approximate λ and L by Eq. (7). Even with very small values of M, i.e. low dislocation densities, the value of $\pi^2\nu^2\lambda^2/v_o^2$ is so small for their 150 kHz transducer, that this is clearly a low ν situation. Assuming CASE I, low ν behavior and that for small grains λ and L are constrained by the grain size, being proportional to $D^{1/2}$, but for large grains by Eq. (7), Eqs. $(21)_{42}$ and $(21)_{32}$ in Table II result. These two equations are compared to the experimental data in Figure 9. The agreement is reasonable with a cross-over in mechanisms at a grain size near 40 µm. If one equates the Eqs. $(21)_{32}$ and $(21)_{42}$ at

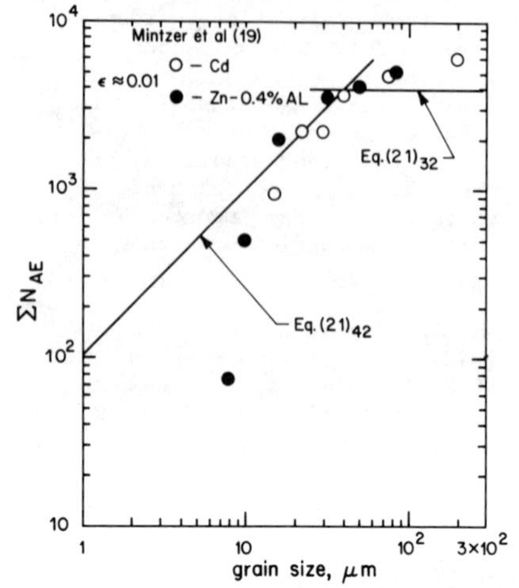

Fig. 9 Effect of grain size on total AE to a given strain in Cd and Zn-0.4%Al.

this crossover point, an M value of about $2.5\times10^9/cm^2$ is calculated which seems about one ot two orders of magnitude lower than what might be expected for many dislocation multiplication coefficients. However, it is in keeping with the very low coefficient observed by Puttick (26) for Cd. Moreover, in a previous paper on the same superplastic alloy (Zn-0.4% Al), Frydman, et al. (27) demonstrated that there was almost a complete absence of dislocations within the grains after deformation but that for other deformation modes at much lower temperatures where no superplasticity was in evidence, acoustic emission was about forty times larger and the dislocation multiplication parameter was much larger.

Finally, reconsider the data of Tandon and Tangri (20) on Fe-3%Si. Since their ΔΣN is essentially 99% of the ΣAE, one can equate this to ΣS(ν,ε). The appropriate situation would be a thermally activated mechanism at high frequency since, with v_o relatively small and ν~480 kHz, $\sin^2\psi \approx 1$. With CASE II controlling dislocation generation, (20) one finds Eq. $(21)_{83}$ to be the solution. Taking typical grain size and dislocation velocity dependencies from the literature for Fe-3Si, $m^*\simeq\partial\ell n\dot{\varepsilon}/\partial\ell n\sigma^*\approx7.5$ and $k_y\simeq0.4MPa\ m^{1/2}$, one can reproduce the data in Figure 7 as well as the curve already presented.

To summarize this section, it is sufficient to make two points. First, acoustic emission depends upon plastic strain and grain size in a large number of ways depending on the output measured, the type of transducer used, and the details of dislocation creation and free flight. Secondly, consider the analysis of the instantaneous acoustic emission output, S(ν,ε), as a function of strain. At small strains for CASE II, S(ν,ε) would increase linearly with strain or the square root of strain as predicted by the high frequency case. However, as strain increased and the mean free path became substantially smaller, the low frequency case would be represented and S(ν,ε) would either not increase or would decrease approximately as $1/\varepsilon^{3/2}$. These two cases probably bracket many AE observations and would explain the peak in acoustic emission as a function of strain as observed in almost all investigations and as postulated by Fleischmann, et al. (10, 12). In principle, this removes any objection of using broad band acoustic emission to measure dislocation processes and provides a basis for understanding emission from relatively clean polycrystalline materials in the absence of second phase particles.

Since this analysis was completed, we received a communication from Wadley (28) that for very coarse grained specimens of aluminum, the ΣAE does increase with increasing grain sizes. This confirms the trend for CASE I in Figure 8. Those specimens also contained relatively few grains in the cross section. The final proof of this argument would be to evaluate very coarse grained specimens of increasing grain size with a series of samples of differing diameters. For example, in the coarsest grained samples, one might want one grain across the diameter in the smallest diameter sample but ten grains in the largest diameter sample.

Particle Fracture Effects

It is now generally accepted that much acoustic emission in complex alloys originates from either surface oxides, (29) dispersed phases (30) or precipitates (5, 31) fracturing. This would certainly be consistent with the irreversibility of much of the acoustic emission on reloading or the so-called Kaiser effect. (32, 33) That is, instead of using up grain boundary sources on the initial loading, one could use up the easily fractured fraction of the particle distribution. The procedure will be to review a few experimental results which highlight some current controversial questions, present a theoretical approach for handling AE from particle fracture and then apply the approach to a few specific cases.

Experimental Observations

The early work (3, 4, 30, 34, 35) on fracture of particles as acoustic emission sources centered on aluminum alloys where it was relatively easy to identify cleavage of large particles with burst-type AE events. Invariably, the acoustic emission detected in tension was greater than that in compression which could be easily understood if the particles were elongated in the direction of the tensile axis. In a very careful study, Hamstad, et al. (30) demonstrated that the same inclusions which failed in tension could also fail in compression but with a fracture plane parallel to the loading direction. The result was that the peak amplitude of the AE burst was at least a factor of five less in compression than in tension since the Poisson stresses and strains responsible for particle fracture in compression are substantially smaller. An example of this is shown in Figure 10 for one of their heats of 2124-T851 aluminum where the RMS voltage is more than an order of magnitude greater in tension than it is in compression. Tensile data of the other orientations give different amounts of emission as interpreted in terms of the number and size of coarse particles (1-20 μm) failing by cleavage. A typical example of such particle fractures during tensile testing of 2000-series aluminum is shown in Figure 11.

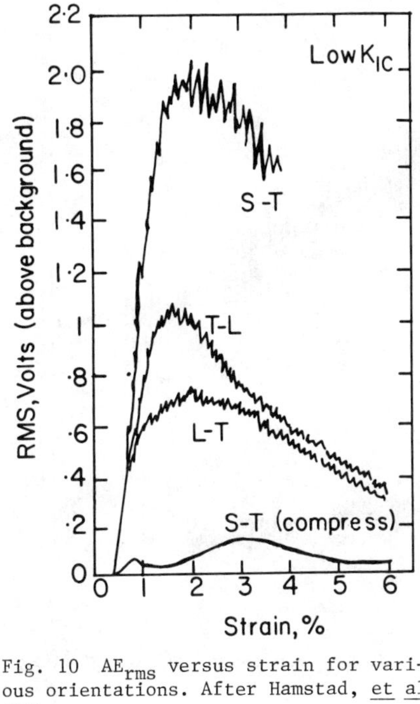

Fig. 10 AE_{rms} versus strain for various orientations. After Hamstad, et al. (30).

With regards to other material systems, Gerberich, et al. (31) and Wadley, et al. (37) have suggested that large carbonitrides, Laves phase precipitates and sulfide inclusions are major sources of emission in superalloys and steels. An example of the former is shown in Figure 12 where count rate versus plastic strain was determined for IN 718 superalloy aged to two conditions. It is clear that there is a bimodal distribution of AE in both aged conditions with burst emission throughout. The first peak

Fig. 11 Fractography of cleaved particles in 2000-series aluminum.

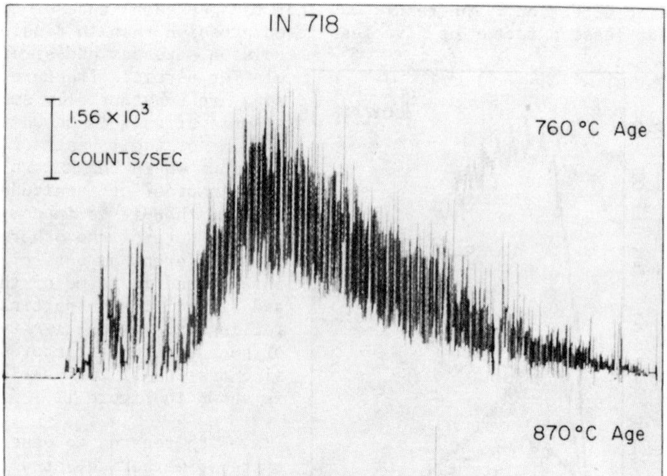

Fig. 12 AE Count rate versus tensile strain for two conditions of IN 718 superalloy. After Gerberich, et al. (31)

Fig. 13 Scanning electron microscopy of fractured NbC or TiN in IN 718 deformed to a strain of 0.05. After Gerberich, et al. (31)

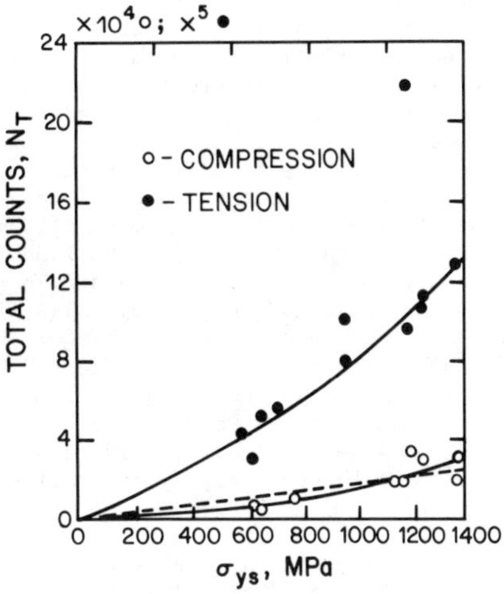

Fig. 14 Effect of yield strength on AE output for compression and tension tests. Note that the tensile scale is ten times the compression scale. After Gerberich, et al. (31)

and the bursts were associated with NbC and TiN particles on the order of 1 to 20 μm in diameter while the second peak was associated with Laves phase (Ni_3Nb) particles on the order of 0.1 to 1 μm in size. One of the large carbide or nitride particles, fractured after a strain of 0.05, is shown in Figure 13. From detailed metallography, it was shown that virtually all of the large particles had failed by the time the strain reached a level where AE ceased. (31) The reason for the bimodal distribution is that the smaller Laves phase particles require a higher fracture stress and hence a larger applied strain. This will be discussed in more detail under Particle Size Fracture Criterion. For this same set of materials, compression tests were run, the results being qualitatively similar to those observed by others. (3, 4, 30, 34, 35) As seen in Figure 14, there is a factor of 50 greater emission in tension compared to compression, attributable to the greater energy release associated with particle fracture.

There appears to be no disagreement in the literature for large particle fractures on the order of 2 to 60 μm being sources of acoustic emission. What is controversial is whether or not fracture of smaller particles on the order of 0.1 to 1 μm give rise to detectable emission and whether fracture events are uniquely separable from dislocation events. (31) With regards to smaller particles, Wadley, et al's review (5) suggests that carbide platelets in steel are doubtful as acoustic emission sources. On the other hand, Gerberich et al's (31) study of 4340 steel heat treated to different strength levels, suggested that the emissions were proportional to the carbide volume fraction extracted from replicas. The emission cannot be related to inclusion fractures since the highest strength condition (where sulfides would fracture most easily) exhibited nearly an order of magnitude lower emission than many of the lower strength conditions. To confuse the issue further, Swindlehurst (37) carefully measured a cumulative density of cracked carbides in a 1.1%C steel as a function of strain and concluded that they could not be the source of acoustic emission. The reason given was that the functional dependence of cumulative crack density was to increase linearly with strain while the functional dependence of AE was to rise sharply then plateau after a strain of about 0.05. What was not reported was any variation in particle size with regards to susceptibility to fracture. One would expect the larger, properly oriented carbides to fracture first at lower strains as suggested by the micrographs presented by Swindlehurst. (37) The remaining particle distribution could have been such that the fracture events beyond the 0.05 strain level were simply below the detection level. Clearly, this question has not been sufficiently settled as to negate the possibility of small carbide particles giving rise to detectable AE.

Finally, it has been suggested that plasticity events triggered by particles (31) could give rise to AE. This could be either due to simultaneous plasticity accomodation around a group of particles, or more likely, a large release of dislocation energy by cascading into an interface or void associated with particle decohesion or fracture. Such doubts about the separability of particle fracture events have been clarified by using a wide band frequency technique to separate out dislocation from particle fracture events. (38) Palmer, Green and Bruchey have been using optical methods with essentially unlimited frequency response and absolute calibration. In complex materials, they have characterized the type of emission as low and high frequency types, with the later exhibiting nearly a 1:1 correlation to intermetallic particle fractures. Thus, with appropriate frequency separation, they have unambiguously shown that there are at least two types of emission events, one associated with particle fracture.

The next step is to describe several possible approaches for quantifying acoustic emission amplitudes from particle fracture events.

Single Particle Event

Considering particle fracture initiation and arrest as accelerating and decelerating acoustic wave sources, in principle one could model the acoustic emission from such a single brittle particle event in a tough metal matrix. We will not do this exactly but will approximate the energy density by smoothing out the local stress and velocity profiles. If one considered a crack front moving across a particle as a line source, then from Eqs. (2) and (3) one can show that, for $\sin(x) \simeq x$, the acoustic emission amplitude may be given by

$$AE_{rms} = K''(\nu) \frac{u_o \Delta A}{\pi r_o} \tag{22}$$

where u_o is the displacement at the source, r_o is the distance the detector is from the source and ΔA is the area swept out by the source. Treating the particle fracture as an analagous event and the displacement associated with an applied stress intensity at the instant the particle fractures, it may be shown that

$$u_o = \frac{K_c^{\,2}}{\sigma_{flow}E} = \frac{2E\gamma_c}{\sigma_{flow}E} = \frac{2\sigma_{flow}^2 q^2_{(\varepsilon)} d_c}{6E\sigma_{flow}} \approx \frac{3\sigma_{flow} d_c}{\mu} \tag{23}$$

where K_c is the particle fracture toughness, γ_c is the specific work of fracture for the particle, σ_{flow} is the macroscopic flow stress at particle fracture, $q_{(\varepsilon)}$ is a strain concentration factor near two, d_c is the cracked particle diameter and E,μ are elastic moduli. Combining Eqs. (22) and (23) gives

$$AE_{rms} \simeq \frac{K''(\nu)\sigma_{flow}\Delta A d_c}{\mu r_o} \tag{24}$$

This indicates the rms amplitude to be proportional to the flow stress and crack area but inversely proportional to the distance from the source. As this is an uncalibrated correlation, we will revert to two previous models which have been calibrated using known fracture events. (40, 41) The first used a compliance calibration technique with known incremental crack advances to define an elastic energy release, ΔU, in terms of load, P, specimen thickness, B, and specimen compliance, C_S, giving

$$\frac{\Delta U}{B} = -\int_{a_o}^{a_1} \frac{\sigma^2 \pi a}{E}\, da = -\frac{\Delta P P C_S}{2B} \tag{25}$$

Writing compliance in terms of specimen geometry and P in terms of applied stress intensity, K, it was found that

$$\Delta P = \frac{3.5 K W^{\frac{1}{2}} \Delta A}{YL} \tag{26}$$

where ΔA is the incremental area swept out by the crack advance, W is the specimen width, Y is the geometry factor for a compact tension sample and L is the distance between load points. In that same study the acoustic emission amplitude was measured for each of a variation of single fracture events giving

$$(AE)_1 = \frac{C''\Delta P}{B} \tag{27}$$

Combining Eqs. (26) and (27) by eliminating ΔP gives for a specific specimen geometry

$$(AE)_1 = C'\sigma\Delta A(\pi a)^{\frac{1}{2}} \tag{28}$$

Thus instead of the rms amplitude, we now have the amplitude of the single event proportional to applied stress and incremental crack area. The second method used a point source calibration to determine the amplitude of the emission arriving from a source a distance, r_o, below the surface. This is given by (41, 5)

$$(AE)_1 = \frac{0.047\sigma\Delta A}{\mu r_o} \tag{29}$$

Note the similarity of this to Eq. (28), specifically that the emission amplitude is proportional to the applied stress times the incremental area advance. As both of these have been calibrated and both give similar results, we will use Eq. (28) since it avoids the necessity of having to specify the depth of the event below the surface. Still, there is one unspecified constant in Eq. (28) since the original constant was calibrated on a system with relatively low gain and frequency response. To allow correlation to low amplitude events an approach similar to that used previously (31) was to take the number of acoustic emission counts, N_{AE}, recorded at a particular gain over a given crack increment. For example, Evans and

Linzer (42) measured emission from the cracking of a glass slide and found that one recorded count represented an incremental crack advance of 60×10^{-10} m^2 for an average stress intensity of 1 MPa $m^{\frac{1}{2}}$. From Eq. (28) this would give the constant to be 1.67×10^8 MPa^{-1} $m^{-3/2}$ for a 170 kHz PZT transducer at 80 dB gain. At 95 dB gain, there is about a factor of 50 increase in sensitivity (43) so that

$$\Sigma AE_{(95dB)} \simeq 8.3 \times 10^9 MPa^{-1} m^{-5/2} K\Delta A \tag{30}$$

This then allows a direct comparison of the AE count rate or sum to cracking events. The simplest comparison is to the cracked particle events in the IN 718 superalloy (31) where it was found that all carbides and nitrides cracked by the time the applied strain in a tensile test reached 0.05. Considering the average particle diameter of 4.3 μm and assuming spherical particles, then for one particle, Eq. (30) gives the number of counts to be

$$N_{AE(1)} \simeq \underbrace{8.3 \times 10^9 MPa^{-1} m^{-5/2}}_{C'} \quad \underbrace{\sigma \left[\pi \times 2.15 \times 10^{-6} m\right]^{\frac{1}{2}}}_{K} \quad \underbrace{\pi/4 \left[4.3 \times 10^{-6} m\right]^2}_{\Delta A} \tag{31}$$

$$\simeq 3.13 \times 10^{-4} MPa^{-1} \sigma$$

Since the typical flow stress for these materials was 1000 MPa, it would say that an average particle would not quite represent one count.

Multiple Particle Analysis

Using the analysis of the previous section, we will next consider the acoustic emission as occurring from a summation of multiple events or as a function of multiple fracture events which in turn depend upon the magnitude of the plastic strain.

Summation of multiple events. First consider the large carbide or nitride particles fracturing in In 718 superalloy. Using Eq. (31) to represent the emission from a single particle, and assuming spherical particles in a gage volume, V_T, it follows that the total emission from particle fracture would be

$$\Sigma AE = \sum_i^{Np} N_{AE(i)} \simeq N_p \cdot \overline{N_{AE(1)}} \simeq \frac{6 f_v V_t}{\pi d^3} \times 3.13 \times 10^{-4} MPa^{-1} \sigma \tag{32}$$

With a volume fraction, f_v, of 3.5×10^{-4} and a gage volume of 0.42 cm^3, this becomes

$$\Sigma AE \simeq 185 MPa^{-1} \sigma \tag{33}$$

for particles with an average diameter of 4.3 μm. The curve represented by Eq. (33) is given in Figure 14 as the dotted line and is seen to represent about 20% of all acoustic emission. Considering the relative magnitude of the first peak in Figure 12 and the frequency of bursts thereafter, this is a reasonable number. Although this gives a rough accountability for the total emission from large second phase particles, it does not explain how AE events from particles might depend upon relative plastic strain and propensity for particle fracture.

Dependency on fraction cracked. Thus far, the assumption has been that all particles in a given population fracture. Clearly, particle fracture occurs over a given strain range and in many instances only a few particles of a given population may fracture. Thus, acoustic emission will depend on the size of the particles being fractured, the stress (or strain) at which these fracture, and the fraction of available particles being cracked. The latter acutely depends on the particle distribution and in the absence of specific information, a log normal distribution will be assumed.

In this section, two sets of data on EN30A and 4340 steels are presented. (36, 31) The first (36) concerns a low alloy steel tempered at 650°C for 6 to 10,080 minutes to produce carbides in the range of 0.1 to 2 μm diameter. For this 0.28 wt% carbon steel, the upper bound volume fraction would be about 0.055. In addition, there is an inclusion distribution of sulfides 2 to 10 μm in diameter which represents a volume fraction of about $5x10^{-4}$. One can first estimate the effects of the inclusions assuming that all fracture and using Eq. (33) with Wadley, et al's (36) reported stress-strain data. As there was no direct way of determining the constant appropriate to the mJ output of their power meter results, the data were scaled to the lowest energy output result. As is seen in Figure 15, the trend of decreased emission with increase tempering is in the right direction but non-predictive. On the other hand, if one assumes that single carbides from the large particle end of a log normal distribution are sufficiently large to produce detectible emission, then a substantially different result is obtained. The treatment must now evaluate what critical particle diameter at what stress (and strain) level will fracture which in turn can give the fraction of particles cracked, f_c. In the simplest sense, the summation of emission at the end of a given deformation level will be given by

$$\Sigma AE \simeq C' f_c N_p \bar{N}_{AE(1)} \tag{34}$$

From Eqs. (31) and (32) it can be shown that the total emission would be

$$\Sigma AE \simeq C f_c f_v \sigma_{flow} \bar{d}^{-\frac{1}{2}} \tag{35}$$

where $\bar{d}$ is the average cracked particle size. Although it appears as though acoustic emission might decrease as particle size increases, that is often not true since f_c may be an even more sensitive function of increasing particle size. In order to use this result, it is necessary to evaluate the minimum particle size that can fracture which in turn is dependent both on the specific work of fracture of the particles and the matrix properties. The specific work of fracture for Fe_3C was taken as the surface energy, γ_s and estimated using an approach suggested by Kelly. (45) For orthorhombic Fe_3C with a=4.825 Å, b=5.087 Å and C=6.743 Å, there is some question as to what to use for the lattice parameter and modulus, but picking average values, Figure 16 suggests γ_s to be about 3800 ergs/cm^2. As will be shown in the section Particle Size Fracture Criterion, this allows an estimate of the critical particle diameter that may be fractured under a given flow stress condition.

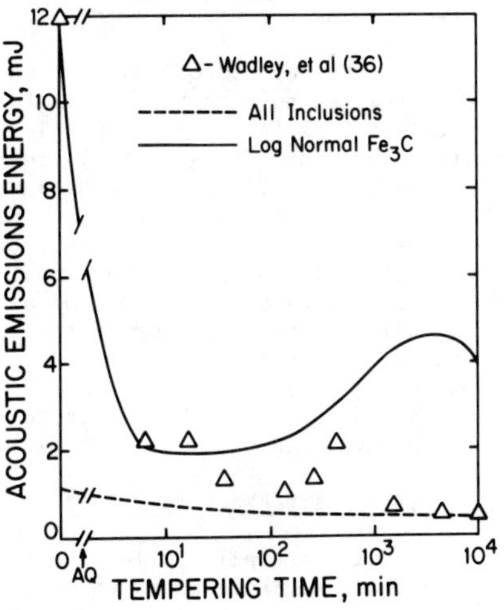

Fig. 15 Prediction of AE during deformation of a low alloy steel in terms of inclusion or carbide fracture.

One additional assumption was how the average particle size varied during tempering at 650°C. As a first order estimate, a 0.20 μm average particle size was estimated at the longest tempering time with others being scaled downward for shorter times using a $t^{1/3}$ growth law. With these, the

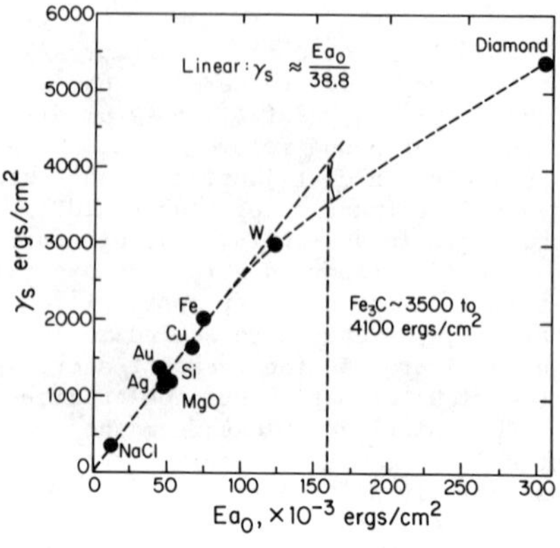

Fig. 16 Prediction of fracture energy for Fe_3C from Kelly's correlation. (44)

resultant estimate was that from three to thirty percent of the carbide fraction cracked which is obviously high. Using the same scaling law corresponding to the MnS fracture estimate, the result in Figure 15 is high but reasonably representative. Although Wadley, et al. (36) suggest that the emission is due to both inclusion fracture and dislocation bursts, this analysis suggests at least some contribution from a third source, that of cracked carbides. Finally, a fourth source for the very high emission level in the as-quenched sample might be quasi-cleavage as suggested by one of their SEM fractographs [their Figure 12(b)].

A second set of data on 4340 steel (31) was also evaluated except now the carbide population was varied by tempering temperature rather than tempering time. In addition, it was somewhat easier to analyze with fewer assumptions since the volume fractions and particle sizes were better known from extraction replicas. For tempering conditions ranging from 200 to 700°C, it was found that the critical particle size for fracture ranged from 0.635 to 2.45 µm. From a log normal distribution of all particles, this allowed estimates of the fraction cracked giving $0.057<f_c<0.111$. In conjunction with the average particle sizes, the volume fraction of particles and the flow stress properties, (31) Eq. (35) gave the result in Figure 17. The comparison to observed ΣAE is very good considering that the constant is determined from sections Single Particle Event and Multiple Particle Analysis to be 7.9×10^3 MPa^{-1}-m$^{\frac{1}{2}}$ and thus there is no adjustible parameter.

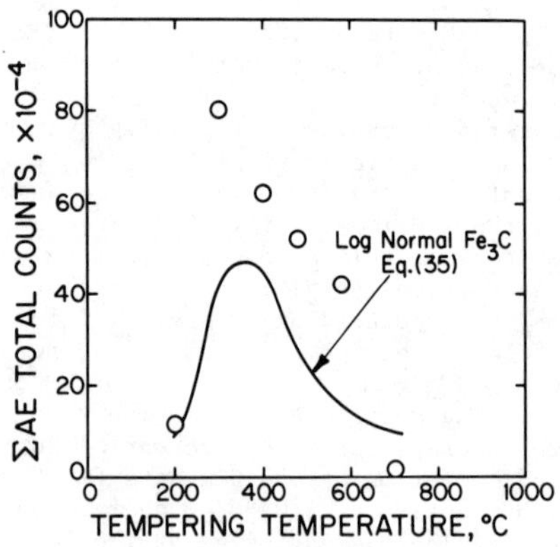

Fig. 17 Prediction of total AE to a strain of 0.01 using a log normal distribution of Fe_3C particles.

From the success of this approach, one is tempted to suggest that most of the emission is from cracked particles in these complex steels. Still, in the absence of definitive measures of numbers of cracked particles, such results must be treated as speculative. The result which seemed to have most information available was that on aluminum alloys with large intermetallic compounds fracturing. Before this is considered, however, the concept of a critical particle size for fracture will be developed.

Particle Size Fracture Criterion

We will generally follow the experimentally verified approach of Van Stone, et al. (45) who carefully quantified the size of particles fracturing in a series of aluminum alloys as a function of plastic strain. Utilizing Gurland and Plateau's concept, (46) it was demonstrated that the flow stress for fracturing a critical particle size, d_c, was given by

$$\sigma_{flow} = \frac{1}{q_{(\varepsilon)}}\left(\frac{6E\gamma_c}{d_c}\right)^{1/2} \quad (36)$$

This result has been slightly modified to account for the fact that $q_{(\varepsilon)}$ would increase with plastic strain rather than being constant. Using Roberts, Lehtinen and Easterling's results, (47) it may be shown that

$$q_{(\varepsilon)} = \frac{1}{\sqrt{3}}\left[\left(\frac{\varepsilon}{\varepsilon_{ys}}\right)^n + \sqrt{3}\left(\frac{\sqrt{6}\ (1+n)}{n\ \bar{M}}\ \frac{\varepsilon}{\varepsilon_{ys}}\right)^{n/1+n}\right] \quad (37)$$

where n is the strain hardening exponent and $\bar{M}$ is the Taylor factor. For the 4340 steel data of the previous section, $q_{(\varepsilon)}$ varied from 1.75 to 1.87. Using Van Stone's estimate of $E\gamma_c$ for particle fracture to be $4.31\times10^{11}N^2/m^3$, Eqs. (36) and (37) were used to determine the flow stress for fracture of particles with given average dimensions. While the calculated flow stresses for particle fracture from Eqs. (36) and (37) were 440-510 MPa, Van Stones's observations (45) ranged from 460-580 MPa. However, the parameter of interest is the critical particle size, which from Eqs. (36) and (37) is seen to be given by

$$d_c = 6E\gamma_c\left\{\sigma_{flow}\ q_{(\varepsilon)}\right\}^{-2} \quad (38)$$

Since both σ_{flow} and $q_{(\varepsilon)}$ increase with plastic strain, it is clear that d_c will decrease with strain. For the intermetallic compounds studied by Van Stone, the result of Eq. (38) is shown for both a constant stress concentration factor of two and the strain concentration factor of Eq. (37) in Figure 18. In doing this, a flow stress relationship for the typical aluminum alloy of Van Stone's investigation (45) was taken as

$$\sigma_{flow} = 620\varepsilon^{0.070}\text{MPa} \quad (39)$$

As the strain increases and d_c decreases in Figure 18, it should be apparent that a larger fraction of the particle distribution will be susceptible to fracture with increasing strain.

Application of Simple Particle Distributions

To complete this approach, some assumption on particle size distributions will be made to demonstrate a range of possibilities. Since Van Stone, et al. (45) found particles generally to be less than five microns in size and that those less than two microns did not fracture, two simple distributions in Figure 19 are suggested. For each of these two cases, the number fraction susceptible to cracking per micron is given by

$$f^N_{III} = 0.333/\mu m \quad ; \quad \text{CASE III}$$
$$f^N_{IV} = \left(0.222/\mu m^2\right)\left(5\mu m - d\right) \quad ; \quad \text{CASE IV} \quad (40)$$

To obtain the fraction cracked for a given critical particle diameter, d_c, it is a simple matter to integrate giving

$$f_c = \int_{d_c}^{d_{max}} \left\{f^N\right\}\, d\, d \quad (41)$$

By determining d_c from Eq. (38) and integrating via Eq. (41), the fraction of cracked particles was determined and compared to Van Stone's observations

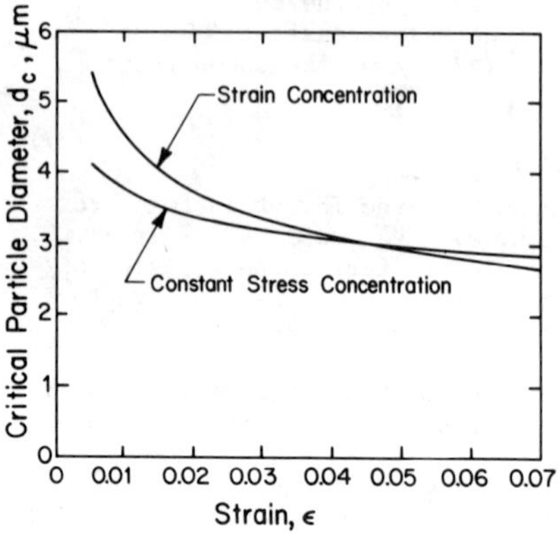

Fig. 18 Critical particle size for fracture using $q_{(\varepsilon)}$ to be a constant of two or to be given by Eq. (37). Curves are predicted from Eqs. (37)-(39).

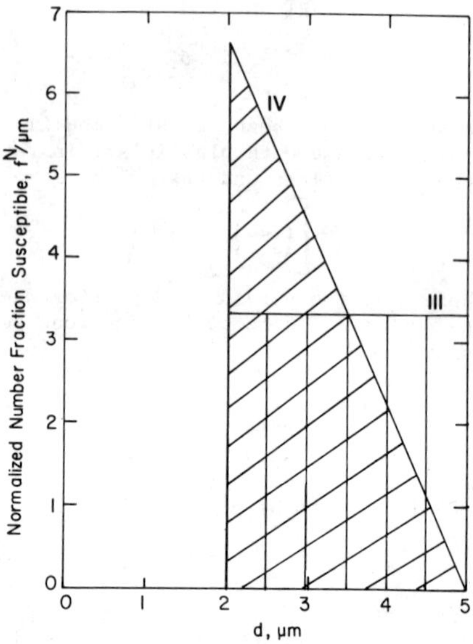

Fig. 19 Two simple particle distributions for large intermetallic compounds.

(45) for 2000 series aluminum alloys in Figure 20. It is seen that CASE III overpredicts the results because of the heavy weighting toward large particles but that CASE IV is a reasonable representation of cracked particles.

With this confirmation, it was possible to proceed to the final step of determining acoustic emission as a function of strain. In terms of acoustic emission rate occurring in a given strain increment, $(AE)_{\Delta\varepsilon}$, this would be given by

$$(AE)_{\Delta\varepsilon} = \left\{ \int_{d_{c2}}^{d_{c1}} f^N \, d d \right\} \left\{ N_p \right\} \left\{ C' \left(\frac{\pi}{2} \right)^{\frac{1}{2}} \bar{\sigma} \, \bar{d}^{\,5/2} \right\} \quad (42)$$

where the last term in brackets represents the emission from an average particle in the strain increment. It should be noted that for the first increment of strain, if $d_c > d_{max}$ then no particles fracture and no emission from particles results. As d_c becomes smaller than d_{max} then an increment of emission obtained by integrating from d_{c1} to d_{max} would result. Subsequent increments would be obtained by integrating from d_{c2} to d_{c1}, d_{c3} to d_{c2}, and so forth. This was accomplished by using 0.005 strain increments and Eqs. (36)-(42). The result seen in Figure 21 is typical of that observed for acoustic emission count rate plotted as a function of strain. First, it is seen that the distribution with the heavy weighting of large particles peaks at a smaller strain. Secondly, the peak emission rate is a factor of three larger. Both of these features are qualitatively in keeping with observations of AE from two 7000 series alloys in Figure 22, where (a) is obtained from a relatively clean alloy with a typical distribution of intermetallic particles while (b) is obtained on a dirty alloy with many 20-60 μm size particles. (48)

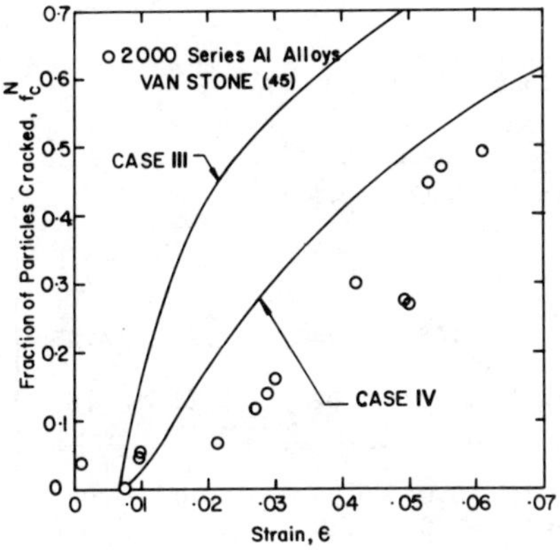

Fig. 20 Prediction of the fraction of particles cracked versus strain by two simple distributions.

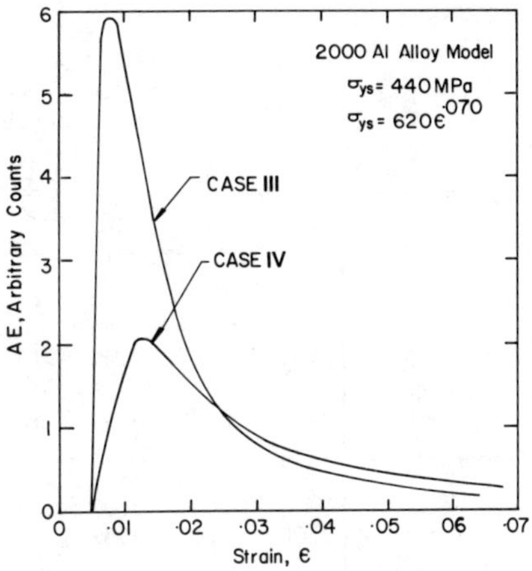

Fig. 21 Acoustic emission count rate versus strain for two simple particle distributions in a model alloy system.

One final comparison is made to Swindlehurst's data (37) on spherodized steel since he carefully measured the cumulative fraction of cracked particles. Since the particle size distribution was not reported, first a CASE III distribution and then a log normal distribution will be assumed. From Eq. (42), it is straightforward to determine the acoustic emission "rate" over an increment of strain, $\Delta\varepsilon$, by using the average particle diameter and the average flow stress in that increment. CASE III was easily determined and gave a good representation of the ΣAE data but did not predict the nearly linear increase in cumulative carbide crack density with strain found by Swindlehurst. (37) On the other hand, the log normal distribution did fit the carbide cracking observations but did not quite predict the shape of the ΣAE curve in Figure 23 quite as well. The pertinent values are given in Table III. Here it is seen that the predicted number of cracked carbides is fairly linear with strain for log normal and in fact by converting volume fraction to number fraction the agreement with SEM observations (37) is good. In addition, if one plots an arbitrary emission rate as a function of strain, it closely matches the observed peak emission rate and the down slope showing AE to disappear near a strain of 0.15. The only characteristic not predicted is the double AE peak which is probably associated with inhomogeneous deformation during Lüders straining. It is very likely that by using a more accurate description of stress-strain and the local stress state of a propagating Lüders band that a closer prediction of AE in this steel could be achieved. This is not to say that factors other than particle fracture are not contributing but is merely emphasized to invalidate the original objection posed by Swindlehurst. (37)

Summary and Conclusions

In the two main sections on dislocation (grain size) and particle fracture processes, it is noted that two very different processes may give rise in the peak emission rate observed in almost all tests. In clean polycrystalline materials it is either due to mean free path arguments or grain boundary source exhaustion and the frequency limitations of most detection

Table III: AE From Theoretical Cracked Carbide Distributions (0.1 to 10 μm)

	ε	0.005	0.01	0.02	0.03	0.04	0.05	0.1	0.15	0.2
	σ_{flow} (MPa)	368	422	485	526	557	583	669	726	768
	$q_{(\varepsilon)}$	2.32	2.62	2.96	3.18	3.35	3.48	3.94	4.23	4.45
	d_c (μm)	9.71	5.79	3.44	2.53	2.03	1.72	1.02	0.75	0.61
	Interval	0–0.005	0.005–0.01	0.01–0.02	0.02–0.03	0.03–.04	0.04–0.05	0.05–0.10	0.10–0.15	0.15–0.20
Case III	f_c	.029	.396	.237	.092	.051	.031	.071 (.014)†	.027 (.0054)†	.015 (.003)†
Case III	$\Sigma\ f_c$	.029	.425	.662	.754	.805	.836	.907	.934	.949
Case III	$\bar{d}_c$ (μm)	9.86	7.75	4.62	2.99	2.28	1.88	1.38	0.88	0.68
Case III	AE††	6.8	63.8	12.0	1.8	0.5	0.2	0.2 (.04)†	.05 (.01)†	.05 (.01)†
Case III	Σ AE††	6.8	70.6	82.6	84.4	84.9	85.1	85.3	85.35	85.4
Log Normal	f_c	.0015	.0098	.0427	.0590	.0660	.061	.250 (.050)†	.154 (.031)†	.098 (.020)†
Log Normal	$\Sigma\ f_c$	.0015	.0113	.054	.113	.179	.240	.490	.644	.742
Log Normal	$\bar{d}_c$ (μm)	9.85	7.26	4.38	2.94	2.27	1.84	1.36	0.84	0.68
Log Normal	AE††	4.2	15.8	22.3	12.7	7.9	4.6	9.8 (1.96)†	1.9 (.38)†	0.8 (.16)†
Log Normal	Σ AE††	4.2	20.0	42.3	55.0	62.9	67.5	77.3	79.2	80

†Parantheses denote interval value appropriate to a 0.01 strain increment.

††Acoustic emission are arbitrary units.

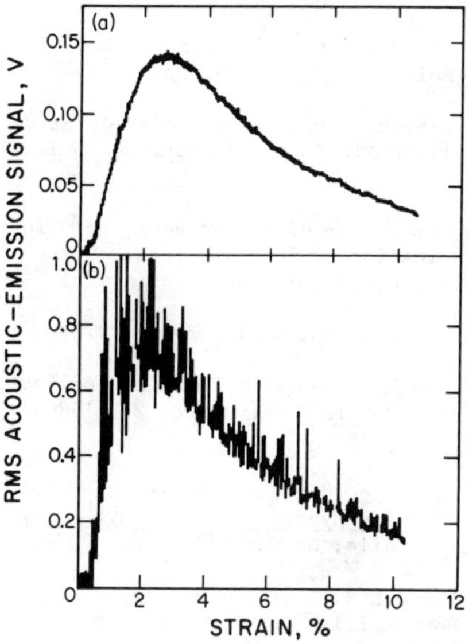

Fig. 22 Acoustic emission during tensile testing of (a) a relatively clean 7075-T6 alloy and (b) a dirty 7075-T6 alloy with large Cr-Al intermetallic compounds giving burst emission (48).

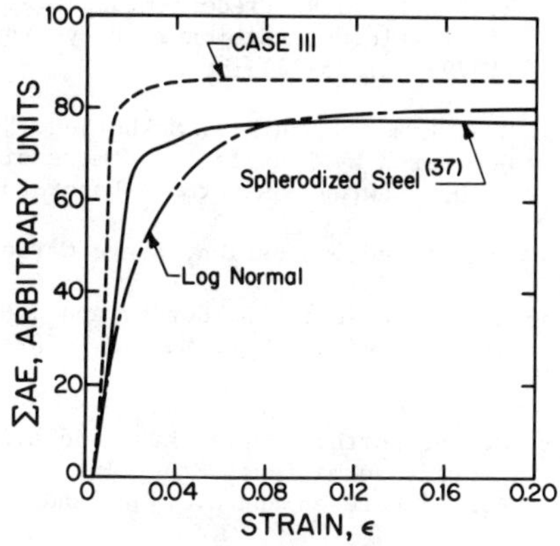

Fig. 23 Prediction of acoustic emission from a spherodized steel using two types of particle size distributions. See Table 3.

equipment. In more complex materials with large dispersoid or precipitate particles, it is probably a combination of the previous process in conjunction with the gradual exhaustion of particle fracture sources. Depending upon the particle sizes, the distribution of those, the strain-intensity and flow stress of the matrix, the specific work of fracture for a particular particle, and the AE instrumentation, a large variety of acoustic emission outputs and concomittant detection are predicted. Such predictions agree reasonably well with observations of AE in complex materials where sulfides, carbides, nitrides and intermetallic compounds are fracturing throughout a tensile test. Proper use of such NDE methods awaits careful analysis of dislocation (grain size), transformation and particle fracture contributions in well-characterized complex materials.

References

1. N. Kiesewetter and P. Schiller, "The Acoustic Emission from Moving Dislocations in Aluminum," Phys. Stat. Sol. (a), 38, (1976) pp. 569-575.

2. R.C. Bill, J.R. Frederick and D.K. Felbeck, "An Acoustic Emission Study of Plastic Deformation in Polycrystalline Aluminum," J. Mat. Sci., 14 (1979) pp. 25-32.

3. M.A. Hamstad and A.K. Mukherjee, "A Comparison of the Acoustic Emission Generated by Tensile and Compression Testing of 7075 Aluminum," UCRL-77502, Lawrence Livermore Laboratories, California, November (1975).

4. Private communication, Steve Carpenter, University of Denver, 1975.

5. H.N.G. Wadley, C.B. Scruby and J.H. Speake, "Acoustic Emission for Physical Examination of Metals," Internat. Metals Reviews, No. 2, (1980) pp. 41-64.

6. C.J.M. Northrup, W.J. Kass and A.G. Beattie, "Acoustic Emission During Hydride Formation," Proc. Int. Symposium on Hydrides for Energy Storage, A.F. Anderesen and A.J. Maeland, eds.; Gelio, Norway (1977) pp. 205-216.

7. C.H. Caceres, W. Arneodo, R. Pascual and H.R. Bertorello, "Acoustic Emission Related to Stress Induced Martensitic Transformation in β Cu-Zn," Scripta Met., 14 (1980) pp. 293-297.

8. G. Cannelli and R. Cantelli, "Acoustic Emission During the Precipitation of β-Phase Hydride from Solid Solution in Vanadium, Scripta Met., 14 (1980) pp. 731-734.

9. F.P. Higgens and S.H. Carpenter, "Sources of Acoustic Emission Generated During the Tensile Deformation of Pure Iron," Acta Met., 26 (1978) pp. 133-139.

10. P. Fleischmann, F. Lakestani and J.C. Baboux and D. Rouby, "Spectral and Energy Analysis of a Moving Ultrasonic Source - Application of Acoustic Emission to Aluminum During Plastic Deformation," Mat. Sci. and Engng. 29 (1977) pp. 205-212.

11. K. Kuribayashi and T. Kishi, "Acoustic Emission Behavior in Al-Mg Solid Solution," Mat. Sci. and Engng. 33 (1977) pp. 159-163.

12. D. Rouby and P. Fleischmann, "Spectral Analysis of Acoustic Emission from Aluminum Single Crystals Undergoing Plastic Deofrmation," Phys. Stat. Sol. (a) 48 (1978) pp. 439-445.

13. A.A. Pollock, "Acoustic Emission - 2, Acoustic Emission Amplitudes," J. Non-Destructive Test., October (1973) pp. 264-269.

14. H.C. Kim and T. Kishi, "The Grain Size and Flow Stress Dependence of Acoustic Emission Energy Release in Polycrystalline Aluminum," Phys. Stat. Sol. (a) 55 (1979) pp. 189-195.

15. D.R. James and S.H. Carpenter, "The Effect of Sample Size on the Acoustic Emission Generated During Tensile Deformation of 7075 Aluminum Alloy," Scripta Met. 10 (1976) pp. 779-781.

16. A.B.L. Agarwal, J.R. Frederick and D.K. Felbeck, Met. Trans., 1 (1970) p. 1069.

17. N.G. Sankar, J.R. Frederick and D.K. Felbeck, "Acoustic Emission from Metals During Unloading and its Relation to the Bauschinger Effect", Met. Trans., 1 (1970) pp. 2979-2980.

18. E. Siegel, "Burst Aocustic Emission During the Bauschinger Effect in FCC and HCP Metals and Alloys," Acta Met., 25 (1977) pp. 383-394.

19. S. Mintzer, R. Pascual and R.M. Volpi, "Acoustic Emission and Grain Size in Plastic Deformation of Metals," Scripta Met., 12 (1978) pp. 531-534.

20. K.N. Tandon and K. Tangri, "Acoustic Emissions During Deformation of Polycrystalline Silicon-Iron," Mat. Sci and Engng., 20 (1975) pp. 47-54

21. L.E. Murr and S.S. Hecker, "Quantitative Evidence for Dislocation Emission from Grain Boundaries," Scripta Met., 13 (1979) pp. 167-171.

22. R.C. Bill, Ph.D. Thesis, The University of Michigan (1970).

23. T. Imanaka, K. Sano and M. Shimizu, "Dislocation Attenuation and Acoustic Emission During Deformation in Copper Single Crystal," Crystal Lattice Defects, 4 (1973) pp. 57-64.

24. M.F. Ashby, Phil. Mag., 21 (1970) p. 399.

25. N. Nagata and T. Vreeland, Jr., "Mobility of Basal Dislocations Interacting with Non-Basal Dislocations In Zinc," in Second Int. Conf. in Strength of Metals and Alloys, III, Am. Soc. for Metals, (1970) pp. 423-427.

26. K.E. Puttick, "The Observation of Dislocations in Cadmium by Optical, X-ray and Electron Microscopy," ibid. pp. 464-467.

27. R, Frydman, R. Pascual and R.M. Volpi, "Acousitc Emission due to Dislocations and Grain Boundaries," Scripta Met., 9 (1975) pp. 1267-1270.

28. Private Communication, H.N.G. Wadley, AERE Harwell, United Kingdom, December 1980.

29. I.G. Palmer, "Acoustic Emission Measurements on Reactor Pressure Vessel Steel," Mater. Sci. and Engng., 11 (1973) pp. 227-236.

30. M.A. Hamstad, R. Bianchetti and A.K. Mukherjee, "A Correlation Between Acoustic Emission and the Fracture Toughness of 2124-T851 Aluminum," Engng. Fract. Mech., 9, (1977) pp. 663-674.

31. W.W. Gerberich, M. Stout, K. Jatavallabuhla and D. Atteridge, "Acoustic Emission Interpretation of Ductile Fracture Processes," Int. J. of Fracture, 15, No. 6 (1979) pp. 491-514.

32. J. Kaiser, Arkiv Für das Eisenhuttenwesen, 24 (1953) p. 43.

33. R.G. Liptai, D.O. Harris, R.B. Engle and and C.A. Tatro, "Acoustic Emission Techniques in Materials Research," Int. J. of Nondestructive Test., 3 (1971) pp. 215-275.

34. R. Bianchetti, M.A. Hamstad and A.K. Mukherjee, J. Test. Eval., 4 (1976) p. 313.

35. L. Graham, "Sources of Acoustic Emission in Aluminum Alloys," Proc. of the ARPA/AFML Review of Quant. NDE Science Center, Rockwell International, Thousand Oaks, CA (1975) p. 631.

36. H.N.G. Wadley, D.C. Furze, C.B. Scruby and B.L. Eyre, "Effect of Isothermal Tempering on Acoustic Emission during Ductile Fracture of Low-Alloy Steel," Metal Science, (August, 1979) pp. 451-462.

37. W.E. Swindlehurst, "On Carbide Cracking as a Source of Acoustic Emission in Steel," J. Mat. Sci., 13 (1978) pp. 209-212.

38. W.J. Bruchey, Jr., "Optical Probing of Acoustic Emission During Deformation of Micro-Tensile Specimens," Ph.D. Thesis, John Hopkins University, Baltimore (1980).

39. C.H. Palmer and R.E. Green, Jr., "Materials Evaluation by Optical Detection of Acoustic Emission Signals," Materials Evaluation, 35, No. 10 (1977) pp. 107-112.

40. J.D. Desai and W.W. Gerberich, "Analysis of Incremental Cracking by the Stress-Wave Emission Technique," Engng. Fracture Mech., 7 (1975) pp. 153-165.

41. C.B. Scruby, J.C. Collingwood and H.N.G. Wadley, J. Phys. D.; (Applied Phys.), 11 (1978) p. 2359.

42. A.G. Evans and M. Linzer, "Failure Prediction in Structural Ceramics Using Acoustic Emission," J. Am. Ceramic Soc., 56, No. 11 (1973) pp. 575-581.

43. K. Ono and H. Ucisik, "Acoustic Emission Behavior of Aluminum Alloys," UCLA ENG-7514, University of California, Los Angeles (February 1975).

44. A. Kelly, Strong Solids, pp. 2-5, Clarendon Press, Oxford, 1966.

45. R.H. Van Stone, R.H. Merchant and J.R. Low, Jr., "Investigation of the Plastic Fracture of High-Strength Aluminum Alloys," in Fatigue and Fracture Toughness-Cryogenic Behavior, ASTM STP 556, Am. Soc. for Test. and Mat'ls, Philadelphia (1974) pp. 93-124.

46. J. Gurland and J. Plateau, Trans. ASM, 56 (1963) p. 442.

47. W. Roberts, B. Lehtinen and K. Easterling, "An In Situ SEM Study of Void Development Around Inclusions in Steel During Plastic Deformation," Acta. Met., 24 (1976) pp. 745-758.

48. R. Bianchetti, M.A. Hamstad and A.K. Mukherjee, "Correlation Between Acoustic Emission and Microstructure in a Disperison Hardened Aluminum Alloy," in Proc. Second Int. Conf. in Mech. Behavior of Mat'ls., ICM II, Boston, (1976) pp. 1625-1630.

MAPPING OF MATERIALS STRESS WITH

ULTRASONIC TOMOGRAPHY

B. P. Hildebrand
Spectron Development Laboratories, Inc.
3303 Harbor Boulevard, Suite G-3, Costa Mesa, California 92626

and

T. P. Harrington
Battelle, Pacific Northwest Laboratories
Battelle Boulevard, Richland, Washington 99352

It is known that internal stress concentrations can give rise to microcracks which then grow when the structure is subjected to external forces. It has also been found that the velocity of sound is altered as it propagates through a region of stress. In this paper we discuss a technique called Computer-Assisted Tomography (CAT) and describe an application that provides pictures of stress fields. We report the results of both simulated and experimental models used to evaluate the technique. We conclude that the CAT approach has great potential for locating and mapping residual stress in metals.

Introduction

One of the outstanding problems of nondestructive testing is the location and measurement of areas of residual stress. In the manufacture of large structures, such as pressure vessels, a number of very large welds are required. The heat-affected zone surrounding the welds will contain residual stresses due to uneven cooling rates. In practice, these stresses are relieved by heating the entire vessel to some appropriate temperature and then carefully controlling the cooling rate. This standard procedure has not always been successful in relieving the residual stress that is induced by the welding process (1). Structural designers realize this, and try to compensate for the possibility of stress concentrations by overdesigning. This leads to penalties in terms of both added weight and cost. Consequently, there is high motivation for finding and delineating such concentrations.

Currently there is no satisfactory test for measuring the success of stress-relief procedures. Thus, it may happen that high residual-stress regions exist in the pressure vessel. If these regions occur in critical areas (such as nozzle-to-vessel joints), cracks may develop, which would require subsequent plant shut-downs and expensive repair. If it were possible to detect and map residual stress, local stress relief could be applied, and future problems could be avoided.

Standard Nondestructive Testing (NDT) examinations consist of radiology and ultrasonic pulse-echo. Neither of these techniques can reveal the presence of residual stress; the former because it shows only density variations, and the latter because stress regions are not sharply defined; hence, they do not reflect much sound. It is known, however, that the velocity of sound in a solid is affected by stress. This phenomenon is a third order effect, and has been used primarily to determine the Lamé and Murnaghan elastic constants for various materials.

A number of studies have shown that it is possible to measure velocities with sufficient accuracy to detect and calculate residual stress (2-4). Typically, for steel, the stress-acoustic constant is approximately 0.197 nsec/cm/6895 kPa. This constant means that a 6895 kPa increment in stress produces a 0.5 nsec change in travel time through a 2.54 cm thick specimen. Since it is relatively easy to measure time-of-flight changes to this accuracy, the sensitivity of the proposed method should be approximately 6895 kPa (= 1 ksi).

At the present time, a number of researchers have made velocity measurements for the computation of stress. This data, however, has been obtained in the form of spot measurements or profiles. No one has yet attempted to take a number of profiles at different angles in order to reconstruct a cross section of the residual stress. The purpose of this paper is to discuss our efforts to develop a method for mapping stress anomalies utilizing velocity information and an image reconstruction technique known as Tomography.

A Tomogram could be described as a picture of a slice. Over the past several years, CAT scanners have revolutionized the field of diagnostic medicine. This technique uses x-rays to obtain visible, thin slices through any section of the human body. We are applying the same reconstruction algorithms used in medical applications, (except for the substitution of ultrasound for the x-rays), to generate velocity maps of cross sections of thick metal sections.

In the following section, we briefly discuss the principles of Tomography, and describe the Algebraic Reconstruction Technique (ART) we are using to construct the velocity profile. The third section describes several computer simulations for testing the ART algorithm. In the third section, we also discuss the effect of varying several parameters and describe the effect they have on the quality of the reconstruction. In the fourth section, we describe an experimental apparatus, which we have developed in our laboratory, for obtaining time-of-flight measurements. It also includes a discussion of reconstructions generated from these experimental data, and a description of the calibration procedure we are using in order to relate the velocity of sound to stress. In our conclusion, we comment on the validity of this method for mapping residual stress and also describe a prototype instrument that we are building to test this technique.

Principles of Image Reconstruction in Computed Tomography

As we noted in the introduction, Tomography has had its greatest impact to date in the field of diagnostic medicine. A good source of references on medical Tomography can be found in a paper by Brooks and Dechiro (5). The principles of image reconstruction using Tomography have also been applied in a number of other disciplines: Electron Microscopy, Radio Astronomy, and Nondestructive Testing (6-11).

To clarify the discussion that follows, a brief digression into the principles and terminology of Tomography will be necessary. Consider Figure 1a, which is a particular cross section of a general three-dimensional object. Define the velocity distribution of the object as v(x,y,z). Assume an ultrasound pulse, transmitted from location t, traverses the object along path ℓ to a point r where it is received. The time-of-flight along path ℓ is a function of the velocity distribution along the path, and is given as the line integral

$$t_\Theta(\rho) = \int_\ell \frac{ds}{v} \quad (1)$$

where ρ and Θ define the path ℓ from t to r. If the variable ρ is allowed to vary continuously over the object at a constant angle, a one-dimensional projection of the velocity distribution will be obtained.

The reconstruction problem is to estimate the velocity of the cross section v(x,y,z) from the projection values. Clearly, this can only be done if the projection values are available for a large number of angles Θ_i. Also, in actual practice the projection values are available only at a discrete number of the ρ_j for each angle Θ_i. Thus, the reconstruction is an approximation to the actual velocity field, with resolution being a function of the spacing between rays $\Delta\rho_j$ and the degrees between angles $\Delta\Theta_i$.

In theory, the velocity field can be reconstructed quite easily from the projection information. To see this, visualize a square grid superimposed over the object of Figure 1a as shown in Figure 1b. If we assume that the velocity within each square (pixel) is constant, then the total travel time through the grid becomes the sum of the travel times in each of the pixels. Each ray that passes through the velocity field will intersect various pixels.

The length of the ray segment in each pixel can be determined quite easily by using geometry. If a ray of length l_k lies in a cell with velocity

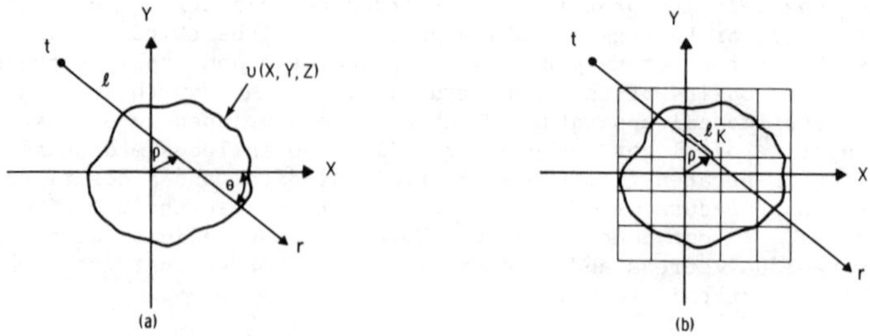

FIGURE 1. Geometry for (a) computation of the time-of-flight profile and (b) the reconstruction of the object from the profiles.

v_k, the travel time through the cell is l_k/v_k. Thus, the total time-of flight can be expressed as a linear equation whose unknowns are the velocities v_k. By using a number of independent rays that is equal to the number of grid cells, a linear system of algebraic equations can be generated and solved for the unknown velocities.

The difficulty with solving for the unknown velocities by inverting a system of algebraic equations is that, even for small grid systems, the size of the resulting matrix is prohibitive. For example, consider a grid system of 20 by 20 pixels. This means that there are 400 unknown velocities to solve for. If we take 20 independent views with 20 rays per view, the result is 400 equations in 400 unknowns, or a 400 by 400 matrix to invert. Furthermore, the matrix will be sparse, since each ray will usually intersect less than 30 pixels.

Other reconstruction algorithms are available that overcome the limitations imposed by the matrix inversion method, some of these alternatives are:

(1) Convolution, which is based on the work of Johann Radonn (who first solved the equations governing image reconstruction in 1917 (12-13).

(2) Fourier Transform, which is the spatial frequency version of convolution (14).

(3) Back Projection, which was used in the first attempts to produce Tomograms of living patients.

(4) Iterative Techniques, which are probably the most widely used reconstruction method.

Iterative Reconstruction Algorithms

All the iterative algorithms start with an "initial guess" as to what the image looks like. Typically, the average value of the projections is divided among all the pixels that make up the image. The algorithms then adjust the pixel values to bring them into better agreement with the measured projections. These algorithms are iterative in that they continually sequence through the set of projection data, updating the pixel values until a stopping criteria is met. An example of a stopping criteria would be a measurement of the change that has been made to the image during an iteration. If the change is less than some prespecified minimum, the iteration is stopped.

The three most popular iterative reconstruction algorithms currently in use are: the Iterative Least Square Technique (ILST) of Goitein (15), the Simultaneous Iterative Reconstruction Technique (SIRT) introduced by Gilbert (16), and the Algebraic Reconstruction Technique (ART) discovered by Gordon, et al. (7, 17). Our preliminary investigation of both ART and SIRT indicates that ART produces better reconstructions than SIRT, given the amount and nature of the data that is available.

As indicated in the discussion earlier in this section that the time-of-flight for a given projection can be approximated as a sum of the time-of-flights in individual pixels. Briefly, the ART algorithm works as follows: for each ray in the data set, ART compares the experimentally obtained time-of-flight with the time-of-flight calculated using the current pixel values. If they are different, ART updates the velocity in each of the pixels as a function of the length of the ray path in the pixel. After the pixels along the ray have been modified, the time-of-flight calculated from the pixel values matches that obtained experimentally. ART then cycles through the remainder of the rays in the data set, repeating the process described above. However, as each new ray is processed, the value of previously updated pixels will be changed. This is the reason for the need to iterate. It has been shown, however, that the image improves after each iteration and converges to the best solution in a least squares sense (17).

The majority of medical CAT scanners in use today employ one of two geometries for obtaining profiles. These two geometries are usually referred to as fan beam and parallel beam. Figure 2 points out the differences between these two approaches. In the parallel beam method, a set of measurements are obtained by scanning the source and detector linearly past the

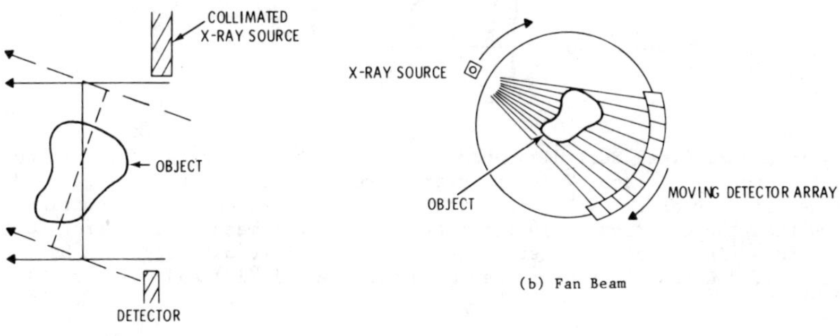

(a) Parallel Beam

(b) Fan Beam

FIGURE 2. Two Reconstruction Geometries

patient. The entire scanner assembly is then rotated by a fixed amount and the scan is repeated. In the fan beam approach, the source rays are formed into a fan of narrow beams that encompass the patient. The rays are received simultaneously by an array of detectors. For this geometry, the source and detector array are also rotated about the patient. Since the fan beam method does not require a linear translation of the source and detector, it is capable of much superior performance.

Simulated Data

To test the feasibility of using time-of-flight information along with the ART algorithm to reconstruct velocity fields, we have simulated several different images on the computer. As mentioned in the previous section, the time-of-flight for a particular ray crossing a velocity field is given by the line integral

$$t_{\Theta}(\rho) = \int_{\ell} \frac{ds}{v} \tag{2}$$

To obtain simultated time-of-flights on the computer, this integral can be discretized to:

$$t_{\Theta_i,\rho_j} = \sum_{k=1}^{K} \ell_{kij}/v_k \tag{3}$$

where ℓ_{kij} is the path length of the ray defined by Θ_i and ρ_j through the cell k. And K is the set of all cells which intersect the line ℓ.

To visually compare the reconstructions with the simulated velocity fields that were used to generate the data, we construct both isometric views and black-and-white gray scale images of the reconstruction. We will use both of these techniques to display results in this paper. To obtain a more qualitative idea of how the reconstructions compare with the original image, we calculate several error parameters at each iteration of the ART algorithm. The discrepancy is defined as:

$$\delta^q = \left\{ \frac{\sum_k (V_k^q - V_k)^2}{\sum_k (V_k - \bar{V})^2} \right\}^{1/2} \tag{4}$$

This is a normalized Euclidian norm where V_k is the velocity value of the kth pixel of the test image, $\bar{V}$ is the average velocity of the test image, and V is the velocity of the kth pixel after the qth iteration. This equation shows that the discrepancy is the ratio of the root-mean-square error to the standard deviation of the test picture. This measure was suggested by Gilbert (16), and has since been used by Herman, et al (17) and Colsher (8).

The second error parameter we calculate is the mean (or average) error which is defined as:

$$E_{av}^q = \frac{\sum_k |V_k^q - V_k|}{\sum_k V_k} \tag{5}$$

This has also been used by Sweeney and Colsher (18, 8). The final error function that we calculate is the residual, which is defined by:

$$R^q = \left\{ \sum_i \sum_j \left[p(\Theta_i, \rho_j) - p^q(\Theta_i, \rho_j) \right]^2 \right\}^{1/2} \tag{6}$$

where $P(\Theta_i,\rho_j)$ is the actual measured time-of-flight for the ray path defined by Θ_i and ρ_j and $p^q(\Theta_i,\rho_j)$ is the computed time-of-flight for the same path using the pixel values after the qth iteration. This measure indicates the degree to which the reconstructed image satisfies the measured time-of-flight data. Notice also that this is the only error measure of the three that is valid for experimentally obtained data. This criterion has been employed by Colsher (8).

An obvious deficiency of the first two error criteria is that the test image must be discretized. Several of the velocity fields that we used to generate our simulated data are circular and thus difficult to discretize. This problem tends to negate the usefulness of these two criteria.

All of our reconstructions were performed on a PDP 11/70 minicomputer. It is not valid to discuss execution time, since this computer operates in a multi-user environment. On the average, however, each iteration of the ART algorithm required approximately 30 sec. of wall clock time and 5 sec. of computer time. All of the isometric and gray scale images presented in this report were post-processed after the ART reconstruction program was completed.

In our initial investigation of the ART algorithm, we simulated parallel beam geometry, as described in the previous section. Fig. 3a represents a velocity field having two islands of 2% velocity increase in a uniform region. Fig. 3b shows the reconstruction when time-of-flight profiles are

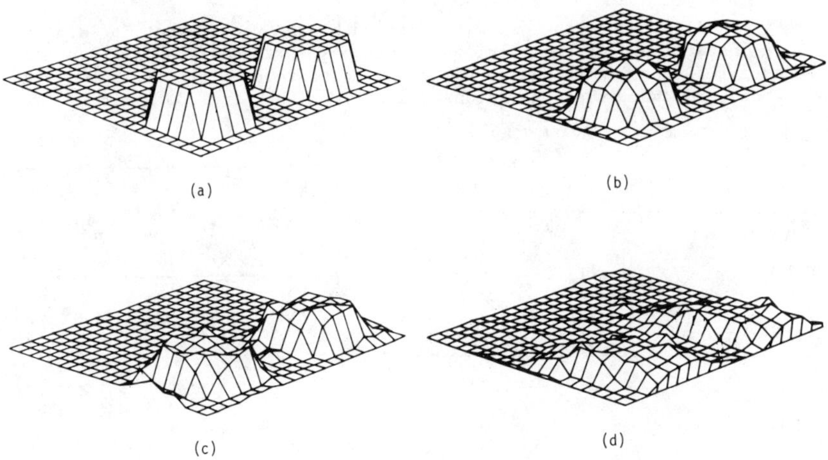

(a) (b) (c) (d)

FIGURE 3. (a) Test Image. Reconstructions: (b) 180° Field of View (c) 90° Field of View (d) 45° Field of View taken over a full 180° field of view. Notice that the tops of the two islands in this reconstruction are fairly flat and the steep sides approximate the test image very well. In Fig. 3c, a 90° field of view was available to the reconstruction algorithm. In this image, the tops of the islands are somewhat irregular and the sides have less slope than the 180° field of view reconstructions. The 90° data base, however, still provides an adequate reconstruction of the test image. Fig. 3d shows the result of using a 45° data base. In this reconstruction, the two islands have been smeared into the uniform velocity region. The tops of the islands are rough, and the height does not represent a 2% velocity difference. The walls of the islands in this reconstruction also have a very shallow slope. The reason for the poor reconstruction with a 45° field of view is that the majority of the rays pass through one of the two islands. As discussed in the last section, the velocity differences are divided among all the pixels in the path of the ray. Since there are few rays which pass completely outside the two velocity islands, the smearing effect is not counteracted.

A major goal of our present work is to develop procedures for mapping residual stress in thick metal sections (typically 10 to 20 cm). In many practical situations, it may not be feasible to obtain time-of-flight profiles by either of the geometries discussed in the previous section. For instance, it may be physically impossible to position a detector inside a pressure vessel.

To overcome this difficulty, we have been investigating the feasibility of locating both the transmitter and the receiver on the same side of the metal section. Figs. 4a and 4b illustrate the differences between this geometry and those discussed in the last section. Fig. 4a illustrates the case in which both sides are accessible. The only way in which different angular profiles can be taken is to launch the waves at different angles by tilting the source. Theoretically, it is possible to launch waves over a $\pm$ 90° field. However, note that the receiver would need to be moved farther away with increasing angle. Hence, a practically obtainable field of view with this method is $\pm$ 45°.

Fig. 4 illustrates the dilemma produced by the requirement for single surface inspection, and the solution. The back surface of the section is used as a reflector, with the receiver placed to receive the reflected signal. Note, however, that the reconstruction will now include a mirror image

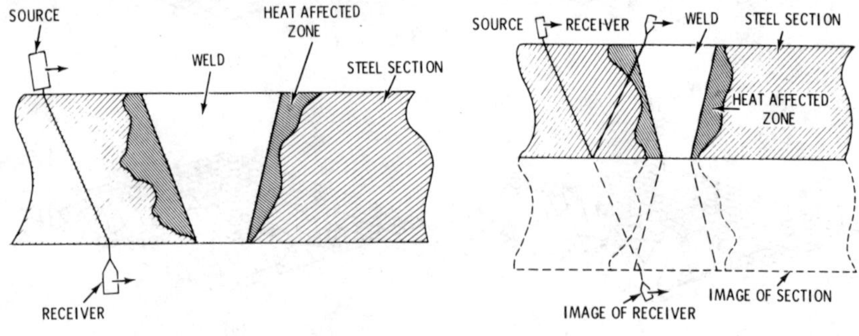

(a) Scanning Thick Sections by Transmission (b) Scanning Thick Sections by Reflection

FIGURE 4. Scanning Thick Sections by Transmission and Reflection.

as well as the object itself. Also note that in both cases, the total geometric path length changes as a function of angle of view. Both of these peculiarities must be taken into account by the reconstruction algorithm.

Experimental Data

Parallel Beam

To demonstrate the ultransonic tomography technique for mapping velocity fields, we developed a model which allows us to obtain time-of-flight profiles in the vicinity of a known stress concentration. This experimental model, diagrammed in Fig. 5, consists of an aluminum specimen measuring 25 mm x 50 mm x 380 mm. At the center of the 50 mm x 380 mm plane, we caused a 6.35 mm

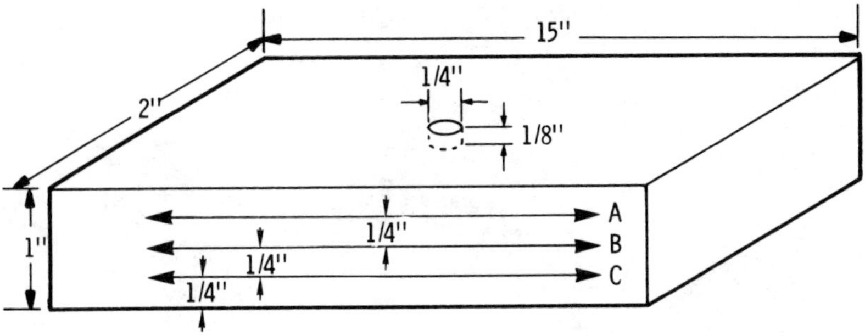

FIGURE 5. Sketch of Aluminum Stress Sample Showing Scans at Levels A, B, and C.

diameter, 3.17 mm deep, flat bottom depression by applying a 6,350 kg force. We then obtained three sets of time-of-flight profiles at Positions A, B, and C as shown in Fig. 5, Position A being 3.17 mm below the bottom of the depression and B and C being 9.52 mm and 15.87 mm below, respectively.

The time-of-flight measurements were made using the parallel beam method. The source and receiver were adjusted to obtain 27 views from -52° to $+52^{\circ}$in steps of 4°.

Figure 6 shows the time-of-flight profiles obtained for the B scan: 6a shows the profiles from 0° to 52°, while 6b shows profiles from the angles -4° to -52°. In this representation, the profiles are all scaled by the same factor but they have been repositioned along the vertical axis so that each set could be displayed on a single plot. One important point to notice in Fig. 6 is that as the angle increases, the resulting profiles deteriorate (contain points which appear to be inconsistent); however, we have found that if a small number of time-of-flights obtained are in error, the resulting reconstruction is not drastically affected.

Figures 7a, 7b, and 7c show gray scale images of the reconstructions from Positions A, B, and C respectively. In generating these images, we have applied the same maximum and minimum velocities to scale each of the reconstructions. Doing this shows how the velocity falls off with distance from the stress region. For these images, we have also positioned the levels of gray so that the middle shade (Level 5, counting from the darkest level of gray) corresponds to the average velocity measured in the aluminum sample.

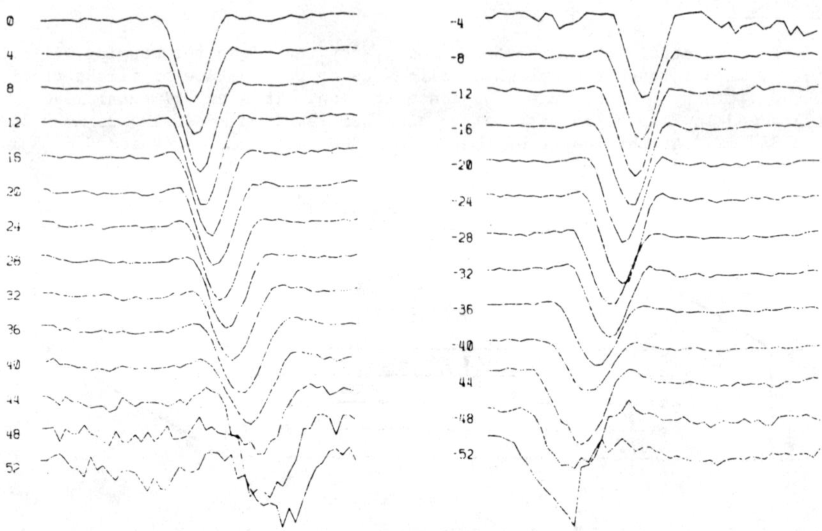

FIGURE 6. Time of Flight Profiles for Aluminum Sample at Level B.

This average velocity was determined by making a series of pulse-echo measurements as far as possible from the induced stress region. Levels lighter than 5 indicate velocity increases and those darker than 5 indicate velocity decreases. Each level represents a 0.3% velocity change. The physical dimensions of the reconstructed region are 38 mm x 38 mm.

Notice that by scaling all three images to the same maximum and minimum velocities, no detail can be observed for the C scan. This is because the velocity change assigned to each shade of gray exceeds the total change in the C scan. The calculated maximum velocity variation was 2.6%, 1.8%, and 0.4% for Scans A, B, and C respectively.

In Figs. 8a, 8b, and 8c, we show the same reconstructions, but here we have scaled each image by its own maximum and minimum. That is, for 8a, the

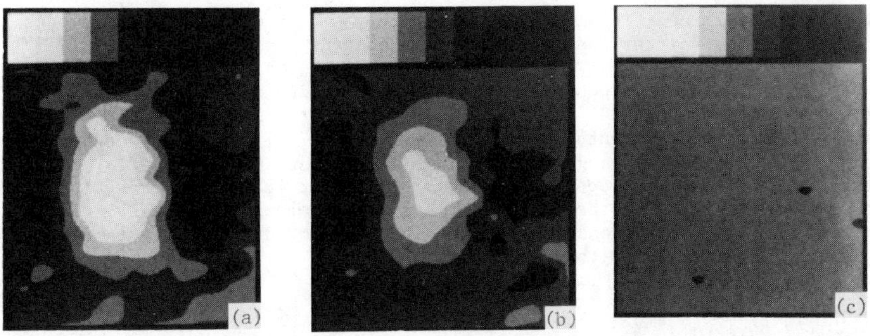

FIGURE 7. Reconstructions at (a) Level A, (b) Level B, and (c) Level C in the aluminum block described in Figure 5. The gray scale for all three is normalized to Level A.

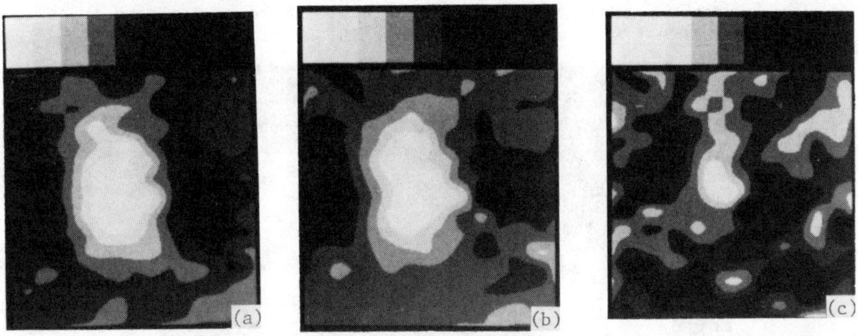

FIGURE 8. Figure 7 repeated with gray scale normalized to each individual scan.

levels represent 0.3% velocity change; for 8b, 0.2%; and for 8c, 0.04%. As in Fig. 7, the middle gray level corresponds to the average velocity in the aluminum section away from the stress region. By using this scheme for displaying reconstructions, we can see detail in the C scan which was hidden in Fig. 7c. However, there is now no convenient way to compare Figs. 8a and 8c.

Reflected Bêam

We have performed several experiments to test the reflected beam techniques. As a model of a region of velocity change in an otherwise uniform metal volume, we chose a rubber glove filled with a water-alcohol mixture surrounded by a large volume of water. The ratio of water to alcohol in the glove was intended to provide a 3% increase in the velocity of sound over the surrounding water bath.

To simulate reflection from a back surface, we placed a transmitter-receiver pair on the same side of the rubber glove, and a mirror on the opposite side of the glove centered between the two transducers. A single profile was obtained by scanning the transmitter-receiver combination linearly past the glove. Additional profiles were obtained by changing the distance between the transmitter and receiver and repeating the linear scan.

In one experiment, we used a data set consisting of 23 profiles taken at angles from 10° to 120° in steps of 5°. Each profile consisted of 31 time-of-flight measurements taken at intervals of 7.37 mm. Fig. 9 shows the reconstruction of a cross section of a single finger of the rubber glove obtained from this data set. The reconstruction indicates a sharp velocity change in the region of the finger. The maximum velocity change calculated for this reconstruction was 2.7%, which is close to our measurements of the actual velocity change within the glove.

Calibration

In order to relate velocity measurements to stress, we have initiated a calibration experiment. The purpose of this experiment is to obtain velocity measurements as a function of stress, both tensile and compression. To accomplish this, we fabricated tensile and compression specimens from 6.45 cm, A516-74A, grade 70, pressure vessel steel plate. These were placed in an MTS 810 material-test system machine located in our laboratory.

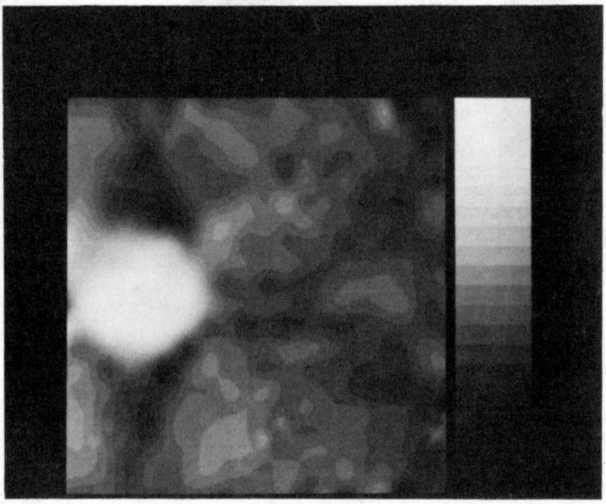

FIGURE 9. Reconstruction of Rubber Glove Model Using Reflected Beam Tomography.

Time-of-flight measurements were made with a system consisting of a Metrotek MP 215 High-energy Pulser (driving a wide-band transducer) and a Metrotek MR 101 Receiver and Metrotek MG 703 Interval Gate (driving an HP 5345A Time Interval Counter). The transducer was coupled directly to one side of the specimen and stress was applied. The MG 703 was set to gate out the first and second echos from the back surface of the specimen. This eliminates any possible error due to the coupling thickness. We have found this measurement to be repeatable to 0.1 nsec. Transverse strain was measured with a Lion Precision Corporation Metri-Gap 300-3 Capacitive Micrometer to a precision of 25 μm. A thermocouple was also attached to the specimen to keep track of the temperature as the specimen was stressed.

The results of these tests are summarized in Figs. 10 and 11. Curves 1, 2, and 3 of Fig. 10 are derived from one tensile specimen and curve 4 is derived from another. Curve 1 represents a test run well below the yield point, curve 2 is a test run of the same specimen taken past the yield limit, and curve 3 is a final run taken to failure. This set of curves is extremely interesting since it seems to indicate that the specimen retains a memory of its last test if taken beyond yield. Curves 1 and 2 follow the expected path of a linear velocity increase with stress. However, after stressing the specimen into yield, as was done in curve 2, the behavior of the velocity (curve 3) is that of a specimen under compressive load. It appears as though a net residual compressive stress is present in the sample. The tension introduced by the machine must first overcome the residual compressive stress before the velocity again increases. This residual stress appears to be about 2.07×10^5 kPa. Note also, that the yield strength of the material increases from about 3.10×10^5 kPa to 4.55×10^5 kPa due to work hardening (as is well known). Curve 4 is the result of a test on a second identical specimen, performed to corroborate the stress acoustic constant derived from the first specimen. This factor, which is the slope of the curve, is $7.33 \times 10^{-4}\%/6.9 \times 10^3$ kPa/2.54 cm.

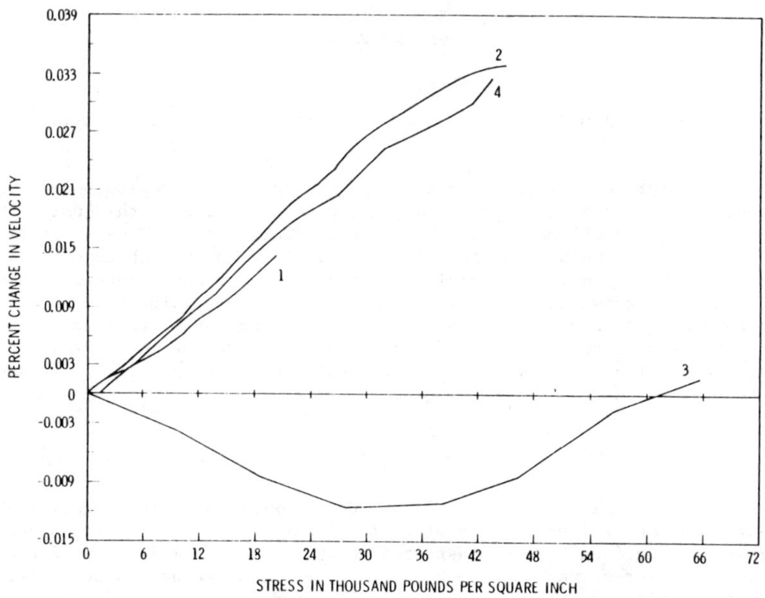

FIGURE 10. Velocity Change as a Function of Tensile Stress

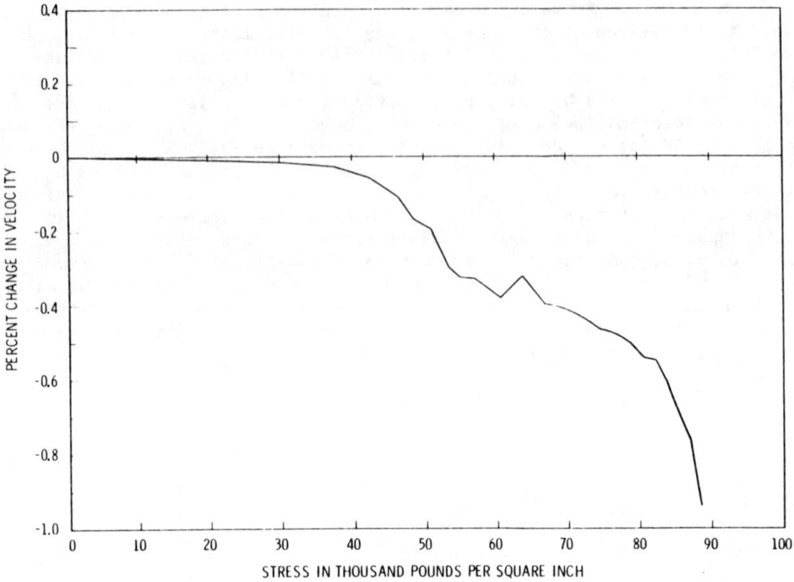

FIGURE 11. Velocity Change as a Function of Compressive Stress

In Fig. 10, the curves are drawn only for the stress region before yield is reched. This is because the readings become erratic past this point. For this material, the velocity change appears to flatten out. However, this may be due to the well known behavior of materials in tension such as dislocation and slip fractures in the material. In compression, a much smoother behavior is revealed, as shown in Fig. 11. In this case, linear velocity decrease (with about the same stress acoustic constant) occurs up to the yield point. After the material begins to yield, the velocity decreases in a highly non-linear manner with the slope increasing precipitately. This, too, is expected from theory.

Conclusion

We have discussed an application of Computer-Assisted Tomography (CAT) for locating and mapping regions of residual stress. The simulations and experiments described have demonstrated that velocity anomalies of 2% can be quite easily resolved. In work not reported here, we have also experimentally mapped velocity anomalies as low as 0.2% and feel that 0.05% is technically feasible. These velocities translate to a sensitivity of 6895 kPa in a 2.54 cm-thick region.

We feel that the reflected beam geometry represents an important advance in the development of methods for inspecting structures which do not physically lend themselves to either the parallel beam or the fan beam geometries.

Thus, the reflected beam geometry could have important applications in the area of in-service inspection.

We are currently testing a prototype instrument for measuring stress in an online production environment. The instrument consists of an array of transducers placed in contact with the metal surface. Two elements of the array are selected to form a pitch-catch arrangement. One is used to transmit a pulse and the other to receive it. The choice of elements defines the angle of the ray to be measured. After a time-of-flight measurement has been made, the selected set of two elements are electronically moved over by one element and second measurement made. In this way, a whole profile can be made very quickly.

After one profile is made, the separation between the two elements is changed by selecting two different transducers and the process is repeated, giving another profile at a different angle. All of the necessary data can thus be taken in a few seconds or less, depending upon the accuracy desired.

The time-of-flight measurements are fed directly into the memory of a microprocessor. When a complete set of data is available, the computer executes the ART algorithm and displays the resulting reconstruction on a CRT display.

An additional problem also needs to be discussed at this time. When sound waves travel through media with different velocities, they are refracted. In the work discussed in this report, we have assumed that the sound travels in a straight line between the source and receiver. The difficulty in trying to include the effect of refraction in our analysis is that we need to know the velocity field to calculate the refraction. However, the velocity field is exactly what we are attempting to determine.

Several solutions to this problem have been suggested. The most accurate method would be to include the refraction equations in a simultaneous solution for the velocities. However, as we discussed earlier, a simultaneous solution to the problem is impossible because of the large number of equations and unknowns involved. A more practical solution would be to first compute the velocity field as we have described. Once we have an estimate of the velocities, we could iterate several additional times using a modified version of the ART algorithm which would take into account the ray path bending.

To date, we have not investigated the problems caused by refraction because the velocity changes which we are interested in are small and will cause only slight bending. However, if in the near future we apply this technique to examine regions containing large velocity variations, we will have to modify our reconstruction method to account for the refraction.

This research is sponsored by the Electric Power Research Institute under Contract RP504-2.

References

1. Gott, K. E., "Residual Stresses in a Weldment of Pressure Vessel Steel", AB Atomenergi Report, NyKoping, Sweden (1978).

2. Hsu, N. H., "Acoustical Birefringence and the Use of Ultrasonic Waves for Experimental Stress Analysis", Experimental Mechanics, May 1974.

3. Smith, R. T., "Stress-Induced Anisotropy in Solids--The Acousto-Elastic Effect", Ultrasonics, July-September 1963.

4. Noronha, P. J. and Wert, J. J., "An Ultrasonic Technique for the Measurement of Residual Stress", J. of Testing and Evaluation, March 1975, Vol. 3.

5. Brooks, R. A. and DiChiro, G., "Theory of Image Reconstruction in Computed Tomography", Radiology, December 1975, Vol. 117, pp. 561-572.

6. DeRosier, D. J. and Klug, A., "Reconstruction of Three-Dimensional Structures from Electron Micrographs", Nature (London), January 1968, No. 217, pp. 130-134, 13.

7. Gordon, R., Bender, R. and Herman, G. T., "Algebraic Reconstruction Techniques (ART) for Three-Dimensional Electron Microscopy and X-Ray Photography", J. Theoretical Biology, December 1970, Vol. 29, pp. 471-481.

8. Colsher, J. G., "Iterative Three-Dimensional Image Reconstruction from Projections: Applications in Electron Microscopy", Ph.D. Thesis, University of California-Livermore, UCRL-52179, December 1976.

9. Bracewell, R. N. and Riddle, A. C., "Inversion of Fan-Beam Scans in Radio Astronomy", The Astrophysics J., November 1967, Vol. 150, pp. 427-434.

10. Kruger, R. P. and Cannon, T. M., "The Application of Computed Tomography, Boundary Detection, and Shaded Graphics Reconstruction to Industrial Inspection", Materials Evaluation, April 1978, Vol. 36, No. 5.

11. Falconer, D. G. and Gates, D. C., "Reactor-Component Inspection with Computed Tomography", EPRI NP-213, Project 610-1, July 1976.

12. Radon, J., "Uber die Bestimmung von Funktionen durch ihre Integralwerte langs gewisser Mannigfaltigkeiten", Ber. Akad. Wiss. (Leipqig), Math. Phys. Klasse, Vol. 69, pp. 262-277, 1917.

13. Rammachandran, G. N. and Lakshminarayanan, A. V., "Three-Dimensional Reconstruction from Radiographs and Electron Micrographs: Applications of Convolutions Instead of Fourier Transforms", Proceedings of National Academy of Science U.S.A., September 1971, Vol. 68, pp. 2236-2240.

14. Mersereau, R. M. and Oppenheim, A. V., "Digital Reconstruction of Multidimensional Signals from their Projections", Proceedings of the IEEE, October 1974, Vol. 62, No. 10, pp. 1319-1338.

15. Goitein, M., "Three-Dimensional Density from a Series of Two-Dimensional Projections", Nucl. Instr. Meth., June 1972, Vol. 101, pp. 509-518.

16. Gilbert, P., "Iterative Methods for the Three-Dimensional Reconstruction of an Object from Projections", J. Theoretical Biology, July 1972, Vol. 36, pp. 105-117.

17. Herman, G. T., Lent, A. and Rowland, S. W., "ART: Mathematics and Applications; A Report on the Mathematical Foundations and the Applicability to Real Data of the Algebraic Reconstruction Techniques", J. Theoretical Biology, 1973, Vol. 42, pp. 1-32.

18. Sweeney, D. W., "Interferometric Measurement of Three-Dimensional Temperature Fields", Ph. D. Thesis, University of Michigan-Ann Arbor, 1972.

EARLY DETECTION OF METAL FATIGUE BY OPTICAL CORRELATION

David R. Sigler and W.L. Haworth

Department of Metallurgical Engineering
and
Research Institute for Engineering Sciences
Wayne State University

The application of an optical correlation technique to the detection of fatigue damage in metals is reviewed in this paper. Measurements of optical correlation intensity (I_c) can be made during fatigue cycling by transmitting light scattered from the metal surface through a holographic filter containing information about the surface topography at an earlier time; the theory underlying this technique is presented. Topographic changes such as slip band development, microcrack formation and crack growth observed during fatigue tests cause corresponding changes in correlation intensity. For both unnotched and notched specimens with either polished or rough-ground surfaces, a three-stage correlation intensity curve (log I_c vs. number of fatigue cycles) is generally observed, region A corresponding to slip band development and microcracking and region C to crack propagation; few metallographic changes are observed during region B of the curve. The loss of correlation intensity from the beginning of fatigue cycling to the development of a crack less than one millimetre in length can be several orders of magnitude: Thus, the technique is an extremely sensitive method of detecting fatigue damage and early crack growth in metals, and no prior knowledge of the presence or precise location of particular flaws or cracks is required. Recent experiments demonstrate that correlation intensity losses early in the fatigue life are very sensitive to the imposed cyclic strain range, and that plastic strains of the order of 10^{-4} can be detected reliably.

Symbols

ξ_m, η_m, ζ_m - Cartesian grid coordinates corresponding to specimen surface at time hologram is taken.

ξ_n, η_n, ζ_n - Cartesian grid coordinates corresponding to specimen surface at a later time.

x_α, y_α - Coordinates of plane containing lens I.

x_β, y_β - Coordinates of plane containing the hologram.

x_μ, y_μ - Coordinates of plane containing lens II.

x_υ, y_υ - Coordinates of plane containing the photodetector.

λ - Wavelength of coherent light.

a_m - Reflectivity of specimen surface at point ξ_m, η_m, ζ_m.

a_n - Reflectivity of specimen surface at point ξ_n, η_n, ζ_n.

ϕ - Angle between illuminating beam and the macroscopic plane of the specimen surface.

θ - Angle between any light ray scattered off the specimen surface and the normal to the specimen surface.

Ω_R - Angle between the reference beam and the plane containing the hologram.

γ - Angle between normal of x_υ, y_υ plane and a reconstructed light ray.

f_1 - Focal length of lens I.

f_2 - Focal length of lens II.

$F_1 = 1/f_1$, $F_2 = 1/f_2$

d_1 = Distance between origin of ξ_m, η_m, ζ_m plane and origin of x_α, y_α plane.

d_2 - Perpendicular distance between x_α, y_α and x_β, y_β plane.

d_3 - Distance between origin of x_μ, y_μ and x_υ, y_υ planes.

d_4 - Perpendicular distance between plane of lens II and the photodetector plane.

$1/d_1 = D_1$, $1/d_2 = D_2$, $1/d_3 = D_3$, $1/d_4 = D_4$.

$u(x,y,z)$ - Complex amplitude of scattered light wave at point x,y,z at time hologram is made.

$v(x,y,z)$ - Complex amplitude of scattered light wave at point x,y,z at some later time.

$w(x,y,z)$ - Complex amplitude of reconstructed reference wave at coordinates x,y,z.

$R(x_\beta,y_\beta)$ - Complex amplitude of reference wave at hologram plane.

$T(x_\beta,y_\beta)$ - Transmittance of hologram. Reconstruction is given by multiplying T(x,y) by a complex wave amplitude.

$r(\xi_m,\eta_m,\zeta_m,x_\alpha,y_\alpha)$ - Distance between points in the ξ_m, η_m, ζ_m and x_α, y_α planes.

$r(x_\beta,y_\beta,x_\mu,y_\mu)$ - Distance between a point in the x_β, y_β and x_μ, y_μ planes.

$S_o, R_o, K_I, K_{II}, K_{III}, K_{IV}, K_V, K_{VI}, K_{VII}$ - constants.

$w_{uv}(x_\upsilon,y_\upsilon)$ - Complex amplitude of reconstructed wave at the point x_υ,y_υ.

$I_{uv}(x_\upsilon,y_\upsilon)$ - Intensity of reconstructed wave at the point x_υ,y_υ.

$\psi(x,y;p) = \exp(-\frac{\pi i}{\lambda}\, p(x^2 + y^2))$

Introduction

The early stages of fatigue damage in metals, including slip line or slip band development, cracking of second-phase particles or inclusions, fatigue crack initiation and early crack growth, generally occur at or near the surface of a specimen or component, where the material is unconstrained and strains are often highest. The optical correlation technique which we will discuss in this paper provides a method of storing information about the surface topography of a specimen in a hologram at time t_1 and, after fatigue cycling, comparing it with the actual surface topography at a later time t_2. If the surfaces are identical, one infers that there is no significant fatigue damage. A change in surface condition, however, may indicate that fatigue damage is occurring, and in principle the optical correlation technique enables us to quantify the damage. In particular, the optical correlation measurements are sensitive to surface changes caused by slip lines, particle cracking, fatigue crack initiation and the growth of fatigue cracks through the metal matrix. Such measurements can be made continuously, in air, as the fatigue test proceeds, independently of any prior knowledge of the thermal or mechanical history of the specimen, and they can be made under a variety of initial surface conditions of the material. Thus, the technique is a very promising one for investigating fatigue damage - and possibly predicting fatigue failure - on an individual component basis rather than by statistical methods based on fatigue life versus stress (or strain) data for large numbers of tests. Since the holographic technique integrates information over an arbitrarily large area of the material surface, it is not necessary to know precisely where damage is occurring in order to detect it.

The initial experiments investigating the application of optical correlation measurements for fatigue detection were carried out by Chuang (1) and by Marom and Mueller (2,3) on aluminum alloys more than ten years ago. Jenkins and McIlwain (4) reported that damage caused by thermal cycling of solder joints could be detected by this technique, and Bond et al. (5) investigated the effect of fatigue crack growth in Fe-3Si on optical correlation measurements. Much of the early work was reviewed by Marom (6), and a reasonably detailed description of the theory underlying the application of optical correlation to nondestructive evaluation of materials can be found in his paper. These experiments established that optical correlation measurements are sensitive to fatigue damage. Chuang (1) and Marom and Mueller (2,3) in particular were able to show that the optical correlation intensity (defined below) varied continuously as a function of the number of fatigue cycles during the fatigue life of a material. Although detailed metallographic observations were not made, these investigators speculated that the initial changes in correlation intensity correspond to cyclic hardening early in the fatigue life, and showed that rapid correlation intensity changes late in the fatigue life correspond to fatigue crack propagation. The measurements could be made on surfaces that were initially either smooth or rough, and were sensitive to the strain amplitude during fatigue. Since the correlation intensity losses could be quite large, it was necessary to re-record holograms several times during a fatigue test and normalize the results back to the initial intensity. Unfortunately, the need for repeated recording of the hologram is a major difficulty common to all of the early investigations of this technique. Problems arise because of the time delay, emulsion shrinkage, and necessity for accurate repositioning after development which are associated with the use of photographic film to record the hologram. This makes the repeated recording of holograms during a fatigue test particularly tedious; furthermore, the test must be interrupted for an extended interval whenever a new hologram is recorded. A further difficulty was the need to interrupt the fatigue test in order to perform the optical correlation measurement.

Experiments carried out in our laboratory over the last few years have been designed to address the problems outlined above: we are now able to make optical correlation measurements in "real-time" during a fatigue test, and to re-record holograms rapidly in situ, without the need for photographic processing, by using a thermoplastic/photoconductor holographic recording device instead of photographic film (7). We have established the general relationships between the optical correlation measurements and the surface changes occurring on the specimen by using in situ metallography, and we are beginning to examine these relationships in detail.

In the following sections, the optical correlation technique is described briefly and a more detailed description is given of the theory underlying its application to nondestructive evaluation of materials. This is followed by a summary of our previously published work on the application of the optical correlation technique to fatigue damage detection, and a review of significant results from work currently in progress.

Optical Correlation

The optical correlation technique, as adapted to the nondestructive testing of metals by Marom and Mueller (2,3) requires two distinct operations which we can describe by referring to Figure 1. The first step is to generate a complex spatial filter, or hologram, by superimposing light scattered from the specimen surface on a coherent plane-wave reference beam. The resulting interference pattern is recorded on light-sensitive film or some alternative recording device. In the second step, the reference beam is blocked. Light from the coherent source which is scattered by the object is then transformed on passing through the filter into a beam with the same phase and amplitude characteristics as the original reference beam. The intensity of the central peak of this reconstructed reference beam can be measured by a photodetector, as shown in Fig. 1, and is termed the correlation intensity, I_c. Changes in the topography of the specimen surface change the phase and amplitude of the object beam, and a corresponding decrease in correlation intensity occurs. However, translation or rotation of the specimen or the holographic filter will also cause the correlation intensity to fall. If the changes caused by translation or rotation are reduced sufficiently, or factored out, then during a fatigue test the I_c values will provide an integrated measure of accumulated fatigue damage over the entire illuminated surface of the specimen.

A more detailed description of the optical reconstruction process and the relationship between surface changes and correlation intensity can be given on the basis of Fig. 2, which shows schematically an off-axis optical correlation system consisting of the specimen, two lenses of focal length, f_1 and f_2 respectively, a holographic recording device, and a light detector. The initial position coordinates of a point on the specimen surface are ξ_m, η_m, ζ_m. To construct the holographic filter, the specimen is illuminated by a plane wave incident at an angle ϕ to the ξ axis and perpendicular to the η axis. If the specimen surface consists of a large number of individual Huygens point radiators, the complex amplitude of the light scattered at each point is

$$u(\xi_m,\eta_m,\zeta_m) = a_m \exp[\frac{2\pi i}{\lambda}(\xi_m \cos\phi + \zeta_m \sin\phi)] \qquad (1)$$

The resulting light field at the collector lens is given by

$$u(x_\alpha,y_\alpha) = \sum_{m=1}^{M} \{\frac{u(\xi_m,\eta_m,\zeta_m)}{|r(\xi_m,\eta_m,\zeta_m,x_\alpha,y_\alpha)|}$$

$$\exp[-\frac{2\pi i}{\lambda} r(\xi_m,\eta_m,\zeta_m,x_\alpha,y_\alpha)]\cos\theta\} \qquad (2)$$

We now assume that $\frac{1}{|r(\xi_m,\eta_m,\zeta_m,x_\alpha,y_\alpha)|}$ can be approximated as $1/d_1$. Also, only a small cone of light emanating from the specimen surface will be analyzed so we assume that θ is less than 10°, hence cos θ is approximately unity. Using the Binomial expansion, $r(\xi_m,\eta_m,\zeta_m,x_\alpha,y_\alpha)$ in the exponential term is given by

$$r(\xi_m,\eta_m,\zeta_m,x_\alpha,y_\alpha) = d_1 + \xi_m \cos\phi - \zeta_m \sin\phi + \frac{(x_\alpha-\xi_m \sin\phi)^2 + (y_\alpha-\eta_m)^2}{2d_1} + \ldots \qquad (3)$$

provided all neglected terms are much less than λ. This puts restrictions on specimen size, lens size, and specimen to lens distance. At the hologram plane (x_β, y_β) the light amplitude is given by

$$u(x_\beta,y_\beta) = K_I \iint_{-\infty}^{\infty} u(x_\alpha,y_\alpha)\, \psi(x_\alpha,y_\alpha;-\frac{1}{f_1})\psi(x_\beta-x_\alpha,y_\beta-y_\alpha;\frac{1}{d_2})dx_\alpha dy_\alpha \qquad (4)$$

The functions $\psi(x_\alpha,y_\alpha;-\frac{1}{f_1})$ and $\psi(x_\beta-x_\alpha,y_\beta-y_\alpha;\frac{1}{d_2})$ account for passage

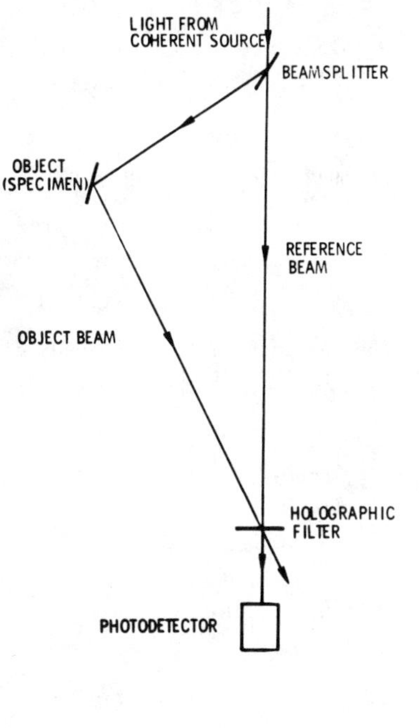

Fig. 1 Simplified schematic of optical correlation system for nondestructive evaluation.

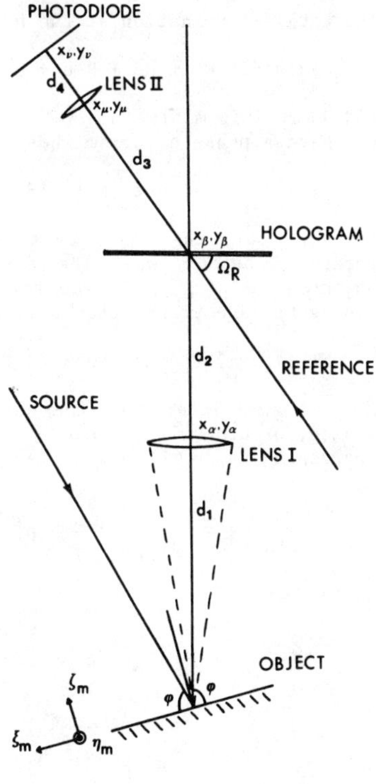

Fig. 2 Schematic of optical system for holographic recording and reconstruction.

of light through lens I, and through the space between lens I and the hologram plane, respectively. Using standard Fourier transforms Eq. (4) can be integrated to yield

$$u(x_\beta,y_\beta) = K_{II}\sum_{m=1}^{M}\{\frac{a_m}{d_1}\exp[-\frac{2\pi i}{\lambda}(d_1-2\zeta_m \sin\phi)]$$

$$\exp[\frac{2\pi i}{\lambda}\{x_\beta \frac{\xi_m \sin\phi\, D_1 D_2}{D_1 - F_1 + D_2} + y_\beta \frac{\eta_m D_1 D_2}{D_1 - F_1 + D_2}\}]$$

$$\psi(\xi_m,\eta_m;D_1 - \frac{D_1^2}{D_1 - F_1 + D_2})\ \psi(x_\beta,y_\beta;D_2 - \frac{D_1^2}{D_1 - F_1 + D_2})\} \qquad (5)$$

If the hologram is located in the focal plane of lens I, then $F_1 = D_2$ and the first ψ term is eliminated. Also incident on the hologram plane is the reference wave R, where

$$R(x_\beta,y_\beta) = R_o \exp(-\frac{2\pi i}{\lambda} x_\beta \cos \Omega_R) \qquad (6)$$

The total illumination at the hologram plane is given by

$$|R+u|^2 = R^2 + u^2 + R^*u + u^*R \qquad (7)$$

and this information is stored when the hologram is recorded. Thus, we now have a phase hologram whose amplitude transmittance is

$$T(x_\beta,y_\beta) = \text{const.}|R + u|^2 \qquad (8)$$

However, only the final term in Eq. (7) leads to reconstruction of the reference beam R when the hologram is subsequently illuminated by the object wave $u(x_\beta,y_\beta)$. The term of interest in Eq. (8) is therefore the conjugate wave transmittance given by

$$T_c(x_\beta,y_\beta) = \text{const. } u^*R \qquad (9)$$

At some later time the surface of the specimen may be strained, displaced, or otherwise changed. The complex amplitude at the hologram plane corresponding to the new surface is

$$v(x_\beta,y_\beta) = K_{III}\sum_{n=1}^{N}\{\frac{a_n}{d_1}\exp[-\frac{2\pi i}{\lambda}(d_1 - 2\zeta_n \sin\phi)]$$

$$\exp[\frac{2\pi i}{\lambda}\{x_\beta \frac{\xi_n \sin\phi\, D_1 D_2}{D_1 - F_1 + D_2} + y_\beta \frac{\eta_n D_1 D_2}{D_1 - F_1 + D_2}\}]$$

$$\psi(x_\beta,y_\beta;D_2 - \frac{D_1^2}{D_1 - F_1 + D_2})\} \qquad (10)$$

The reconstructed wave at the plane x_μ,y_μ produced by $v(x_\beta,y_\beta)$ illuminating the hologram is

$$w(x_\mu,y_\mu) = K_{IV} \iint_{-\infty}^{\infty} \frac{u^*(x_\beta,y_\beta)v(x_\beta,y_\beta)R(x_\beta,y_\beta)}{|r(x_\beta,y_\beta,x_\mu,y_\mu)|}$$

$$\exp[-\frac{2\pi i}{\lambda} r(x_\beta,y_\beta,x_\mu,y_\mu)] \cos\gamma \, dx_\beta \, dy_\beta \qquad (12)$$

The angle γ is less than 10° so cos $\gamma \cong 1$ and $|r(x_\beta,y_\beta,x_\mu,y_\mu)|$ in the denominator is approximately equal to d_3. Using the Binomial expansion $r(x_\beta,y_\beta,x_\mu,y_\mu)$ can be written as

$$r(x_\beta,y_\beta,x_\mu,y_\mu) \cong d_3 - x_\beta \cos \Omega_R$$

$$+ \frac{(x_\mu - x_\beta \sin \Omega_R)^2 - (y_\mu - y_\beta)^2}{2d_3} + \ldots \qquad (13)$$

All neglected terms must be much less than λ. At the plane x_υ,y_υ the reconstructed wave is given by

$$w(x_\upsilon,y_\upsilon) = K_V \iiiint_{-\infty}^{\infty} w(x_\mu,y_\mu) \exp(-\frac{2\pi i}{\lambda} r(x_\beta,y_\beta,x_\mu,y_\mu))$$

$$\psi(x_\mu,y_\mu;-\frac{1}{f_2}) \, \psi(x_\upsilon - x_\mu, \, y_\upsilon - y_\mu;\frac{1}{d_4}) \, dx_\beta \, dy_\beta \, dx_\mu \, dy_\mu \qquad (14)$$

where the ψ functions account for the passage of the wave through lens II and space d_4. The photodetector is placed in the focal plane of lens II, so that $d_4 = f_2$. Equation (14) can be integrated with respect to x_μ,y_μ and x_β,y_β using Fourier transforms to yield two delta functions. The reconstructed wave is finally written as

$$w(x_\upsilon,y_\upsilon) = K_{VI} \exp[-\frac{2\pi i}{\lambda} d_3] \, \psi(x_\upsilon,y_\upsilon;D_4 - \frac{D_4{}^2}{D_3})$$

$$\sum_{m=1}^{M} \frac{a_m}{d_1} \exp[\frac{2\pi i}{\lambda}(d_1 - 2\zeta_m \sin\phi - x_\beta \xi_m \frac{\sin\phi}{d_2} - y_\beta \frac{\eta_m}{d_2})]$$

$$\sum_{n=1}^{N} \frac{a_n}{d_1} \exp[-\frac{2\pi i}{\lambda}(d_1 - 2\zeta_n \sin\phi - x_\beta \xi_n \frac{\sin\phi}{d_2} - y_\beta \frac{\eta_n}{d_2})]$$

$$\delta[\frac{1}{\lambda}(\xi_n \frac{\sin\phi}{d_2} - \xi_m \frac{\sin\phi}{d_2} + \sin \Omega_R \frac{x_\upsilon}{d_4})]$$

$$\delta[\frac{1}{\lambda}(\frac{\eta_n}{d_2} - \frac{\eta_m}{d_2} + \frac{y_\upsilon}{d_4})] \qquad (15)$$

Since the number of point radiators used to represent either the original specimen surface or its modifed surface is arbitrary, we can set N=M. The photodetector is placed at $x_\upsilon = y_\upsilon = 0$. If the delta functions are not to be zero, then in Eq. (15) $\xi_m^\upsilon = \xi_n^\upsilon$ and $\eta_m = \eta_n$. Hence $w(x_\upsilon,y_\upsilon)$ reduces to

$$w(x_\upsilon,y_\upsilon) = w_{uv}(0,0) = K_{VI} \sum_{m=1}^{M} \frac{a_m a_m^d}{d_1^2} \exp[\frac{4\pi i}{\lambda} (\zeta_m^d - \zeta_m)] \qquad (16)$$

where a_m^d denotes a_n when $\xi_m = \xi_n$ and $\eta_m = \eta_n$, and ζ_m^d denotes ζ_n when $\xi_m = \xi_n$ and $\eta_m = \eta_n$. This does not imply that no surface changes have occurred, but only that the same set of coordinates (ξ_m,η_m) is used as a reference grid for either surface.

The photodetector measures energy incident per unit area per unit time, or intensity, at the point $x_\upsilon = y_\upsilon = 0$. This intensity is given by

$$I_c \equiv I_{uv}(0,0) \propto w^*_{uv}(0,0)\, w_{uv}(0,0) \qquad (17)$$

Thus the intensity of light at the focal point of lens II provides a quantitative measure of the similarity between the original and changed specimen surface. Any surface change will make the phase terms in equation (17) non-zero and produce destructive interference, which will reduce $I_{uv}(0,0)$, the correlation intensity.

Fatigue Tests on Sheet Material

In a series of experiments over the past few years, we have investigated the relationship between optical correlation intensity and the development of fatigue damage in sheet specimens of various materials (8-12). Typical specimen geometry is shown in Fig. 3, together with the area of the specimen illuminated by the laser beam. The fatigue tests were conducted in air with a superimposed tensile stress (R=0.5 approximately) to avoid buckling, at a frequency of about 15 Hz, using a mechanically-driven machine which operates essentially at constant load amplitude with a sinusoidal load waveform. The loads were adjusted to give fatigue lives between about 10^4 and $2x10^5$ cycles to failure. The hologram was recorded with the specimen under a static tensile load. Correlation intensity measurements were then made by monitoring the output of the photodetector (Fig. 1) during fatigue cycling, either continuously as the test proceeded, or intermittently by interrupting the test periodically. The dynamic signal from the photodetector during fatigue cycling, which shows reconstruction of the reference beam occurring each time the specimen passes through its original position close to minimum load, is shown in Fig. 4. The scatter in individual peak heights seen in Fig. 4 is largely caused by cycle-to-cycle specimen displacement due to system vibration: several successive peaks can be averaged to give consistent results, or the test can be stopped during the correlation measurement to eliminate vibration. More recent tests in strain control on a closed-loop machine greatly reduce this scatter (see below).

Correlation Intensity Curves

Typical results of these tests on sheet materials are shown in Fig. 5, where correlation intensity is plotted as a function of the number of fatigue cycles for notched and unnotched specimens of a 2024-T3 aluminum alloy polished to a mirror finish prior to testing. Correlation intensity losses over the first hundred or so fatigue cycles may amount to several orders of magnitude: the initial I_c value was recorded after twenty fatigue cycles for both specimens in Fig. 5, and subsequent I_c values are normalized to this initial reading. A graph of log (I_c) vs. number of

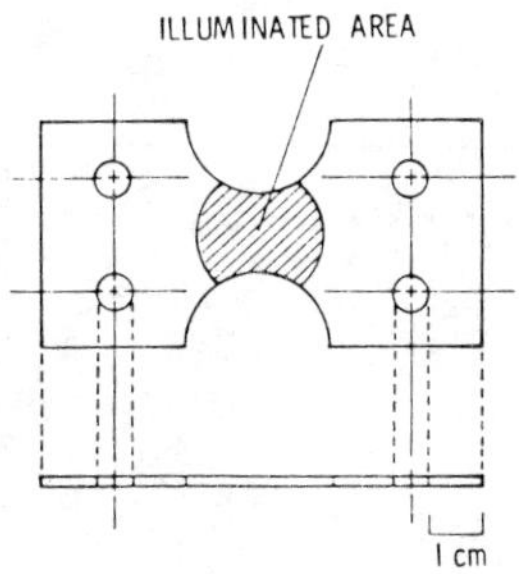

Fig. 3 Specimen geometry for fatigue tests on sheet material.

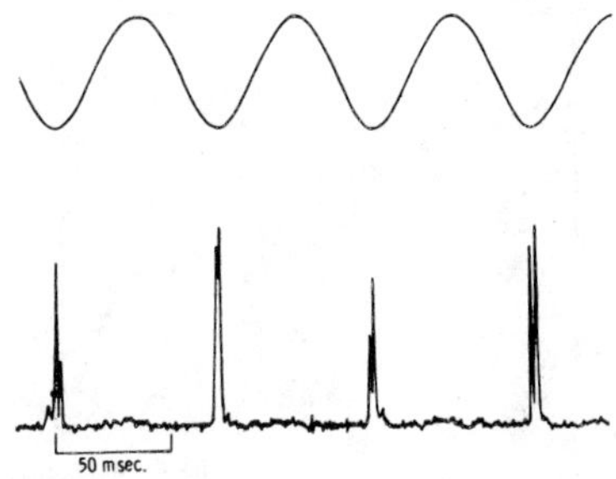

Fig. 4. Correlation intensity during fatigue cycling (lower curve). The upper curve shows axial load on specimen versus time.

fatigue cycles (N) generally falls into the three stages shown in Fig. 5. The slope of the correlation intensity curve increases (becoming less negative) during region A, remains essentially constant in region B, and decreases in an accelerating manner during region C until failure occurs. The transitions from regions A to B and B to C occur at various fractions of the total fatigue life, depending on the material and test parameters. While most of our work has been carried out using 2024-T3 aluminum alloy, tests on titanium and low-carbon steel sheet gave essentially similar results (13). Although factors such as initial surface condition, the presence of notches, or the choice of material affect the precise I_c losses observed during a given test, the general division of the correlation intensity curve into three successive stages is not affected. As shown in Fig. 5, for example, the effect of a notch is to shift the transitions from regions A to B and B to C of the correlation curve back to earlier fractions of the total fatigue life: region A occupies a proportionally smaller fraction of the total fatigue life than for unnotched specimens, and the fraction of life corresponding to region C is greater.

The effect of initial surface finish on the correlation intensity curve is illustrated in Fig. 6 for tests under essentially identical stress conditions where the initial I_c reading corresponds to one thousand fatigue cycles, on 2024-T3 aluminum specimens finished on 600 grit SiC paper or by polishing with 0.05 micron alumina abrasive, respectively. Although variations from specimen to specimen may mask general trends in some cases, the results shown in Fig. 6 suggest that the correlation intensity curve is not strongly affected by the initial surface finish under otherwise similar test conditions.

Metallography

Topographic changes such as slip band development, crack initiation and propagation were investigated either by removing the specimen from the fatigue machine at a given point on the correlation curve and inspecting the surface by optical metallography (9), or by using a metallurgical microscope mounted on a travelling stage to obtain in situ photomicrographs of the surface at selected intervals when the fatigue test was interrupted to make a correlation measurement (11,12). These observations were made on 2024-T3 aluminum specimens polished to a finish of either 0.3

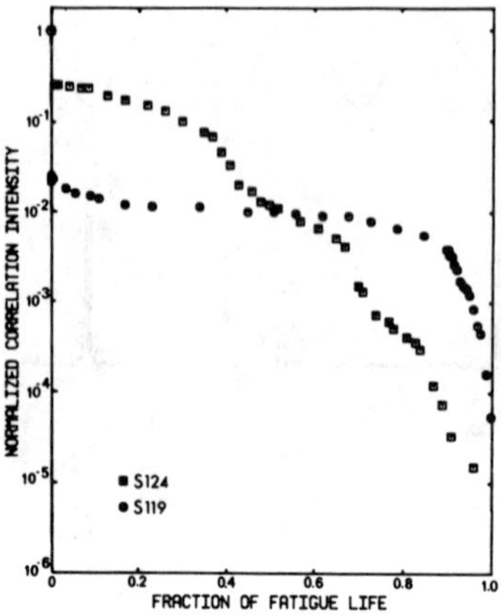

Fig. 5 Correlation intensity curves for unnotched (S119) and notched (S124) specimens of 2024-T3 aluminum alloy. Initial correlation data measured after 20 fatigue cycles.

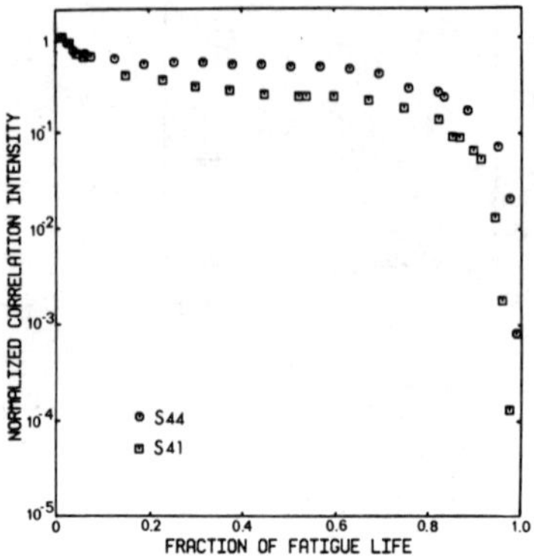

Fig. 6 Correlation intensity for 2024-T3 aluminum alloy finished with 0.05 micron abrasive (S44) or no. 600 emery paper (S41). Initial correlation data measured after 1000 fatigue cycles.

or 0.05 microns.

Using the travelling microscope technique, Haworth, Singh and Mueller (12) observed the extensive development of coarse slip bands over the specimen surface very early in the fatigue life of unnotched specimens, as shown in Fig. 7. The points on the correlation intensity curve corresponding to these photomicrographs are indicated in Fig. 8. Prior to fatigue cycling, the specimen surface contains constituent particles and also voids where particles are lost by abrasion during grinding or polishing, but there is no evidence of slip lines or microcracks. During the first few fatigue cycles, a substantial amount of deformation occurs, as shown in Fig. 7(a), and there is some evidence of microcracking and severe local deformation associated with constituent particles. The microcracks are predominantly perpendicular to the tensile axis and seldom exceed about 10 microns in length. These changes are accompanied by a substantial loss in correlation intensity, but the loss can also be attributed in part to overall extension of the specimen, since the test is stress-controlled with a superimposed mean stress. After measuring grip displacements and comparing notched and unnotched specimens, Haworth, Singh and Mueller suggested that the contribution from specimen extension may predominate over the first 1 pct of fatigue life, but that later correlation losses are caused primarily by the topographic changes. Our more recent work (see below) indicates that topographic changes themselves cause substantial correlation losses even over the first few fatigue cycles.

During the first 1 pct of fatigue life approximately, a rapid intensification of slip band density is observed, and the number of microcracks

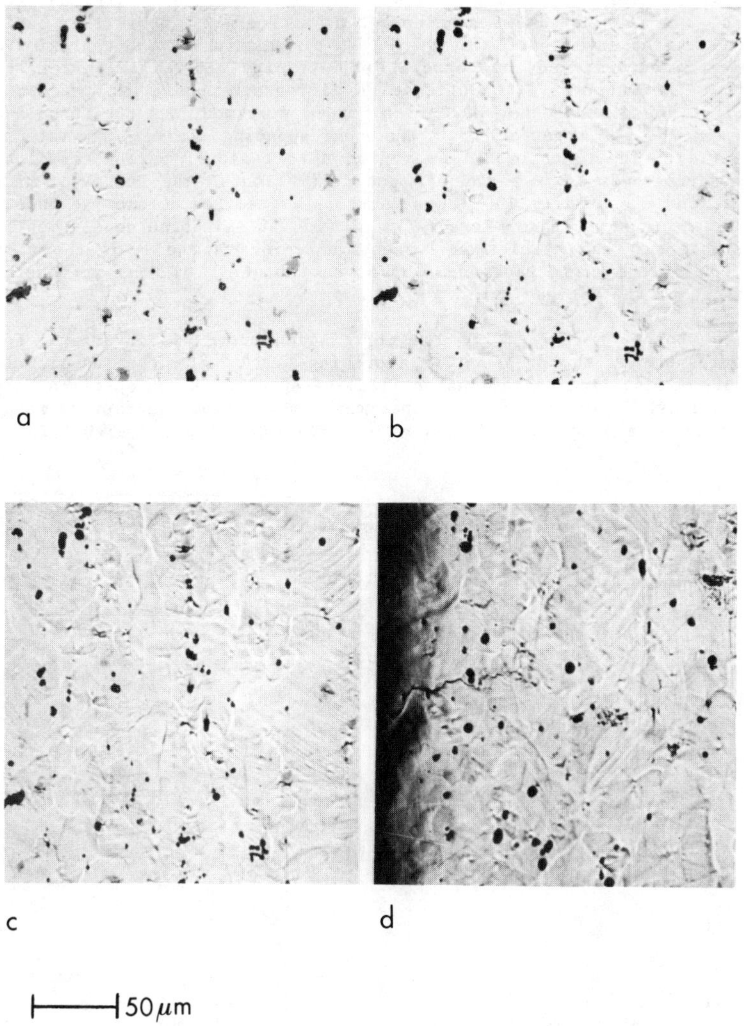

Fig. 7. Progressive development of fatigue damage in unnotched 2024-T3 aluminum (S114). Number of cycles elapsed: (a) 25, (b) 500 (same area), (c) 20,000 (same area), (d) 78,200 (specimen edge).

increases rapidly. These changes are shown in Fig. 7(b) after 500 fatigue cycles, for the same surface area as in Fig. 7(a). The same area is shown in Fig. 7(c) after 20,000 cycles. The overall surface roughness increases significantly between 25 and 500 cycles, but the changes between 500 and 20,000 cycles are somewhat less pronounced. The corresponding correlation intensity losses, in region A of the correlation curve, decay out with increasing number of fatigue cycles as shown in Fig. 8. With the exception of a small increase in the number of microcracks, no further surface changes are observed until fatigue crack propagation beyond approximately 20 microns occurs, and the correlation intensity remains relatively constant in the interim. The increase in microcrack density was determined quantitatively for another 2024-T3 aluminum specimen, and the results are shown in Fig. 9 together with the corresponding correlation intensity curve (12). The transition from region A to region B of the correlation curve corresponds to more than 75 pct saturation in the number of microcracks, and essentially 100 pct saturation is reached before the onset of region C. Thus, surface microcracks as well as slip band development may contribute to the correlation losses in regions A and B of the curve. Very few of the microcracks develop beyond about 20 microns in length at any stage of the fatigue life.

The onset of region C of the correlation curve was invariably found to correspond to the presence of a growing crack or cracks, usually less than 100 microns in length, in the illuminated area of the specimen. Such a crack is shown in Fig. 10 in a specimen removed from the fatigue machine soon after the accelerating loss of correlation intensity signifying the

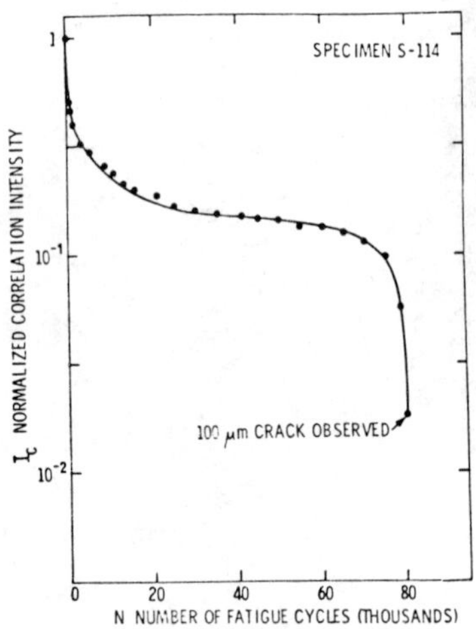

Fig. 8 Correlation intensity curve for the specimen shown in Fig. 7.

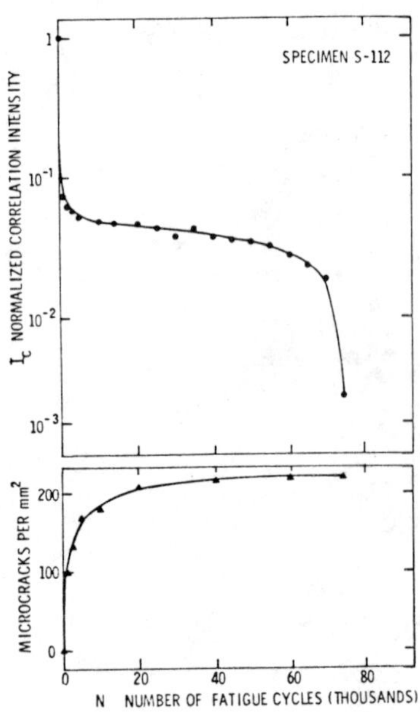

Fig. 9 Microcrack density and corresponding correlation intensity curve for aluminum alloy specimen S112.

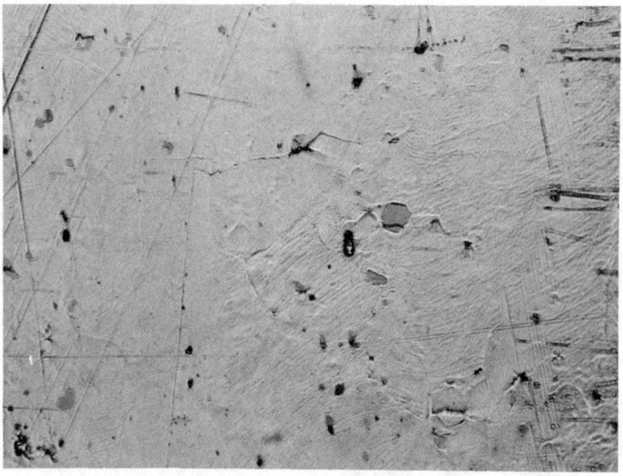

50 μm

Fig. 10 Fatigue crack associated with cracked constituent particle in 2024-T3 Al, observed early in region C of the correlation intensity curve.

transition from region B to C was observed (10). In this case the crack is associated with a constituent particle in the upper center of the field of view. More typical is the crack shown in Fig. 7(d) propagating from one edge of the specimen. At a length of only 60 microns in the illuminated area, this crack caused a significant loss in correlation intensity, as shown in Fig. 8. With the specimen geometry shown in Fig. 3, however, it was not possible to make correlation measurements on the edge of the specimen where the crack shown in Fig. 8 presumably originated: it is possible that region B of the correlation curve corresponds to a period of early crack growth in a region of the specimen that is not illuminated. Further experiments on specimens with a suitably modified geometry are required. Nevertheless, we have shown (11,12) that region C of the correlation curve can be related directly to early crack propagation in both unnotched and notched specimens, and to crack length in particular. For example, in situ photomicrographs of a propagating fatigue crack in 2024-T3 aluminum are shown in Fig. 11, and corresponding correlation intensity data are plotted in Fig. 12 as a function of crack length during region C of the I_c curve (12). The 60 micron crack shown in Fig. 11a was first observed following a correlation intensity loss of more than an order of magnitude after the beginning of region C. Successive crack advances shown in Fig. 11a through 11d caused the corresponding changes in correlation intensity indicated by points P through S in Fig. 12. Crack length, which directly affects both the size of the crack-tip plastic zone and the magnitude of the elastic displacements around the crack, is probably the most important factor controlling correlation intensity losses in region C of the correlation curve. The data in Fig. 12 show that crack length and correlation intensity are closely related, but there is some deviation from a smooth curve. For example, over an interval of 100 fatigue cycles there is a sharp loss in correlation intensity from point Q to point R in Fig. 12, but little change in crack length. Comparison of the crack topography in Fig. 11b and 11c suggests that crack-tip branching may be largely responsible for the correlation loss over this fatigue interval. The rapidly accelerating losses after the crack length exceeds about 500 microns corresponds to increasingly severe cracktip damage and

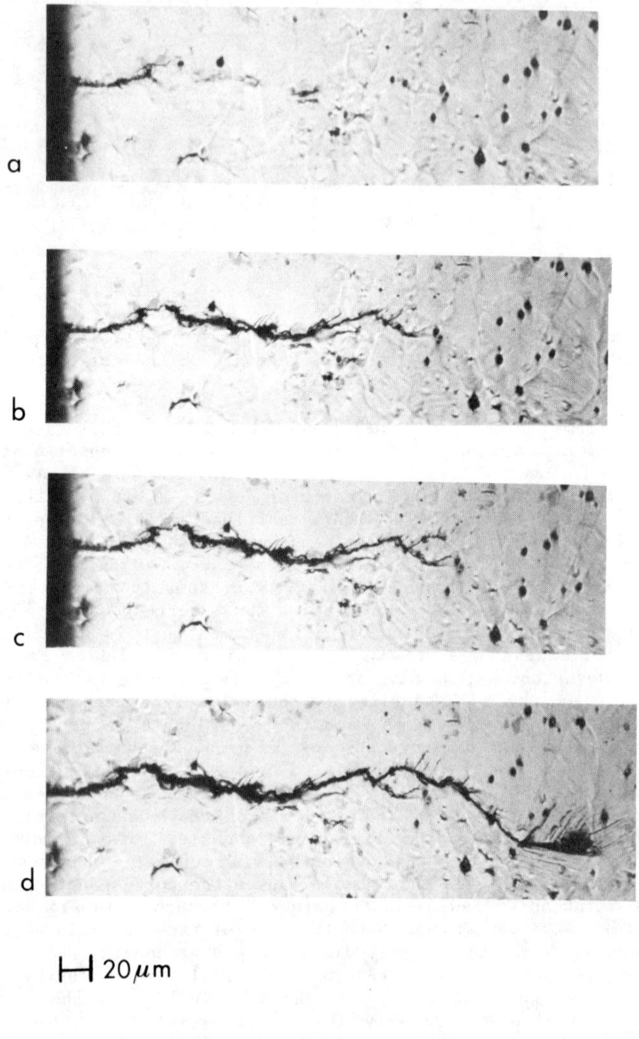

Fig. 11 Fatigue crack growth in unnotched 2024-T3 Al specimen (S117). Micrographs correspond to points P, Q, R, S in Fig. 12. Number of cycles elapsed: (a) 96,000; (b) 96,300; (c) 96,400; (d) 96,800.

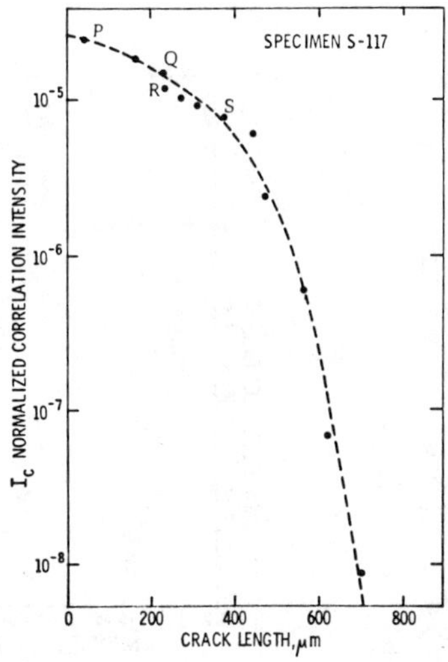

Fig. 12 Correlation intensity in region C as a function of crack length in specimen S117. Points P, Q, R, S correspond to the photomicrographs in Fig. 11.

to a rapid increase in crack width. The loss of correlation intensity caused by crack growth from about 20 microns to 700 microns is well over three orders of magnitude, and the total change from the beginning of the fatigue test to a 700 micron crack is eight orders of magnitude. These observations demonstrate that the optical correlation technique is an extremely sensitive way of detecting fatigue damage and monitoring the initiation and early growth of fatigue cracks, with no prior knowledge of the presence or precise location of a particular flaw or crack.

Strain-Controlled Tests

We are now conducting experiments in strain control on specimens fabricated from 2024-T4 aluminum round bar stock of 12.7 mm diameter. The specimens are machined with a 0.32 mm wide flat in the 0.63 mm diameter gauge section, and this flat surface is illuminated for the optical correlation measurements. The tests are conducted in a closed-loop machine in axial tension-compression, and strain is controlled by means of a clip-on electrical resistance strain gauge. In situ optical observations are made using a travelling metallurgical microscope equipped for Nomarski interference contrast. (See footnote.)

Preliminary results are shown in Figs. 13-17. The hologram is exposed initially at zero load, and stress, strain and correlation intensity are monitored continuously during a test, in this case at a frequency of 0.1 Hz. As shown in Fig. 13b, for a polished specimen subjected to

Notation: In strain control more than one correlation experiment may be conducted on a given specimen. Specimens are numbered consecutively with the prefix N, for example N1, N2. More than one test on the same specimen is indicated by a second digit, for example, N2.1, N2.2 and so on.

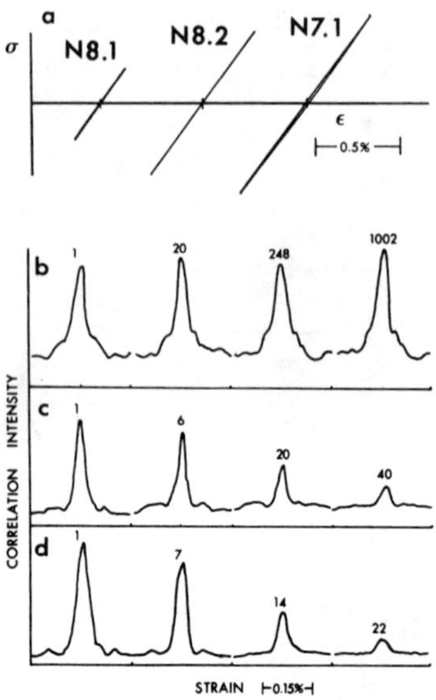

Fig. 13 Effect of strain cycling on correlation intensity in 2024-T4 aluminum bar specimens for the number of strain reversals indicated. (a) Stress-strain relationships for specimens N8 (two tests) and N7 (one test) (b) Correlation intensity peaks for test N8.1, total strain range $\pm 1.5 \times 10^{-3}$ (c) Test N8.2, total strain range $\pm 3.0 \times 10^{-3}$ (d) Test N7.1, total strain range $\pm 4.0 \times 10^{-3}$, prior strain 2% after polishing.

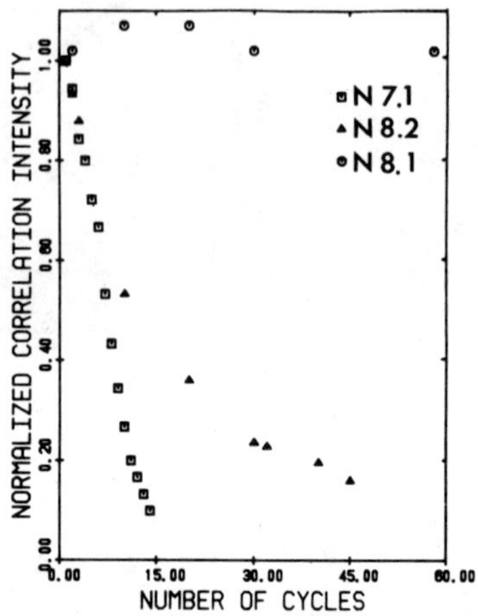

Fig. 14 Correlation intensity data (peak height) during strain-controlled experiments. Total strain ranges are shown in Fig. 13.

strain cycling through a total strain range $\pm 1.5 \times 10^{-3}$ for over 500 cycles, no loss in peak correlation intensity is observed: the correlation intensity is a reversible function of strain, and thus the maximum value of the correlation intensity is reproduced each time the specimen moves through its original position. At this total strain range, plastic strain is undetectable by means of the clip-on gauge, and the plastic strain range must therefore be less than $\pm 5 \times 10^{-5}$. With this experimental configuration, the correlation intensity measurements are sensitive to elastic strains (or strain changes) of the order of 1×10^{-5}. Furthermore, the scatter in peak intensity encountered in our earlier experiments is greatly reduced (compare Fig. 4).

The results presented in Fig. 13c show that irreversible changes in correlation intensity occur when the same specimen is cycled over a slightly higher total strain range, $\pm 3 \times 10^{-3}$. The peak correlation intensity falls more than 50 pct over the first 10 cycles, and decays essentially to background level over about 50 cycles. In spite of the correlation loss, the plastic strain is still too small to be detected directly by the clip-on gauge. Results from a second specimen, shown in Fig. 13d, indicate that the peak correlation intensity decays to background level after only about 15 cycles at a still higher total strain range of $\pm 4 \times 10^{-3}$, and in this case a plastic strain range of $\pm 0.6 \times 10^{-4}$ can be estimated from the width of the stress-strain hyteresis loop. Since this specimen was strained plastically in tension about 2 pct after polishing (but prior to recording the hologram), care must be exercised in comparing the data of Fig. 13d with those of Fig. 13b and 13c. Nevertheless, when the peak correlation intensity is plotted as a function of the number of strain cycles as in Fig. 14, it seems reasonable to conclude that the irreversible losses in correlation intensity are caused by fatigue-induced surface deformation, and that increasing the cyclic strain range causes a more rapid loss of intensity. In terms of the correlation intensity curves presented previously (Fig. 5, for example) the data in Fig. 14 represent a very early stage in region A of the overall curve, and suggest that even in our earlier (tension-tension) tests a substantial proportion of the I_c loss in region A may be attributed to topographic changes alone.

The irreversible loss in correlation intensity can be related to the magnitude of the plastic strain applied during a single half-cycle of tensile strain. Experiments are conducted on specimens that have been

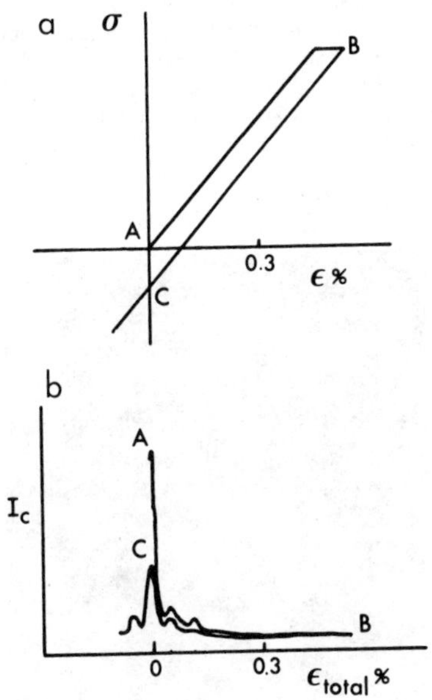

Fig. 15 Effect of tensile strain on correlation intensity in 2024-T4 aluminum. (a) Schematic half-cycle of strain follows path ABC (b) Corresponding correlation intensity.

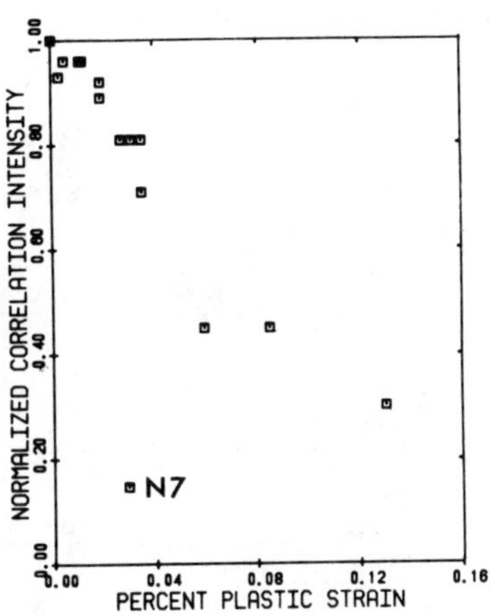

Fig. 16 Correlation intensity (peak height) as function of applied tensile strain in 2024-T4 aluminum. Each point represents one half-cycle of plastic strain for specimen N7 as shown in Fig. 15. A new hologram is recorded prior to each test.

strained slightly after polishing so that the illuminated surface is initially "slipped" rather than "as-polished." A typical strain half-cycle for 2024-T4 aluminum is shown in Fig. 15a. The hologram is made at zero load (point A), the specimen is strained in tension to a predetermined limit (point B), and the total strain for this cycle is then reduced to zero (point C) by applying a small compressive load. The plastic strain determined from the width of the hysteresis loop in this case is $8.5x10^{-4}$. The associated correlation intensity (Fig. 15b) shows an irreversible loss corresponding to the changes in surface topography introduced during the strain cycle. Correlation intensity loss is shown as a function of plastic strain for a series of such single tension-cycle experiments on 2024-T4 Al in Fig. 16. Typical surface changes (slip bands) caused by a plastic strain of even four times this magnitude on a polished surface are very difficult to detect by optical metallography using conventional illumination (Fig. 17a) but they can be resolved by a suitable interference contrast technique (Fig. 17b). Using the optical microscope, we detect no change on the previously slipped surface of specimens subjected to incremental plastic strains of 10^{-3} or less: however, the loss in correlation intensity shown in Fig. 15b strongly suggests that such changes are occurring, and that the correlation technique is extremely sensitive to small changes in surface topography. These results demonstrate that the optical correlation measurements are sensitive to incremental plastic strains of the order of 10^{-4} under conditions where direct visual observation is complicated by prior slip or surface roughness, and where the total incremental strain is zero. The loss in correlation intensity increases with increasing plastic strain, but further experiments are needed to establish a quantitative relationship between correlation intensity and plastic strain.

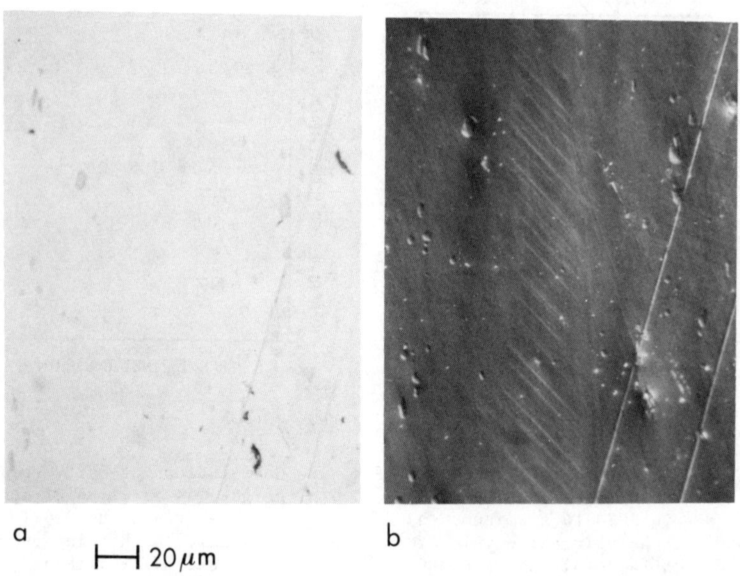

Fig. 17 Surface changes caused by plastic strain of $3.4x10^{-3}$ in 2024-T4 aluminum. (a) Conventional bright-field illumination (b) Nomarski interference contrast micrograph of the same area.

Conclusions

1. Analysis of the wavefront reconstruction process shows that local surface changes in metals, such as those arising from slip or the early development of fatigue cracks, will cause corresponding changes in the optical correlation intensity (I_c) produced when coherent light scattered from the metal surface passes through a holographic filter containing information about the surface topography at an earlier time.

2. In experiments on high-strength aluminum alloy sheet (2024-T3), a three-stage correlation intensity curve (log I_c vs. number of fatigue cycles) is observed for unnotched and notched specimens either polished or rough-ground prior to fatigue testing. A rapid loss of correlation intensity early in the fatigue life (region A) is followed by a more gradual loss (region B) and finally by an accelerating loss (region C) prior to failure. Similar results are observed for titanium and low-carbon steel.

3. Metallographic observations show that in 2024-Al, region A of the correlation intensity curve corresponds to slip band development and microcrack formation, and region C corresponds to the propagation of cracks beyond about 20 microns in length. The overall loss in correlation intensity may be several orders of magnitude before the largest crack reaches a length of one millmetre. Few metallographic changes are observed during region B of the curve.

4. Correlation intensity losses early in the fatigue life in strain-controlled tests are very sensitive to the imposed cyclic strain range. Elastic strains lead to reversible changes in correlation intensity, whereas plastic strains cause irreversible losses in I_c. In tensile tests over a single half-cycle, plastic strains of the order of 10^{-4} can be detected reliably.

Acknowledgement

Andrew Hieber and Vijay Singh made major contributions to the work reviewed in this paper. Financial support from the National Science Foundation, and from International Harvester Corporation in the form of a fellowship to David Sigler, is gratefully acknowledged.

References

1. K.C. Chuang, "Application of the Optical Correlation Measurement to Detection of Fatigue Damage," Materials Evaluation 26 (1968), pp. 116-119.

2. E. Marom and R.K. Mueller, "Nondestructive Early Fatigue Detection," p. 79 in Proceedings of the Sixth Symposium on Nondestructive Evaluation of Aerospace Components and Materials, Western Periodicals Co., North Hollywood, California (1967).

3. E.Marom and R.K. Mueller, section 5.5 "Optical Strain Measurement" in Applications de l'Holographie, J-Ch. Vionot et al., eds., University of Besancon, France, July 1970.

4. R.W. Jenkins and M.C. McIlwain, "Solder Joint Fatigue Measurement by Optical Cross-Correlation," pp. 183-192 in Holography and Optical Filtering, NASA SP-299 (1973).

5. R.E. Bond, R.E. Beissner, James Lankford, Jr., and W.W. Bradshaw, "Optical Correlation Evaluation of Surface Deformation Due to Fatigue," pp. 177-182 in Holography and Optical Filtering, NASA SP-299 (1973).

6. E. Marom, "Holographic Correlation," pp. 149-180 in Holographic Nondestructive Testing, Robert K. Erf, ed., Academic Press, New York (1974).

7. W.S. Colburn and E.N. Tompkins, "Improved Thermoplastic-Photoconductor Devices for Holographic Recording," Applied Optics 13 (1974), pp. 2934-2941.

8. W.L. Haworth, T. Cheng, A.F. Hieber, F.T.S. Yu and R.K. Mueller, "A Dynamic Optical Correlation Technique for Fatigue Failure Prediction," J. Engg. Materials and Technology, ASME Trans., 99 (1977) series H, pp. 229-233.

9. W.L. Haworth, A.F. Hieber and R.K. Mueller, "Fatigue Damage Detection in 2024 Aluminum Alloy by Optical Correlation," Met. Trans. 8A (1977), pp. 1597-1604.

10. W.L. Haworth, V.K. Singh and R.K. Mueller, "Fatigue Damage Detection in 2024-T3 Aluminum, Titanium and Low-Carbon Steel by Optical Correlation," J. Engg. Materials and Technology, ASME Trans., 99 (1977) series H, pp. 319-323.

11. W.L. Haworth, "Holographic Study of Fatigue Deformation and Crack Growth in Metals," Fatigue Engg. Materials and Structures 1 (1979), pp. 351-361.

12. W.L. Haworth, V.K. Singh and R.K. Mueller, "Holographic Detection of Fatigue-Induced Surface Deformation and Crack Growth in a High-Strength Aluminum Alloy," Met. Trans. 11A (1980), pp. 219-229.

REMAINING FATIGUE LIFETIME PREDICTION FOR RETIREMENT-FOR-CAUSE

IN ALUMINUM ALLOYS

W. L. Morris and M. R. James
Rockwell International Science Center
Thousand Oaks, CA 91360
&
O. Buck
Ames Laboratory, Iowa State University
Ames, Iowa 50011

Abstract

A methodology to predict the probability of specimen failure with subsequent fatigue, after a short surface crack has been detected in an Al 2219-T851 alloy is evaluated. Cracks are detected and tracked to failure using optical microscopy. Predictions of the distribution in remaining lifetimes, for comparison with measured values, are made with a Monte Carlo procedure in conjunction with crack growth laws that model the effect of grains of differing size, shape and crystallographic orientation on the propagation rate. A slower average rate of crack growth for cracks formed later during fatigue is observed and predicted. We attribute this to a progressive, fatigue-induced change in the mechanical properties of the surface. The change in surface properties produces a shift toward longer average remaining fatigue lifetimes after crack detection, but with more scatter about the mean life, for those specimens in which more cycles are required to initiate cracks. Our methodology successfully predicts these two effects. Implications of this behavior to the retirement-for-cause concept are discussed.

Introduction

The purpose of retirement-for-cause (RFC) is to extend the service of expensive, high-performance components beyond their intended design life. For components whose useful lifetime is fatigue limited, infrequent, very early failures may occur substantially before the mean lifetime. This forces the design to be extremely conservative, and hence the mean fatigue lifetime is typically much longer than the design life. If the few components destined to fail early are identified and eliminated by careful non-destructive evaluation, the remainder can safely be used well beyond their intended design lifetime. Development of such an RFC strategy requires assessment of the factors affecting the statistics of material failure, the sensitivity and reliability of the methodology used to inspect components for fatigue cracks, and the acceptability and cost of in-service failures.

Prediction of early failure is a problem, particularly if the critical crack size is small and the alloy grain size is large. In the latter case, non-continuum processes related to the formation and early growth of microcracks can have a substantial effect on lifetime (1,2). Variations in service loads and environment may dominate the statistics of early failure and can be treated with traditional statistical procedures. However, even if the service loads are identical from component to component, the statistics of the failure process governed by microstructural variations may lead to failures that are too early to be ignored. These failures occur at too low a probability to be observed in the laboratory and, consequently, a method is required to determine such probabilities directly from the alloy's mechanical properties and microstructure.

Research that addresses this problem is extremely limited and has principally been directed at the problem of predicting early fatigue failure probability in the absence of inspection for cracks. Bloomer and Roylance (3) and Korbacher (4) have tested the applicability of Weibull and log-normal distributions to describe the statistics of fatigue failure for polycrystalline aluminum and copper, respectively. Their results indicate that no single distribution can be used to extrapolate very early failure probabilities from small numbers of fatigue tests. The major problem, and one that must be faced with RFC, is that the mechanism which controls an early failure event may differ from those which determine failures near the mean lifetime.

A key requirement for accurate prediction of the probability of early failure is that the failures must not be anomalous events. For instance, if a foreign object is dropped into the melt during casting and remains undetected in the final product, little can be done to predict the probability of such an event or its effect on lifetime. Early failures that result from unusual but predictable variations in alloy microstructure and mechanical properties can, however, be treated so that accurate fatigue lifetime predictions may be made. Approaches such as given by Provan (5), Ghonem and Provan (6) and McCartney (7) have emphasized statistical methods that recognize these constraints, but do not correctly identify the mechanisms of early fatigue failure important for the aluminium alloy of interest here.

Practical RFC type predictions cannot be made for short cracks in large grain alloys unless the effect of alloy microstructure on crack growth is understood quantitatively. Short crack growth typically involves crack propagation below the threshold stress intensity range required for long crack growth. To relate growth rates to alloy microstructure, conventional fracture mechanics must be replaced by non-continuum models of the effect of alloy microstructure on plastic deformation at the crack front. We have previously described (1) Monte Carlo procedures for predicting early failure probabilities for aluminum alloys based upon such models. In this paper we consider the application of such models to an RFC situation in which surface cracks are detected and tracked to failure during constant deflection fatigue of an Al 2219-T851 alloy. Cycles to failure are calculated for failure probabilities as low as 10^{-4}, and predicted and experimentally measured failure probabilities are compared for probabilities in the range of 0.03 to 0.5. In particular, the effect of prior fatigue on the probability of failure with additional fatigue after a crack has been detected is assessed.

Experimental Procedure

Flexural fatigue specimens (Fig. 1) of Al 2219-T851, with a yield strength $\sigma_{yield} = 360$ MN/m^2, were prepared by machining with progressively decreasing cutting depths (ending with 20 μm) to minimize residual surface stresses. 80 μm of additional material were then removed from each surface, using 600 grit emery paper, and a final polishing sequence with Al_2O_3 powder was used to give specimens a mirror-like surface finish for ease of crack length measurement. Repeated thickness measurements were made during polishing to maintain flatness, and final specimen thickness was 0.1422 cm ± .0003. The resulting residual surface stress, measured by X-ray diffraction, was less than ±20 MN/m^2 from zero.

Specimens were fatigued in deflection control in laboratory air (50% relative humidity) with fully reversed loading at a frequency of 5Hz. Two maximum stress amplitudes (σ_{max}) were used with $\sigma_{max} = 0.63\ \sigma_{yield}$ or $\sigma_{max} = 0.9\ \sigma_{yield}$. Specimens were transferred at intervals to an optical microscope to measure crack lengths. A bending jig was used to load the specimen surface in tension for improved crack resolution during measurement.

The distribution in remaining fatigue lifetimes after crack detection was determined by tracking many cracks on each fatigue specimen. Statistical variations in alloy microstructure affect growth rate most strongly when the crack size is less than five grain diameters. We measured the number of cycles required for cracks to grow to 500 μm in length (approximately 8 grain diameters) to assess failure statistics. Our rationalization for this procedure is that while growth beyond 500 μm in flexural fatigue may take many thousands of cycles, the scatter in lifetime introduced in this period is small.

Characterization of early failures requires that the growth of a large number of cracks be determined quantitatively. Unfortunately, at low stress amplitudes, the density of grain-sized surface cracks initiated in Al 2219-T851 is extremely small. To give an example: observed values of cracking densities are approximately 0.01/cm^2 after 1.5×10^4 cycles and 0.2/cm^2 after $4\text{x}10^4$ fatigue cycles at a cyclic stress amplitude $\sigma_{max} = 0.63\ \sigma_{yield}$. We alleviated this problem with the discovery that a small drop of ink from a felt tip pen placed on the specimen surface during fatigue greatly accelerates crack initiation underneath the ink, provided the ink remains liquid during fatigue. Since the mode of crack initiation remained unchanged by this procedure, we were able to seed cracks at any point during fatigue by delaying application of the ink until the specimen had been prefatigued a desired amount.

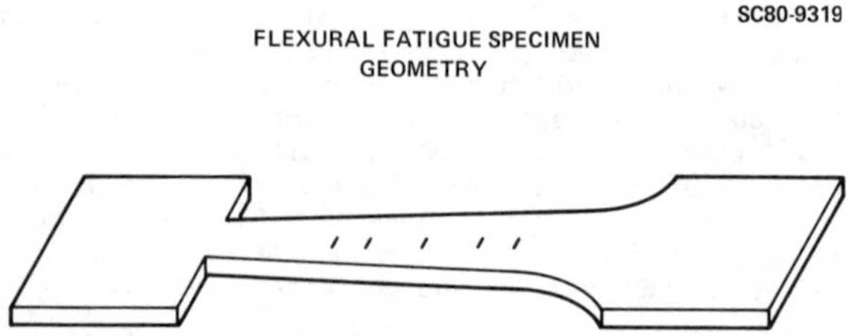

Fig. 1 - Flexural fatigue specimen, illustrating location of surface cracks which have been seeded using ink dots.

Our procedure was to apply ink dots of approximately 50 μm in diameter along a line parallel with the stress axis, with separation between dots of several millimeters. After 5×10^3 fatigue cycles, the ink was removed with an acetone solvent and 5×10^3 additional cycles were applied to grow the cracks to grain size. Growth rates after removal of the ink were found to be the same as on un-inked specimens, indicating that this procedure leaves subsequent growth behavior unaltered.

By using this technique, we were able to initiate numerous microcracks within each desired range in fatigue cycles, and then follow their subsequent growth. Ideally, we would have preferred to select and follow only cracks initially the same length (80 μm). However, in Al 2219-T851, early crack growth is slowed by grain boundaries (2), with the result that a distribution in crack lengths is observed after prefatigue, governed by the distribution in grain sizes. To obtain sufficient data for reasonable statistics, we analyzed the growth of cracks with initial lengths of 80±30 μm, and constructed from these data an effective distribution in remaining lifetime for cracks initially 80 μm long. The new length of each crack after each progressive fatigue increment was plotted against its initial length at the first fatigue increment (10^4 cycles) (Fig. 2). A linear interpolation of these data was then used to construct an effective distribution in lengths for each fatigue increment that approximates the distribution expected if all the cracks were initially 80 μm in length (Fig. 3). This procedure compensates for the tendency of the average growth rate to increase slightly with a larger initial crack length.

Only the faster growing cracks can be characterized with our technique, because shadowing of short cracks by neighboring long cracks further suppresses the rate of growth of the shorter cracks. Furthermore, the specimen ultimately fails before all the cracks that have been tracked can reach 500 μm. The experimental method, therefore, is useful to determine fractional probabilities of failure slightly larger than the reciprocal of the total number of cracks initially seeded.

Results

Remaining specimen fatigue lifetime is frequently determined by the largest crack on a component. This is not always the case, however, as statistical fluctuations in growth rate can lead to failure from propagation of a smaller crack initiated at a site unrelated to the main crack. The probability of occurrence of such independent initiations increases with both specimen surface area and with cyclic stress amplitude. Our

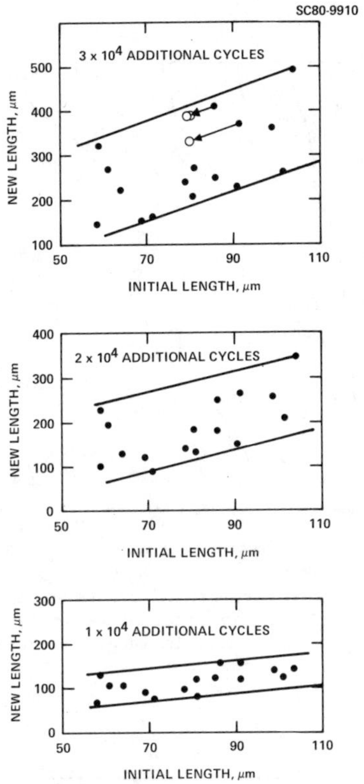

Fig. 2 - New lengths (solid dots) of 16 cracks are plotted for 1,2 and $3x10^4$ cycles after detection at $1.0x10^4$ cycles, against their initial length. An interpolation procedure illustrated (open circles) in top figure is used to construct an effective distribution in crack lengths corresponding to cracks initially 80μm in length.

experimental procedure determines probable remaining lifetime where growth of a single crack determines the lifetime. We call this situation a zero-surface-area limiting case. For comparison, later we calculate the probability of remaining lifetime for a sufficiently large surface area in which cracks unrelated to the main crack may participate in specimen failure.

Experimentally measured distributions in probable remaining fatigue lifetime (zero-surface-area limit) are shown in Fig. 4 for the $\sigma_{max} = 0.63\ \sigma_{yield}$. Three cases are illustrated corresponding to 80 μm cracks initiated at 10K, 25K or 40K cycles ($K=10^3$). The probability of failure is plotted as a function of additional cycles to produce a 500 μm crack. The curves are theoretical predictions obtained from growth models discussed later. The observed trend is that cracks initiated earlier in fatigue

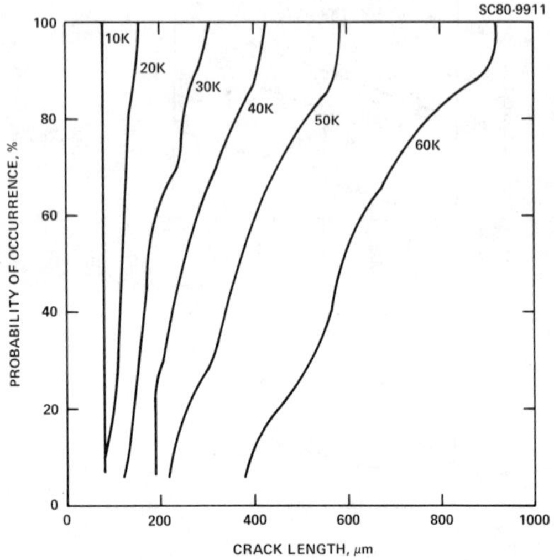

Fig. 3 - Probability of occurrence of cracks of length smaller than specified for five increments in fatigue, after detection of cracks 80μm in length at $1.0x10^4$ fatigue cycles (calculated curves).

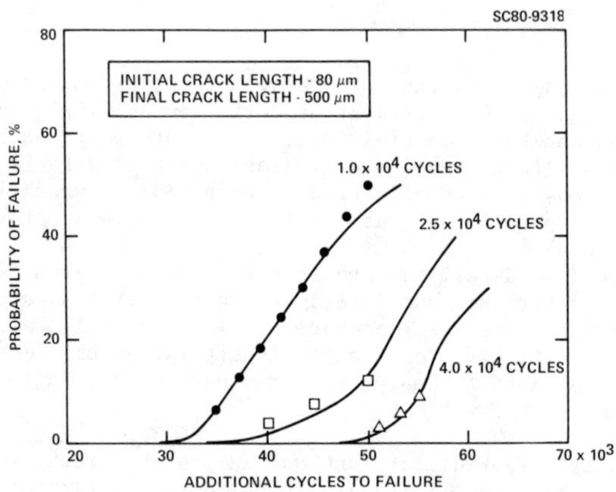

Fig. 4 - Additional cycles to "failure" after an 80μm long crack has been detected at 1.0, 2.5, or $4.0x10^4$ cycles, for a peak stress amplitude $\sigma_{max} = 0.63\ \sigma_{yield}$. Increased mean in remaining lifetime for later initiated cracks is due to cyclic hardening of the surface. Symbols are experimental data, curves are computer predictions.

lifetime grow more rapidly on an average. This is the result of a progressive hardening of the surface which makes it more difficult for the short surface cracks to propagate across grain boundaries later in the lifetime.

Experimentally measured distributions in probable remaining fatigue lifetime (zero-surface-area limit) are illustrated in Fig. 5 for $\sigma_{max} = 0.9\ \sigma_{yield}$. The additional cycles to failure are plotted for cracks initially of 200 μm in length detected at 5K or 10K cycles. Again, "failure" is defined to occur as the crack length reaches 500 μm. In this case, crack coalescence dominates the growth of cracks initiated at 10K cycles leading to average growth rates faster than those for cracks formed at 5K cycles.

Modeling and Discussion

The rate of propagation of short surface cracks in Al 2219-T851 is controlled by the size, shape and crystallographic orientation of grains in the crack path (8, 9). Stage I (crystallographic), Stage II (transgranular, non-crystallographic) and intergranular cracks are all observed in this material. In addition to their characteristic growth habit, Stage I and Stage II cracks are clearly distinguished by measurement of the shape of the plastic zone at the surface crack tips (2). For Stage I cracks, the plastic deformation lies in the plane of the crack and the tensile (Mode I) opening normal to the crack is elastic. Stage II cracks exhibit plastic deformation normal to the fracture plane as well as in front of it, attesting to the homogeneity of slip from the crack tip. The

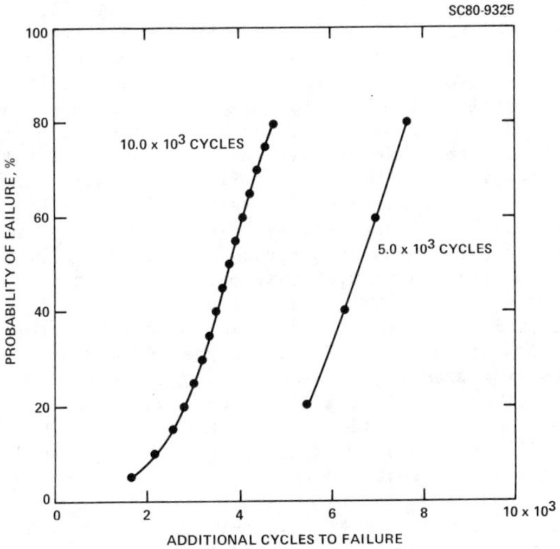

Fig. 5 Probability of failure with additional cycles after a 200μm crack has been detected for high cyclic stress amplitude. Crack coalescence for $\sigma_{max} = 0.90\ \sigma_{yield}$ accelerates average growth rate and is important for later (1.0×10^4 cycles) initiated cracks, for which the density of surface cracks producing coalescence is larger.

plastic deformation normal to the crack plane leads to the development of a tensile Mode I closure stress which slows and may stop the growth of Stage II cracks (2).

For both Stage I and Stage II microcracks, the plastic zone size is constrained at the surface by the grain boundaries, because slip is not continuous across the boundaries (2). The plastic zone size and the closure stress is larger as cracking proceeds into a large grain where the slip distance to the next boundary is greatest. For Al 2219-T851, the closure stress for Stage II cracks reaches a magnitude equal to the peak applied tensile stress if the length of the crack at the surface is the same length or shorter than the slip distance. Thus short cracks in large grains cannot grow in Stage II. Crack closure stress, therefore, is one reason why Stage I growth is preferred for cracks with lengths less than the mean grain size. For the very short cracks, Stage II growth can only occur into comparatively small grains.

As noted by Lynch (10), the character of slip at the surface also affects the mode of crack propagation. For Al 2219-T851, the microplastic deformation behavior of the surface and its effect on crack growth changes with fatigue. The propensity for a Stage I crack to propagate in this mode into a new grain decreases with increased fatigue cycles as slip at the surface becomes more homogeneous. Thus, cracks initiated later during fatigue or at a lower cyclic stress amplitude are more likely to propagate in Stage II (9), but at low growth rates because of the large closures stresses developed in this growth mode.

Also, just before a surface crack tip enters a new grain, propagation at that point on the crack front ceases as slip begins across the grain boundary. If the new growth is subsequently to be Stage II, a progressive increase in the maximum Mode I component of the surface crack tip opening displacement occurs (and has been measured (9)) as the number of dislocations increases in the grain with fatigue. Propagation begins across the boundary only when a critical crack tip opening displacement has been reached. This period of no growth we refer to as incubation, (8); we find incubation to be longer the shorter the crack and the smaller the grain at the crack tip, as predicted by Chang et al (11). The same incubation phenomenon is observed for crystallographic cracks, although the effect cannot be measured via Mode I crack tip opening displacement measurements. Also, the later that a surface crack tip reaches a boundary, the longer is the period of incubation at that point on the crack front. We attribute this phenomenon to a progressive increase in the friction stress at the surface which must be exceeded for microplastic deformation to occur. The mechanism of this change has not been determined, but we believe it results from trapping of dislocations below the surface oxide, as has been suggested by Grosskreutz (12) from studies on Al 1100. This "hardening" of the surface is evident in crack nucleation (9) and residual stress relaxation (13) processes in Al 2219-T851, as well as in the crack propagation behavior of interest here.

With sufficient fatigue, the friction stress that must be exceeded for microplastic deformation to occur at the surface can increase and effectively reach the peak cyclic stress. Therefore, growth modes which depend upon slip near and parallel to the surface cease later during fatigue. This apparently occurs most rapidly in those grains oriented with (111) slip planes favoring slip in the surface plane. In particular, Stage II propagation and Mode II Stage I growth across grain boundaries (at the surface) becomes difficult when the friction stress increases. If the cracks are short and shallow, the deformation properties of the surface completely dominate crack growth which occurs by these modes, and propagation across grain boundaries never occurs later in the fatigue life. For surface cracks longer than about 200μm in Al 2219-T851, our models which mathematically describe this behavior are not rigorously obeyed. We believe this is caused by the subsurface deformation behavior, which cannot be ignored. Growth across the boundaries for longer surface cracks is greatly retarded but not stopped. Additionally, intergranular crack propagation or Mode III Stage I growth becomes a preferred propagation mechanism, because these growth modes are less sensitive to the mechanical properties of the surface. By measuring the opening displacement of surface crack tips we have found that intergranular cracking ahead of the main crack may occur even if hardening at the surface becomes sufficient to prevent slip at the surface from the crack tip into the next grain. In this case, we attribute the grain boundary cracking to an enhancement of the strain amplitude at the boundaries due to the elastic strain field associated with the crack.

The mathematical details of the growth laws outlined above have been described elsewhere (2, 8, 9). This includes growth models for Stage I and Stage II-transgranular growth, which relate the magnitude of the closure stress, incubation period and rates of crack coalescence, and hence the resulting crack growth rates, to alloy microstructure. The general approach using a Monte Carlo simulation of crack growth to determine the statistical effects of microstructure on growth rate has also been discussed elsewhere (1). Empirical relationships have been used to simulate intergranular crack propagation for the low stress amplitude case since we do not yet have a sufficiently detailed theory to predict the effect of cyclic stress amplitude on the rate of growth of intergranular cracks.

Remaining Lifetime Prediction

A computer simulation of the growth process is used to determine the cycles to failure for a series of 80 μm long cracks. For each crack, the size, shape and crystallographic orientation of grains in the path of the crack at the surface are selected for the growth simulation by using distributions in size and shape measured for the alloy and by assuming that the distribution of crystallographic orientations is random. A simplifying assumption is made in the initial conditions for propagation which makes the calculation substantially easier. We assume that each 80 μm crack is in an 80 μm grain with the surface tips touching the grain boundaries. We further assume that the cracks arrive at the boundary at a random time during an interval up to 5,000 fatigue cycles before they are detected. This is consistent with experimental observation. More accurate calculations would require modeling and simulation of the crack initiation processes.

The predictions of distribution in remaining fatigue lifetimes shown in Fig. 4 were obtained by simulation of the growth of 10^4 cracks for each of the three initial conditions of prefatigue prior to crack detection. Figure 4 is replotted in Fig. 6, using a log scale to emphasize the very early failure characteristics. Some disagreement exists between prediction and experimental data for the case in which 2.5×10^4 cycles were required in initiation. We attribute this to the simplicity of our assumptions regarding the initial location of each crack with respect to the grain boundaries. In Fig. 7, remaining lifetime predictions for $\sigma_{max} = 0.63\ \sigma_{yield}$ have been extended to probabilities of failure of one chance in ten million by calculating the effect which crack coalescence

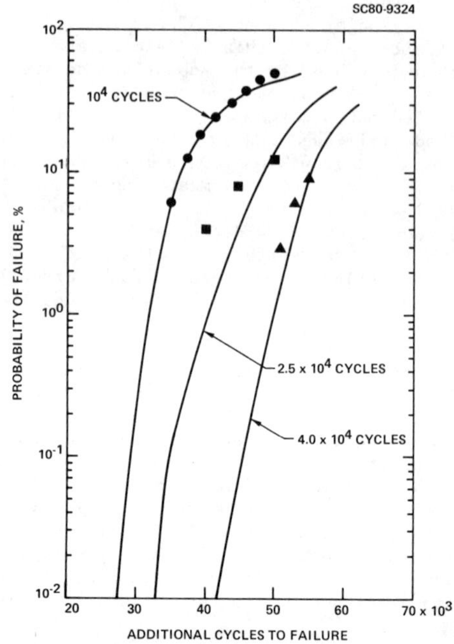

Fig. 6 - Results of Fig. 4, shown using a log scale to emphasize the early failure behavior.

will have on remaining lifetime. In Fig. 8 we compare the calculated remaining lifetime for a zero-surface-area to a 100 cm^2 surface area case for $\sigma_{max} = 0.63\ \sigma_{yield}$ and 10^4 cycles to initiation. The shorter remaining lifetime for the larger surface area is due to initiation and propagation of cracks other then the main detected crack. Both the coalescence and non-zero surface area calculations were made with experimental values of the average numbers of cracks/cm^2 developed during fatigue.

Comments on RFC Strategy

For Al 2219-T851, four separate growth mechanisms govern the statistics of early failure for propagation of short surface cracks. The relative importance of these are a function of the cyclic stress amplitude and of the cycles at which the main crack is formed. For the case of $\sigma_{max} = 0.63\ \sigma_{yield}$, with 10^4 cycles to initiation, specimens with the shortest remaining lifetimes fail by crack coalescence, and at longer lifetimes fail by crystallographic crack propagation with growth rates controlled by incubation. Specimens of even longer lifetime fail due to transgranular non-crystallographic propagation with rates controlled by crack closure stress and by incubation, and still longer life specimens fail predominantly by intergranular crack propagation. Each separate growth mechanism is sensitive to different features of the alloy microstructure and mechanical properties. Thus, early failure statistics are governed both by material features and by growth laws different from those which determine the average lifetime. The probability of very early failure of

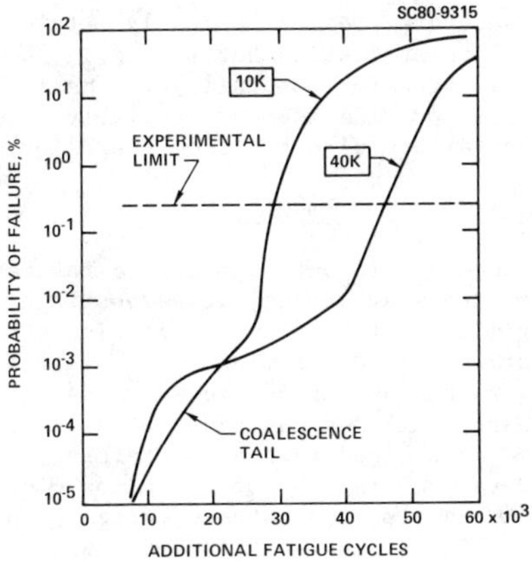

Fig. 7 - Probability of failure with additional fatigue cycles for extremely low probabilities of failure is altered by crack coalescence for $\sigma_{max} = 0.63\ \sigma_{yield}$. Coalescence is more important for cracks initiated at 4.0×10^4 cycles, because the cracking density in higher. Dashed line is approximately the smallest failure probability that can practically be evaluated in the laboratory.

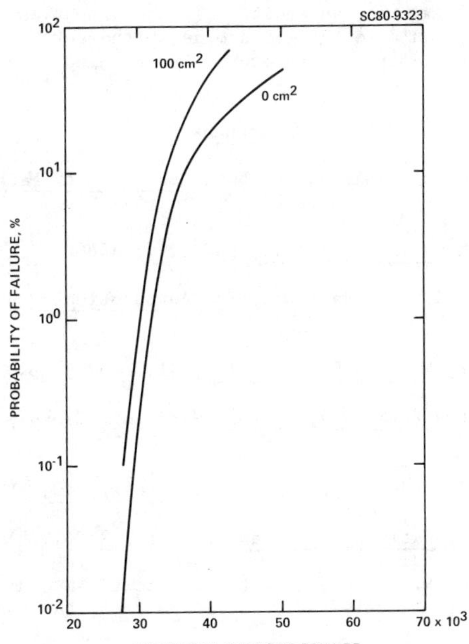

Fig. 8 - Probability of failure distribution shifts to smaller numbers of additional cycles for specimens with larger surface area, because of crack initiation and propagation at sites away from the main crack.

Al 2219-T851 cannot, therefore, be realistically obtained by extrapolation from probability of failure statistics determined in the laboratory. We believe this is generally true for most alloys. The methodology we describe may in the future provide means to reliably calculate the probability of early fatigue failure directly from the alloy microstructure.

Summary

Although this methodology is only applied to laboratory specimens fatigued at a constant stress amplitude, it serves to highlight problems which may be encountered in early failure prediction of real components. Once a short surface crack of predetermined size has been detected by NDE, statistical variations in the number of cycles to failure can be substantial. Exceptionally early failures can occur at a lifetime 1/10 of the mean remaining lifetime determined in the laboratory. These early failures result from rare combinations of the alloy microstructure in the crack path which lead to accelerated rates of crack propagation. Because of their rarity it is unlikely that such failures will be observed during laboratory testing, and even if a single anomolous event is recorded it is likely to be discounted. We have described a method to calculate the probable remaining fatigue lifetime from alloy microstructure for small probabilities of failure. Such predictions are an essential element in the development of a quantitatively based RFC strategy for aluminum alloys. A further complexity of the early failure process, which can place additional demands upon NDE of components, is that the remaining fatigue lifetime after crack detection is a function of the prior loading history. If service records on the component are inadequate to extract loading history information, an additional requirement is that NDE provide a measure of the accumulated fatigue damage of the surface, as well as detect and size surface cracks, before reliable calculations of probable remaining lifetime can be made.

References

1. W. L. Morris, M. R. James and O. Buck, Eng. Fract. Mech., 13, (1980) pp. 213.

2. W. L. Morris, Met. Trans. A., 11, pp. 1117, 1980.

3. N. T. Bloomer and T. F. Roylance, The Aeronautical Quarterly, 16, (1975) pp. 307.

4. G. K. Korbacher, Experiment Mechanics, 11, (1971) pp. 540.

5. J. W. Provan, Fracture 1977, Vol. 2, Waterloo, Canada, June 19-24, (1977) pp. 1170.

6. H. Ghonem and J. W. Provan, Eng. Fract. Mechs., 13, (1980) pp. 963.

7. L. N. McCartney, Fat. Eng. Mat. and Struct., 2, (1980) pp. 387.

8. W. L. Morris, M. R. James and O. Buck, "Growth Rate Models for Short Surface Cracks in Al 2219-T851," Met. Trans. A, in press.

9. M. R. James and W. L. Morris, "Quantative Modeling of Fatigue Crack Initiation," accepted by Mat. Sci. and Eng.

10. S. P. Lynch, Metal. Sci., 9, (1975) pp. 401.

11. R. Chang, W. L. Morris and O. Buck, Scripta Met. 13, (1979) pp. 191.

12. J. C. Grosskreutz, Surface Sci., 8, (1967), pp. 173.

13. M. R. James and W. L. Morris, "The Relaxation of Machining Stresses in Aluminum Alloys During Fatigue," in Residual Stress For Designers and Metallurgists," ASM Conf. April 9-10, 1980, Chicago, Ill, in press.

Subject Index

Author Index